Test Bank

for
Kotz, Treichel, and Weaver's

Chemistry & Chemical Reactivity

Sixth Edition

David Treichel
Nebraska Wesleyan University

THOMSON
BROOKS/COLE
Australia • Canada • Mexico • Singapore • Spain • United Kingdom • United States

Printed in the United States of America
3 4 5 6 7 09 08 07 06 05

Printer: Thomson/West

0-534-99850-X

For more information about our products, contact us at:
Thomson Learning Academic Resource Center
1-800-423-0563

For permission to use material from this text or product, submit a request online at
http://www.thomsonrights.com.
Any additional questions about permissions can be submitted by email to **thomsonrights@thomson.com.**

Thomson Higher Education
10 Davis Drive
Belmont, CA 94002-3098
USA

Asia (including India)
Thomson Learning
5 Shenton Way
#01-01 UIC Building
Singapore 068808

Australia/New Zealand
Thomson Learning Australia
102 Dodds Street
Southbank, Victoria 3006
Australia

Canada
Thomson Nelson
1120 Birchmount Road
Toronto, Ontario M1K 5G4
Canada

UK/Europe/Middle East/Africa
Thomson Learning
High Holborn House
50–51 Bedford Road
London WC1R 4LR
United Kingdom

Latin America
Thomson Learning
Seneca, 53
Colonia Polanco
11560 Mexico
D.F. Mexico

Spain (including Portugal)
Thomson Paraninfo
Calle Magallanes, 25
28015 Madrid, Spain

Table of Contents

Preface

This test bank contains printed tests to accompany Kotz and Treichel's CHEMISTRY AND CHEMICAL REACTIVITY, 5th edition. The test bank contains of over 1250 questions in multiple choice, fill-in-the-blank, and short answer formats. Over 66% of the questions are new to this edition.

I have tried to write questions that span a wide range of difficulties. Some questions require a single step and others require multiple steps to solve. I have paid careful attention to rules governing the handling of significant figures, both for questions containing a mixture of addition and multiplication and for questions involving logarithms.

I would like to thank Jack Kotz and Paul Treichel for the comments, which helped improve this test bank. In addition, I gratefully acknowledge the help provided by Julie Ealy, who checked the manuscript for accuracy.

CHAPTER 1: Matter & Measurement

1.1 Classifying Matter

1. According to the kinetic-molecular theory of matter, particles in a gas
   a) are packed closely together in a regular array.
   b) are packed closely together like solids, but they are not confined to specific positions.
   c) expand to fill their container.
   d) vibrate back and forth about an average position.
   e) move slower as the temperature increases.

   Answer: c

2. According to the kinetic-molecular theory of matter, particles in a liquid
   a) are packed closely together in a regular array.
   b) move slower as the temperature increases.
   c) expand to fill their container.
   d) vibrate back and forth about an average position.
   e) are packed closely together, but are not confined to specific positions.

   Answer: e

3. Which one of the following is a homogeneous mixture?
   a) baked beans
   b) sugar dissolved in water
   c) chocolate chip ice cream
   d) chicken noodle soup
   e) a jar of chunky peanut butter

   Answer: b

4. Which of the following is a heterogeneous mixture?
   a) mortar (a mixture of calcium carbonate and sand)
   b) gasoline
   c) antifreeze (a mixture of water and ethylene glycol)
   d) sodium chloride dissolved in water
   e) the air trapped inside a balloon (not including the balloon)

   Answer: a

5. A pure substance composed of two or more different elements is
   a) a homogeneous mixture.
   b) a heterogeneous mixture.
   c) a polyelement.
   d) an atom.
   e) a chemical compound.

   Answer: e

6. A(n) ________ is a pure substance that is composed of only one type of atom?
   a) element
   b) homogeneous mixture
   c) heterogeneous mixture
   d) macroscopic particle
   e) particulate

   Answer: a

<u>1.2 Elements and Atoms</u>

7. Of the 113 known elements, about _____ are found in nature?
   a) 17 b) 26 c) 60 d) 90 e) 110

   Answer: d

8. What is the correct symbol for lead?
   a) L b) Ld c) Pb d) Sn e) Pd

   Answer: c

9. What is the correct symbol for iron?
   a) Sn b) Fe c) Ir d) I e) Pb

   Answer: b

10. What is the correct symbol for phosphorus?
    a) Pm b) P c) Po d) Pt e) Pd

    Answer: b

11. What is the name of the element with the symbol B?
    a) boron b) beryllium c) bismuth d) bromine e) barium

    Answer: a

12. What is the name of the element with the symbol Ca?
    a) cobalt b) carbon c) calcium d) cadmium e) californium

    Answer: c

13. Which of the following substances is an element?
    a) CO b) S_8 c) NaCl d) $C_6H_{12}O_6$ e) CO_2

    Answer: b

1.3 Compounds and Molecules

14. Which term best describes carbon dioxide, CO_2?
    a) homogeneous mixture
    b) heterogeneous mixture
    c) element
    d) chemical compound
    e) none of the above

    Answer: d

15. Which statement concerning solid NaCl is true?
    a) NaCl is a chemical compound.
    b) NaCl is a homogeneous mixture.
    c) NaCl is a heterogeneous mixture.
    d) The percentage of Cl in NaCl is dependent on where the sample is obtained.
    e) NaCl has properties similar to it component elements, sodium metal and chlorine gas.

    Answer: a

16. The molecular model below depicts a molecule composed of carbon (black), oxygen (gray), and hydrogen (white) atoms. What is the correct molecular formula?

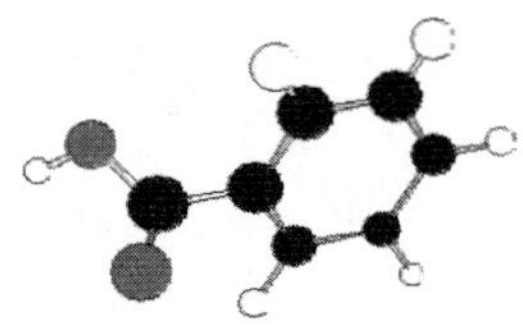

    a) CHO b) $C_6H_6O_2$ c) C_6H_7O d) C_7H_6O e) $C_7H_6O_2$

    Answer: e

1.4 Physical Properties

17. All of the following are examples of physical properties EXCEPT
    a) the boiling point of ethanol.
    b) the conductivity of aluminum wire.
    c) the frying of an egg.
    d) the density of neon gas.
    e) the melting point of gold.

    Answer: c

18. All of the following are examples of physical properties EXCEPT
    a) gold is yellow in appearance.
    b) aluminum oxidizes to form Al_2O_3.
    c) copper is ductile and malleable.
    d) mercury is a liquid at room temperature.
    e) the density of silver is 10.45 g/cm^3.

    Answer: b

19. An intensive property of a substance is
    a) only visible with magnification.
    b) dependent on its volume, but not its mass.
    c) independent of the amount present.
    d) dependent only on its temperature.
    e) dependent only on its mass and volume.

    Answer: c

20. Which of the following is an extensive property?
    a) volume
    b) density
    c) melting point
    d) boiling point
    e) both melting and boiling point

    Answer: a

21. The density of acetic acid is 1.05 g/mL. What is the volume of 327 g of acetic acid?
    a) 2.91×10^{-3} mL
    b) 3.21×10^{-3} mL
    c) 2.62×10^{2} mL
    d) 3.11×10^{2} mL
    e) 3.43×10^{2} mL

    Answer: d

22. The density of liquid mercury is 13.5 g/cm^3. What mass of mercury will fill a 0.500 L container?
a) 3.70×10^{-2} g b) 6.75 g c) 27.0 g d) 37.0 g e) 6.75×10^3 g

Answer: e

23. When 24 copper pennies are submerged in water, the pennies displace 8.26 cm^3 of water. If the combined mass of the pennies is 73.86 g, what is the density of copper?
a) 0.112 g/cm^3 b) 1.34 g/cm^3 c) 8.94 g/cm^3 d) 32.8 g/cm^3 e) 107 g/cm^3

Answer: c

24. You can identify a metal by carefully determining its density. A 8.44 g piece of an unknown metal is 1.25 cm long, 2.50 cm wide, and 1.00 cm thick. What is a possible identity of the element?
a) nickel, 8.90 g/cm^3
b) aluminum, 2.70 g/cm^3
c) titanium, 4.54 g/cm^3
d) zirconium, 6.51 g/cm^3
e) chromium, 7.20 g/cm^3

Answer: b

25. You can identify a metal by carefully determining its density. A 2.166 g piece of an unknown metal is 1.50 cm long, 0.550 cm wide, and 0.250 cm thick. What is a possible identity of the element?
a) silver, 10.5 g/cm^3
b) iridium, 22.4 g/cm^3
c) platinum, 21.4 g/cm^3
d) palladium, 12.0 g/cm^3
e) none of the above

Answer: a

26. A cube of iron has a mass of 29.31 g. If each side of the cube has dimensions of 1.55 cm, what is the density of iron?
a) 0.127 g/cm^3 b) 5.92 g/cm^3 c) 7.87 g/cm^3 d) 12.2 g/cm^3 e) 18.9 g/cm^3

Answer: c

27. At 0 °C, a bottle contains 325 mL of water in its liquid state. What is the volume of the water after it freezes (at 0 °C)? The densities of liquid water and ice at 0 °C are 1.000 g/mL and 0.917 g/mL, respectively.
a) 27.0 mL b) 298 mL c) 325 mL d) 354 mL e) 391 mL

Answer: d

28. All of the following are examples of intensive properties EXCEPT
    a) melting point.
    b) color.
    c) volume.
    d) density.
    e) boiling point.

    Answer: c

1.5 Physical and Chemical Change

29. All of the following are examples of chemical change EXCEPT
    a) the tarnishing of silver.
    b) the rusting of iron.
    c) the combustion of butane gas.
    d) the condensation of steam.
    e) the decomposition of water to hydrogen gas and oxygen gas.

    Answer: d

30. Which one of the following is an example of physical change?
    a) the evaporation of water
    b) the rusting of iron
    c) the combustion of hydrogen
    d) the tarnishing of silver
    e) both the rusting of iron and the tarnishing of silver

    Answer: a

31. Which of the following observations are examples of chemical changes to aluminum?
    1. When aluminum reacts with hydrochloric acid, hydrogen gas is evolved.
    2. Aluminum melts at 660 °C.
    3. A cube of aluminum with a volume of 4.0 cm^3 has a mass of 10.8 g.
    4. Aluminum is a good electrical conductor.

    a) 1 only   b) 2 only   c) 1 and 2   d) 3 and 4   e) 1, 2 and 4

    Answer: a

1.6 Units of Measurement

32. Which is a correct method for converting Fahrenheit to Celsius?

a) $T(^\circ C) = \left(\frac{5\ ^\circ C}{9\ ^\circ F}\right) T(^\circ F) + 32$

b) $T(^\circ C) = \left(\frac{9\ ^\circ C}{5\ ^\circ F}\right) T(^\circ F) + 32$

c) $T(^\circ C) = \frac{5\ ^\circ C}{9\ ^\circ F}\left(T(^\circ F) + 32\right)$

d) $T(^\circ C) = \frac{9\ ^\circ C}{5\ ^\circ F}\left(T(^\circ F) - 32\right)$

e) $T(^\circ C) = \frac{5\ ^\circ C}{9\ ^\circ F}\left(T(^\circ F) - 32\right)$

Answer: e

33. Thermostats are often set to 68 °F. What is this temperature in Celsius?
a) 2.0×10^1 °C b) 25 °C c) 38 °C d) 56 °C e) 7.0×10^2 °C

Answer: a

34. The boiling point of liquid helium is 4 K. What is this temperature in Celsius?
a) -294 °C b) -277 °C c) -269 °C d) -255 °C e) -169 °C

Answer: c

35. The temperature required to melt NaCl is 528 °C. What is this temperature in kelvin?
a) 231 K b) 255 K c) 528 K d) 801 K e) 977 K

Answer: d

36. Chemical reactions are often studied at 25 °C. What is this temperature in Fahrenheit?
a) 45 °F b) 57 °F c) 68 °F d) 77 °F e) 103 °F

Answer: d

37. Which one of the following is not a SI base unit?
a) kilogram b) Fahrenheit c) kelvin d) meter e) second

Answer: b

38. What is a correct method for converting Celsius to Kelvin?

a) $T\ (K) = \frac{1\ K}{1\ ^\circ C}\left(T(^\circ C) - 273.15\ ^\circ C\right)$

b) $T\ (K) = \frac{1\ K}{1\ ^\circ C}\left(T(^\circ C) + 273.15\ ^\circ C\right)$

c) $T\ (K) = \frac{9\ K}{5\ ^\circ C}\left(T(^\circ C) - 273.15\ ^\circ C\right)$

d) $T\ (K) = \frac{5\ K}{9\ ^\circ C}\left(T(^\circ C) + 273.15\ ^\circ C\right)$

e) $T\ (K) = \frac{5\ K}{9\ ^\circ C}\left(T(^\circ C)\right) + 273.15\ ^\circ C$

Answer: b

39. Which of the following is the greatest mass?

a) 2.3×10^9 pg b) 1.5×10^8 ng c) 6.6×10^5 μg d) 8.4×10^2 mg e) 3.0×10^{-1} g

Answer: d

40. The radius of a helium atom is 31 pm. What is the radius in nanometers?

a) 3.1×10^{-9} nm b) 3.1×10^{-6} nm c) 3.1×10^{-5} nm d) 3.1×10^{-3} nm e) 3.1×10^{-2} nm

Answer: e

1.7 Making Measurements: Precision, Accuracy, and Experimental Error

41. Two electronic balances are tested using a standard weight. The true mass of the standard is 1.3111 g. The results of 5 individual measurements on each balance are recorded below.

	Balance A	Balance B
	0.8888 g	1.3110 g
	0.9959 g	1.3109 g
	1.1182 g	1.3111 g
	1.0033 g	1.3110 g
	0.9938 g	1.3110 g
average mass =	1.0000 g	1.3110 g

Which statement best describes the results?

a) A: good precision, good accuracy. B: good precision, good accuracy.
b) A: good precision, good accuracy. B: poor precision, poor accuracy.
c) A: poor precision, good accuracy. B: good precision, good accuracy.
d) A: poor precision, poor accuracy. B: good precision, good accuracy.
e) A: poor precision, poor accuracy. B: good precision, poor accuracy.

Answer: d

1.8 Mathematics in Chemistry

42. The wavelength of light emitted from a green laser pointer is 5.32×10^2 nm. What is the wavelength in meters?
a) 5.32×10^{-9} m b) 5.32×10^{-7} m c) 5.32×10^{-5} m d) 5.32×10^{-3} m e) 5.32×10^{-1} m

Answer: b

43. A cylinder has a radius of 5.08 cm and a height of 125 cm. Calculate the volume of the cylinder in liters.
a) 0.313 L b) 0.625 L c) 10.1 L d) 2.53×10^3 L e) 1.01×10^4 L

Answer: c

44. The dimensions of a box are 12 inch by 11 inch by 5.5 inch. Calculate the volume of the box in cm^3. There is 2.54 cm per inch.
a) 44 cm^3 b) 2.8×10^2 cm^3 c) 1.8×10^3 cm^3 d) 4.7×10^3 cm^3 e) 1.2×10^4 cm^3

Answer: e

45. If the fuel efficiency of an automobile is 27 miles per gallon, what is its fuel efficiency in kilometers per liter? (1 km = 0.621 mile, 1.000 L = 1.057 quarts, 4 quarts = 1 gallon)
a) 11 km/L b) 14 km/L c) 17 km/L d) 63 km/L e) 180 km/L

Answer: a

46. A gallon (3.78 L) of latex paint can cover 385 ft^2 of the surface of a wall. What is the average thickness of one coat of paint (in micrometers)?
a) 0.781 μm b) 3.22 μm c) 9.82 μm d) 32.2 μm e) 106 μm

Answer: e

47. Barium sulfate ($BaSO_4$) is relatively insoluble in water. At 25 °C, only 244 mg will dissolve in 1.0 liter of water. What volume of water is needed to dissolve 2.0 g of barium sulfate?
a) 3.5×10^{-2} L b) 0.12 L c) 8.2 L d) 1.2×10^2 L e) 4.9×10^2 L

Answer: c

48. At 25 °C, the density of oxygen in air is 0.275 g/L. What volume is occupied by 25 kg of oxygen at 25 °C?
a) 1.1×10^{-5} L b) 6.9 L c) 2.8×10^2 L d) 6.9×10^3 L e) 9.1×10^4 L

Answer: e

49. Which method is correct for determining the liters of gas required to fill an automobile's 15 gallon tank? (1.000 L = 1.057 quarts, 4 quarts = 1 gallon)

a) 15 gallons $\left(\frac{1\text{ gallon}}{4\text{ quarts}}\right)\left(\frac{1.057\text{ quarts}}{1\text{ liter}}\right) =$

b) 15 gallons $\left(\frac{1\text{ gallon}}{4\text{ quarts}}\right)\left(\frac{1\text{ liter}}{1.057\text{ quarts}}\right) =$

c) 15 gallons $\left(\frac{4\text{ quarts}}{1\text{ gallon}}\right)\left(\frac{1\text{ liter}}{1.057\text{ quarts}}\right) =$

d) 15 gallons $\left(\frac{4\text{ quarts}}{1\text{ gallon}}\right)\left(\frac{1.057\text{ quarts}}{1\text{ liter}}\right) =$

e) none of the above

Answer: c

50. An electronic balance is used to determine that a sample has a mass of 25.7171 g. If the balance's precision is ±0.1 mg, what is the correct number of significant figures for this measurement?

a) 2 b) 3 c) 4 d) 5 e) 6

Answer: e

51. Convert 5.000×10^{-2} meters to millimeters and express the answer in standard notation using the correct number of significant figures.

a) 50 mm b) 50.0 mm c) 50.00 mm d) 500 nm e) 5000 mm

Answer: c

52. What is the correct answer to the following expression: $(72.61 - 68.59) \times 18.76$? Carry out the subtraction operation first.

a) 7×10^1 b) 75 c) 75.4 d) 74.542 e) 75.5415

Answer: c

53. What is the correct answer to the following expression: $(18 + 95) \times 0.077351$? Carry out the addition operation first.

a) 8.7 b) 8.74 c) 8.741 d) 8.7407 e) 8.74066

Answer: b

54. What is the correct answer to the following expression: $2.97 \times 10^{-5} + 6.13 \times 10^{-3}$?

a) $6.1 \text{ x } 10^{-3}$ b) 6.16×10^{-3} c) 6.160×10^{-3} d) 6.1597×10^{-3} e) 6.15970×10^{-5}

Answer: b

55. Express 0.38300 in exponential notation.
a) 3.8300×10^{-1} b) 3.83×10^{-1} c) 3.830×10^{-2} d) 3.83×10^{1} e) 3.8300×10^{1}

Answer: a

56. What mass of zinc (in kg) is in a 3.50 kg roll of tin foil that is 92.5% tin and 7.5% zinc?
a) 0.26 kg b) 0.47 kg c) 0.90 kg d) 2.1 kg e) 3.2 kg

Answer: a

57. The radius of a lithium atom is 152 pm. What is the volume of a lithium atom is cubic centimeters? The volume of a sphere is $(4/3)\pi r^3$.
a) 1.47×10^{-27} cm^3
b) 1.47×10^{-23} cm^3
c) 1.47×10^{-20} cm3
d) 1.47×10^{-14} cm^3
e) 1.47×10^{7} cm^3

Answer: b

Short Answer Questions

58. There are __________ cubic centimeters (cm^3) in a cubic meter (m^3).

Answer: 1×10^6

59. A mass of 10 g of sugar dissolves in water to form a ________ mixture (i.e., a mixture that is uniform throughout).

Answer: homogeneous

60. A(n) ________ is the smallest particle of an element that retains the characteristic chemical properties of that element.

Answer: atom

61. The ________ of a substance is defined as its mass per unit volume.

Answer: density

62. The separation of a mixture of soil and water is best accomplished by ________.

Answer: filtration

63. What is the difference between the accuracy of measurements and the precision of measurements?

Answer: Accuracy is an indicator of how close a measurement, or an average of a set of measurements, is to the correct value. Precision is an indicator of how close two or more measurements are to each other, or how much numerical spread is present in a set of measurements.

64. Significant figures allow us to estimate uncertainty in calculated values. In some circumstances, following significant figure rules can lead to estimates of uncertainty that are too high or low. This is the case for the mathematical expression below.

$$99 \times 1.02 = 11.98$$

Following the rules governing significant figures in multiplication, the answer can be rounded to 1.0×10^2. What is wrong with rounding this answer to two significant figures?

Answer: Assumed that absolute uncertainty in a value is equal to ±1 in the last place. An absolute uncertainty of ±1 in the value 99 is approximately a 1% uncertainty. Likewise, the absolute and relative uncertainty of 102 is ±1 and 1%, respectively. If both values have uncertainties of 1%, then their product ought to have an uncertainty close to 1%. But, by rounding to 2 significant figures, we have estimated the uncertainty to be 10%.

Diagnostic Quiz Questions

Goal #1: Classify Matter

1. Which of the following properties are consistent with liquids?
   1. The volume of a liquid is determined by the size of its container.
   2. A liquid has a rigid shape and fixed volume.
   3. A liquid has a fixed volume that varies little with temperature or pressure changes.

   a) 1 only
   b) 2 only
   c) 3 only
   d) 1, 2, and 3

   Answer: c

2. A pure substance composed of two or more elements is
   a) a heterogeneous mixture.
   b) a homogeneous mixture.
   c) an element.
   d) a chemical compound.

   Answer: d

3. A tablespoon of sugar will completely dissolve in a glass of pure water to give
   a) a chemical compound.
   b) a homogeneous mixture.
   c) a heterogeneous mixture.
   d) an element.

   Answer: b

4. Concrete is composed of sand, gravel, and calcium oxide. Concrete is best described as
   a) a heterogeneous mixture.
   b) a homogenous mixture.
   c) a pure substance.
   d) a chemical compound.

   Answer: a

Goal #2: Apply the kinetic-molecular theory to the properties of matter.

5. Which of the following statements concerning the kinetic-molecular theory are correct?
   1. Gas particles move faster when they are heated.
   2. Particles in a liquid are closely spaced, but are not confined to specific positions.
   3. Particles in a solid are closely spaced and are confined to specific positions.

   a) 1 only
   b) 2 only
   c) 3 only
   d) 1, 2, and 3

   Answer: d

Goal #3: Recognize elements, atoms, compounds, and molecules.

6. Nitrogen, N_2, is a(n) ________ that is composed of two nitrogen ________.
   a) atom, molecules
   b) compound, molecules
   c) element, atoms
   d) atom, elements

   Answer: c

7. What is the name of the element having the symbol S?
   a) silicon
   b) sulfur
   c) selenium
   d) scandium

   Answer: b

8. The symbol for cesium is _____.
   a) C
   b) Ce
   c) Cs
   d) Cm

   Answer: c

Goal #4: Identify physical and chemical properties and changes

9. Which of the following are chemical properties of iodine?
   1. Iodine is a purple solid at 25 °C.
   2. Iodine reacts with sodium metal to form sodium iodide.
   3. The density of iodine is 4.93 g/cm^3.

   a) 1 only
   b) 2 only
   c) 3 only
   d) 1 and 3

   Answer: b

10. Which of the following are physical properties of potassium?
    1. Potassium reacts with water, producing hydrogen gas and aqueous potassium hydroxide.
    2. Potassium conducts electricity.
    3. Potassium is malleable at room temperature.

    a) 1 only
    b) 2 only
    c) 3 only
    d) 2 and 3

    Answer: d

11. You can identify a metal by carefully determining its density. A 23.1 g piece of an unknown metal is 1.23 cm long, 2.11 cm wide, and 1.00 cm thick. What is the identity of the element?
    a) nickel, 8.90 g/cm^3
    b) aluminum, 2.70 g/cm^3
    c) zirconium, 6.51 g/cm^3
    d) chromium 7.20 g/cm^3

    Answer: a

12. The density of liquid mercury is 13.5 g/cm^3. What mass of mercury (in kg) is required to fill a hollow cylinder having an inner diameter of 2.00 cm to a height of 25.0 cm?
   a) 1.06 kg
   b) 4.24 kg
   c) 0.171 kg
   d) 1.71×10^{-4} kg

   Answer: a

13. Which one of the following lists contains only extensive properties?
   a) melting point, density and color
   b) electrical conductivity and mass
   c) density, boiling point and volume
   d) volume and mass

   Answer: d

Goal #5: Use metric units correctly

14. Thermostats are often set to 22 °C. What is this temperature in kelvin?
   a) 251 K
   b) 284 K
   c) 295 K
   d) 321 K

   Answer: c

15. Helium boils at 4.3 K. What is this temperature in Celsius?
   a) 268.9 °C
   b) 277.4 °C
   c) -277.4 °C
   d) -268.9 °C

   Answer: d

16. The radius of a carbon atom is 7.7×10^{-11} m. What is the radius in picometers?
   a) 7.7 pm
   b) 77 pm
   c) 7.7×10^2 pm
   d) 7.7×10^3 pm

   Answer: b

17. A typical volumetric flask holds a volume of 0.250 L. What is this volume in cubic centimeters?
    a) $0.25\ cm^3$
    b) $2.50\ cm^3$
    c) $2.50 \times 10^2\ cm^3$
    d) $2.50 \times 10^3\ cm^3$

Answer: c

Goal #6: Understand and use the mathematics of chemistry

18. A student determines the density of a bar of silver by measuring its dimensions (2.00 cm by 1.15 cm by 1.00 cm) and determining its mass (25.3 g). If the true density of silver is 10.5 g/cm^3, what is the percent error in the student's measurement?
    a) 1%
    b) 3%
    c) 5%
    d) 10%

Answer: c

19. You and your lab partner are asked to determine the mass of a bar of silver. You use an analytical balance that measures mass to four decimal places (Method A). Your partner uses a top-loader balance that measures mass to two decimal places (Method B). The results are tabulated below.

	Method A (g)	Method B (g)
Measurement #1	3.3682	3.41
Measurement #2	3.3684	3.71
Measurement #3	3.3682	3.35
Measurement #4	3.3681	3.92
Average mass	3.3682	3.60
Percent Error	6.495%	0.1%

The actual mass of the silver bar is 3.6022 g. Which statements best describe the results?
    a) Method A has good precision and poor accuracy. Method B has poor precision and good accuracy.
    b) Method A has poor precision and good accuracy. Method B has good precision and poor accuracy.
    c) Method A has poor precision and poor accuracy. Method B has good precision and good accuracy.
    d) Method A has good precision and good accuracy. Method B has good precision and poor accuracy.

Answer: a

CHAPTER 2: Atoms & Elements

2.1 Protons, Electrons, and Neutrons: Development of Atomic Structure

1. Which of the following postulates of Dalton's atomic theory are now known to be incorrect?
   1. Matter is made up of atoms.
   2. All atoms of a given element are identical.
   3. Atoms are indivisible and indestructible.

   a) 1 only b) 2 only c) 3 only d) 2 and 3 e) 1, 2, and 3

   Answer: d

2. Which of the following statements concerning Marie Curie are correct?
   1. Marie Curie isolated polonium and radium.
   2. Marie Curie discovered that uranium emits neutrons when it decays.
   3. Marie Curie suggested that certain atoms disintegrate, emitting unusual rays. She named the phenomenon radioactivity.

   a) 1 only b) 2 only c) 3 only d) 1 and 2 e) 1 and 3

   Answer: e

3. The Millikan oil drop experiment determined
   a) the mass of a drop of oil.
   b) the charge on an electron.
   c) the charge to mass ratio of a proton.
   d) the mass of a neutron.
   e) the number of electrons in a drop of oil.

   Answer: b

4. Which of the following particles will not be deflected by charged plates?
   a) γ particles
   b) β particles
   c) α particles
   d) protons
   e) α and β particles

   Answer: a

5. J. J. Thomson determined the charge to mass ratio of electrons by
   a) creating a charge on the amber by rubbing it with a cloth.
   b) deflecting cathode rays with magnetic and electric fields.
   c) exposing photographic plates to radioactive uranium ores.
   d) striking a beryllium target with particles emitted from radioactive polonium.
   e) bombarding gold foil with alpha particles.

   Answer: b

6. From the results of his gold foil experiment, Ernest Rutherford concluded that
   a) electrons have a charge of -1.602×10^{-19} C.
   b) atoms contain equal numbers of protons and electrons.
   c) uranium ores emit a form of radiation that affect photographic plates.
   d) alpha particles are helium nuclei.
   e) atoms are composed of a small, dense nucleus surrounded by a cloud of electrons.

   Answer: e

7. Alpha particles (α) are
   a) identical to electrons.
   b) high energy electromagnetic radiation.
   c) helium nuclei.
   d) identical to canal rays.
   e) positively charged electrons.

   Answer: c

8. Beta (β) particles are identical to _________.
   a) neutrons b) electrons c) helium nuclei d) light e) protons

   Answer: b

<u>2.2 Atomic Number and the Mass of an Atom</u>

9. Rank the subatomic particles from lowest to highest mass.
   a) electrons = protons < neutrons
   b) electrons < neutrons < protons
   c) electrons < protons < neutrons
   d) neutrons < electrons < protons
   e) electrons < protons = neutrons

   Answer: c

10. An atomic mass unit (u) is defined as
    a) the mass of one hydrogen-1 atom.
    b) 1/8 the mass of one oxygen-16 atom.
    c) 1/12 the mass of one carbon-12 atom.
    d) 1.99×10^{-23} g.
    e) the sum of the masses of one proton, one neutron, and one electron.

    Answer: c

11. Which of the following atoms contains the largest number of protons?
a) ^{231}Pa b) ^{227}Ac c) ^{232}Th d) ^{226}Ra e) ^{222}Rn

Answer: a

12. How many protons, neutrons, and electrons are in an oxygen-18 atom?
a) 6 protons, 8 neutrons, 4 electron
b) 6 protons, 10 neutrons, 8 electrons
c) 8 protons, 8 neutrons, 8 electrons
d) 8 protons, 10 neutrons, 8 electrons
e) 8 protons, 10 neutrons, 18 electrons

Answer: d

13. What is the atomic symbol for an element with 28 protons and 31 neutrons?
a) $^{59}_{28}Ni$ b) $^{31}_{28}Ni$ c) $^{59}_{31}Ga$ d) $^{28}_{59}Pr$ e) $^{31}_{3}Li$

Answer: a

14. What is the identity of $^{72}_{32}X$?
a) Hf b) Zr c) Rf d) Ge e) Cu

Answer: d

15. What is the atomic symbol for an element that has 30 neutrons and a mass number of 55?
a) Cs b) At c) Zn d) Hg e) Mn

Answer: e

16. How many neutrons are in cobalt-59?
a) 18 b) 27 c) 28 d) 32 e) 59

Answer: d

17. Which of the following atoms contains the largest number of protons?
a) ^{158}Gd b) ^{157}Tb c) ^{127}I d) ^{138}Ba e) ^{144}Nd

Answer: b

18. Which of the following atoms contains the largest number of neutrons?
a) $^{31}_{15}P$ b) $^{30}_{14}Si$ c) $^{37}_{17}Cl$ d) $^{32}_{16}S$ e) $^{34}_{16}S$

Answer: c

19. Which two of the atoms below have the same number of neutrons?

$${}^{64}_{28}\mathrm{Ni}\,,\ {}^{63}_{29}\mathrm{Cu}\,,\ {}^{64}_{30}\mathrm{Zn}\,,\ {}^{68}_{30}\mathrm{Zn}\ ?$$

a) ${}^{63}_{29}\mathrm{Cu}$ and ${}^{64}_{30}\mathrm{Zn}$
b) ${}^{64}_{28}\mathrm{Ni}$ and ${}^{63}_{29}\mathrm{Cu}$
c) ${}^{64}_{28}\mathrm{Ni}$ and ${}^{64}_{30}\mathrm{Zn}$
d) ${}^{63}_{29}\mathrm{Cu}$ and ${}^{68}_{30}\mathrm{Zn}$
e) ${}^{64}_{30}\mathrm{Zn}$ and ${}^{68}_{30}\mathrm{Zn}$

Answer: a

20. Give the mass number of a bromine atom with 46 neutrons.
a) 11 b) 35 c) 46 d) 80 e) 81

Answer: e

2.3 Isotopes

21. Isotopes of a given element have the same number of ________, but different numbers of ________ in their nucleus.
a) neutrons, electrons
b) electrons, protons
c) protons, electrons
d) neutrons, protons
e) protons, neutrons

Answer: e

22. All of the following statements are true EXCEPT
a) for any neutral element, the number of protons and electrons are equal.
b) electrons and protons have equal mass, but opposite charges.
c) the mass number is the sum of the number of protons and neutrons.
d) the atomic number equals the number of protons.
e) isotopes of an element have identical atomic numbers.

Answer: b

23. What experimental method is used to precisely determine the exact masses of isotopes and their relative abundances?
a) determination of density by water displacement
b) distillation
c) bombardment by alpha particles
d) mass spectrometry
e) electron microscopy

Answer: d

24. Which two of the following atoms are isotopes?

$$^{45}_{21}\mathrm{Sc},\ ^{48}_{22}\mathrm{Ti},\ ^{50}_{22}\mathrm{Ti},\ ^{50}_{23}\mathrm{V}$$

a) $^{45}_{21}\mathrm{Sc}$ and $^{50}_{23}\mathrm{V}$
b) $^{48}_{22}\mathrm{Ti}$ and $^{50}_{22}\mathrm{Ti}$
c) $^{50}_{22}\mathrm{Ti}$ and $^{50}_{23}\mathrm{V}$
d) $^{45}_{21}\mathrm{Sc}$ and $^{50}_{22}\mathrm{Ti}$
e) $^{48}_{22}\mathrm{Ti}$ and $^{50}_{23}\mathrm{V}$

Answer: b

2.4 Atomic Weight

25. Bromine has two naturally occurring isotopes. The average mass of bromine is 79.904 u. If 50.54% of bromine is found as bromine-79 (78.9183 u), what is the mass of the other isotope?
a) 79.82 u b) 79.97 u c) 80.91 u d) 81.93 u e) 82.91 u

Answer: c

26. An element consists of two isotopes. The abundance of one isotope is 95.72% and its atomic mass is 114.9041 u. The atomic mass of the second isotope is 112.9043 u. What is the average atomic mass of the element?
a) 113.9 u b) 113.0 u c) 113.9 u d) 114.8 u e) 115.1 u

Answer: d

27. Silver has an average atomic mass of 107.87 u. If 48.18% of Ag exists as Ag-109 (108.9047 u), what is the identity and the atomic mass of the other isotope?
a) Ag-106; 106.9 u
b) Ag-107; 106.9 u
c) Ag-107; 107.9 u
d) Ag-108; 107.9 u
e) Ag-108; 108.9 u

Answer: b

28. Lithium has two stable isotopes with masses of 6.01512 u and 7.01600 u. The average molar mass of Li is 6.941 u. What is the percent abundance of each isotope?
a) 7.49% Li-6 and 92.51% Li-7
b) 9.18% Li-6 and 90.813% Li-7
c) 12.2% Li-6 and 87.78% Li-7
d) 50.00% Li-6 and 50.00% Li-7
e) 62.99% Li-6 and 37.01% Li-7

Answer: a

2.5 Atoms and the Mole

29. You have 0.330 mole of each of the following elements: Be, B, Br, Ba, and Bi. Which sample has the largest mass?
a) Be b) B c) Br d) Ba e) Bi

Answer: e

30. You have 4.15 g of each of the following elements: Ca, Cu, Ce, Cs, Cf. Which sample contains the largest number of atoms?
a) Ca b) Cu c) Ce d) Cs e) Cf

Answer: a

31. Calculate the number of moles in 0.41 g titanium.
a) 9.1×10^{-4} mol b) 8.6×10^{-3} mol c) 0.051 mol d) 2.0×10^{1} mol e) 1.2×10^{2} mol

Answer: b

32. Calculate the number of moles in 39 g silicon.
a) 9.1×10^{-4} mol b) 0.72 mol c) 1.4 mol d) 11 mol e) 1.1×10^{3} mol

Answer: c

33. What is the mass of 5.1 mol P?
a) 6.3×10^{-3} g   b) 1.6×10^{-1} g   c) 6.1 g   d) 1.6×10^{2} g   e) 2.0×10^{2} g

Answer: d

34. What is the mass of 0.25 mol Zn?
a) 3.8×10^{-3} g   b) 0.061 g   c) 7.3 g   d) 16 g   e) 2.6×10^{2} g

Answer: d

35. A 3.6 g sample of lithium contains ________ atoms.
a) 5.2×10^{-1} b) 1.7×10^{23} c) 3.1×10^{23} d) 1.2×10^{24} e) 1.5×10^{25}

Answer: c

36. What is a correct method for calculating the mass of 1.9×10^{23} potassium atoms?

a) $6.02 \times 10^{23} \text{ atoms K}\left(\dfrac{1 \text{ mol K}}{1.9 \times 10^{23} \text{ atoms}}\right)\left(\dfrac{39.1 \text{ g K}}{1 \text{ mol K}}\right) =$

b) $1.9 \times 10^{23} \text{ atoms K}\left(\dfrac{6.02 \times 10^{23} \text{ atoms}}{1 \text{ mol K}}\right)\left(\dfrac{39.1 \text{ g K}}{1 \text{ mol K}}\right) =$

c) $1.9 \times 10^{23} \text{ atoms K}\left(\dfrac{6.02 \times 10^{23} \text{ atoms}}{1 \text{ mol K}}\right)\left(\dfrac{1 \text{ mol K}}{39.1 \text{ g K}}\right) =$

d) $1.9 \times 10^{23} \text{ atoms K}\left(\dfrac{1 \text{ mol K}}{6.02 \times 10^{23} \text{ atoms}}\right)\left(\dfrac{1 \text{ mol K}}{39.1 \text{ g K}}\right) =$

e) $1.9 \times 10^{23} \text{ atoms K}\left(\dfrac{1 \text{ mol K}}{6.02 \times 10^{23} \text{ atoms}}\right)\left(\dfrac{39.1 \text{ g K}}{1 \text{ mol K}}\right) =$

Answer: e

37. The molar mass of carbon is 12.01 g/mole. What is the mass of a 1.0×10^{9} carbon atoms?
a) 1.4×10^{-16} g
b) 2.0×10^{-14} g
c) 1.2×10^{-8} g
d) 8.3×10^{7} g
e) 7.2×10^{15} g

Answer: b

38. What is a correct method for calculating the number of atoms in 3.5×10^{-3} mole Cr?

a) $3.5 \times 10^{-3} \text{ mol Cr}\left(\dfrac{6.02 \times 10^{23} \text{ atoms Cr}}{1 \text{ mole Cr}}\right) =$

b) $3.5 \times 10^{-3} \text{ mol Cr}\left(\dfrac{1 \text{ mol Cr}}{52.0 \text{ g Cr}}\right)\left(\dfrac{6.02 \times 10^{23} \text{ atoms Cr}}{1 \text{ mole Cr}}\right) =$

c) $3.5 \times 10^{-3} \text{ mol Cr}\left(\dfrac{52.0 \text{ g Cr}}{1 \text{ mol Cr}}\right)\left(\dfrac{6.02 \times 10^{23} \text{ atoms Cr}}{1 \text{ mole Cr}}\right) =$

d) $3.5 \times 10^{-3} \text{ mol Cr}\left(\dfrac{1 \text{ mole Cr}}{6.02 \times 10^{23} \text{ atoms Cr}}\right) =$

e) $3.5 \times 10^{-3} \text{ mol Cr}\left(\dfrac{52.0 \text{ atoms Cr}}{1 \text{ mole Cr}}\right) =$

Answer: a

39. Pennies minted after 1983 are composed of 97% zinc and 3.0% copper and have a mass of 2.46 g. How many moles of copper are in a penny?
a) 0.0012 mol b) 0.014 mol c) 0.038 mol d) 0.040 mol e) 25 mol

Answer: a

40. The molar mass of silver is 107.9 g/mol. What is the mass of a single silver atom?
a) 1.539×10^{-26} g
b) 1.792×10^{-22} g
c) 1.079×10^{-20} g
d) 1.901×10^{-18} g
e) 5.581×10^{21} g

Answer: b

41. What mass of He contains the same number of atoms as 5.0 g Kr?
a) 0.24 g b) 0.80 g c) 1.2 g d) 5.0 g e) 1.0×10^{2} g

Answer: a

42. All of the following are possible quantities EXCEPT
a) a gold coin with a mass of 0.0031 kg.
b) aluminum foil that is 2.0×10^{-2} cm thick.
c) 2.0×10^{-22} grams He.
d) 1.0×10^{-24} mol C.
e) 22 atoms of Bi.

Answer: d

43. The density of lithium is 0.546 g/cm^3. What volume is occupied by 1.96×10^{23} atoms of lithium?
a) 0.0859 cm^3 b) 0.596 cm^3 c) 4.14 cm^3 d) 5.63 cm^3 e) 39.0 cm^3

Answer: c

44. A nail is coated with a 0.042 cm thick layer of zinc. The surface area of the nail is 9.17 cm^2. The density of zinc is 7.13 g/cm^3. How many zinc atoms are used in the coating?
a) 5.0×10^{20} atoms
b) 2.5×10^{22} atoms
c) 3.3×10^{22} atoms
d) 1..7 $\times 10^{23}$ atoms
e) 1.7×10^{24} atoms

Answer: b

45. An average sample of coal contains 3.0% sulfur by mass. Calculate the moles of sulfur present in 2.40×10^3 kg of coal.
a) 9.4×10^2 mol b) 2.2×10^3 mol c) 6.0×10^3 mol d) 7.2×10^4 mol e) 7.5×10^4 mol

Answer: b

46. The density of iron is 7.87 g/cm^3. Calculate the number of iron atoms present in a cube that has an edge of 3.00 cm.
a) 1.75×10^{21} atoms
b) 3.69×10^{22} atoms
c) 1.75×10^{23} atoms
d) 2.29×10^{24} atoms
e) 1.28×10^{26} atoms

Answer: d

2.6 The Periodic Table

47. The most abundant element in the universe is
a) hydrogen. b) carbon. c) helium. d) nitrogen. e) silicon.

Answer: a

48. Which gaseous element comprises over three quarters of the Earth's atmosphere?
a) nitrogen b) helium c) oxygen d) argon e) hydrogen

Answer: a

49. How many elements are in the fifth period of the periodic table?
a) 8 b) 14 c) 18 d) 32 e) 40

Answer: c

50. How many elements are in the first period of the periodic table?
a) 2 b) 7 c) 8 d) 10 e) 18

Answer: a

51. What element is in the fifth period in Group 3A?
a) As b) Nb c) P d) In e) Tl

Answer: d

52. What alkali metal is in the fourth period?
a) Ca b) K c) Na d) Sr e) Sc

Answer: b

53. What chalcogen is in the second period?
a) N b) O c) F d) Ne e) As

Answer: b

2.7 An Overview of the Elements, their Chemistry, and the Periodic Table

54. Which three elements are likely to have similar chemical and physical properties?
a) sodium, lithium, and potassium
b) sodium, magnesium, and aluminum
c) nitrogen, oxygen, and neon
d) nickel, copper, and zinc
e) uranium, plutonium, and americium

Answer: a

2.8 Essential Elements

55. Four elements constitute over 95% of the human body. These four elements are
a) iron, calcium, carbon, and sodium.
b) phosphorus, carbon, nitrogen, and sodium.
c) sodium, calcium, carbon, and iron.
d) magnesium, oxygen, nitrogen, and hydrogen.
e) oxygen, carbon, hydrogen, and nitrogen.

Answer: e

Short Answer Questions

56. Marie Curie was the first person to isolate the element ________, which she named after her native country.

Answer: polonium

57. Pure carbon can exist in several forms, or _______. Examples include diamond, graphite, and buckminsterfullerene.

Answer: allotropes

58. Group 2A elements are also known as ________ metals.

Answer: alkaline earth

59. Mendeleev's original periodic table placed the elements in order of increasing ________.

Answer: mass

60. The modern periodic table places the elements in order of increasing ________.

Answer: atomic number

61. Brimstone or burning stone are ancient names for the element ________.

Answer: sulfur

62. What are the names of four metalloids?

Answer: boron, silicon, germanium, arsenic (antimony, and tellurium)

Diagnostic Quiz Questions

Goal #1: Describe the atomic structure and define atomic number and mass number

1. J. J. Thomson determined the charge to mass ratio of electrons by
   a) suspending charged oil droplets between magnetic plates.
   b) exposing photographic plates to a radioactive uranium ore.
   c) deflecting cathode rays with magnetic and electric fields.
   d) striking a beryllium target with particles emitted from radioactive polonium.

   Answer: c

2. Which of the following statements are correct?
   1. Electrons and protons have identical masses but opposite charges.
   2. Most of an atom's mass is concentrated in a small, positively charged, nucleus.
   3. Atoms have equal numbers of protons and neutrons.

   a) 1 only
   b) 2 only
   c) 3 only
   d) 1 and 2

   Answer: b

Goal #2: Understand the nature of isotopes and calculate atomic weight from isotope abundances and exact atomic masses.

3. Which of the following statements are correct?
   1. Atomic number equals number of protons plus neutrons.
   2. Mass number equals the number of neutrons.
   3. An atomic mass unit equals 1/12 the mass of a carbon-12 atom.

   a) 1 only
   b) 2 only
   c) 3 only
   d) 2 and 3

   Answer: c

4. How many protons, neutrons and electrons are in a fluorine-19 atom?
   a) 9 protons, 9 neutrons, 10 electron
   b) 9 protons, 10 neutrons, 9 electrons
   c) 10 protons, 9 neutrons, 9 electrons
   d) 10 protons, 9 neutrons, 10 electrons

   Answer: b

5. Which of the following atoms has the greatest number of protons?
   a) ^{12}C
   b) ^{15}O
   c) ^{14}C
   d) ^{15}N

   Answer: b

6. Which of the following atoms contains the greatest number of neutrons?
   a) $^{28}_{14}Si$
   b) $^{31}_{15}P$
   c) $^{36}_{16}S$
   d) $^{35}_{17}Cl$

   Answer: c

7. What is the identity of an atom that has 28 neutrons and a mass number of 52?
   a) Cr
   b) Te
   c) Ni
   d) Hg

   Answer: a

8. Silver has an average atomic mass of 107.9 u and is known to have only two naturally occurring isotopes. If 51.84% of Ag exists as Ag-107 (106.9051 u), what is the identity and the atomic mass of the other isotope?
   a) Ag-110; 110.1 u
   b) Ag-110; 109.9 u
   c) Ag-108; 107.9 u
   d) Ag-109; 109.0 u

   Answer: d

9. An element consists of two isotopes. The abundance of one isotope is 60.40% and its atomic mass is 68.9257 u. The atomic mass of the second isotope is 70.9249 u. What is the average atomic mass of the element?
   a) 69.72 u
   b) 69.93 u
   c) 70.13 u
   d) 139.9 u

   Answer: a

Goal #3: Understand the relationship between atomic weight, isotopic mass and isotopic abundance

10. You have 0.500 g of the following elements: He, Ne, Ar, and Kr. Which sample contains the largest number of atoms?
    a) He
    b) Ne
    c) Ar
    d) Kr

    Answer: a

11. What is the mass of 0.442 mol Fe?
    a) 7.91×10^{-3} g
    b) 24.7 g
    c) 4.05×10^{-2} g
    d) 126 g

    Answer: b

12. Calculate the moles of Na in a 4.15 mg sample.
    a) 1.81×10^{-4} mol
    b) 9.54×10^{-2} mol
    c) 1.05×10^{-2} mol
    d) 5.54×10^{3} mol

    Answer: a

13. The molar mass of cesium is 132.9 g/mol. What is the mass of a single Cs atom?
    a) 2.207×10^{-22} g
    b) 1.249×10^{-26} g
    c) 2.763×10^{-23} g
    d) 4.531×10^{21} g

    Answer: a

14. What is the mass of 5.8×10^{18} atoms of Ne?
    a) 9.6×10^{-6} g
    b) 1.9×10^{-4} g
    c) 4.8×10^{-7} g
    d) 5.1×10^{3} g

    Answer: b

15. The density of silver is 10.5 g/cm^3. What is the volume of a piece of Ag that contains 2.8×10^{22} atoms?
    a) 0.48 cm^3
    b) 53 cm^3
    c) 4.8×10^{-2} cm^3
    d) 2.1 cm^3

    Answer: a

Goal #4: Know the terminology of the periodic table

16. What alkaline earth metal is located in the fourth period?
    a) K
    b) Ca
    c) Ga
    d) Ge

    Answer: b

17. What halogen is located in the fourth period?
    a) Se
    b) Kr
    c) Cl
    d) Br

    Answer: d

18. Which chalcogen is located in the second period?
    a) P
    b) N
    c) O
    d) S

    Answer: c

19. Which grouping of elements is composed entirely of nonmetals?
    a) iodine, indium, and xenon
    b) aluminum, silicon, and phosphorus
    c) sulfur, neon, and bromine
    d) gallium, argon, and oxygen

    Answer: c

20. Which grouping of elements is composed entirely of metalloids?
    a) B, As, and Sb
    b) Si, P, and Ge
    c) As, Ge, and Pb
    d) In, Sn, and Ge

    Answer: a

21. The following lists of elements are all found in the human body. Which three elements are found in the highest concentrations?
    a) sodium, oxygen, and magnesium
    b) oxygen, carbon, and hydrogen
    c) selenium, oxygen, and potassium
    d) carbon, oxygen, and iron

    Answer: b

CHAPTER 3: Molecules & Compounds

3.1 Molecules, Compounds, and Formulas

1. The smallest unit into which a pure substance such as sugar or water can be divided, while still retaining its composition and chemical properties is a(n) ________.
   a) ion b) atom c) isotope d) mixture e) molecule

   Answer: e

2. The formula for oxalic acid, HO_2CCO_2H, is an example of a(n)
   a) ionic compound formula.
   b) empirical formula.
   c) structural formula.
   d) condensed formula.
   e) ball and stick model.

   Answer: d

3. $C_2H_4Cl_2$ is the formula for two possible molecules. $C_2H_4Cl_2$ is an example of a(n)
   a) molecular formula.
   b) empirical formula.
   c) condensed formula.
   d) ionic compound formula.
   e) ball and stick model.

   Answer: a

4. Chalcopyrite is a copper containing ore. This ore contains 1 iron atom and 2 sulfur atoms for each copper atom. What is the formula of chalcopyrite?
   a) CuFeS b) CuFe(2S) c) $CuFeS_2$ d) 2SCuFe e) $CuFeS^2$

   Answer: c

5. Acetylsalicylic acid, commonly known as aspirin, has 9 carbon atoms, 8 hydrogen atoms, and 4 oxygen atoms per molecule. What is the molecular formula of aspirin?
   a) $C_9He_8O_4$ b) $C_9H_8O_4$ c) $C_9H_8Ox_4$ d) $Ca_9H_8O_2$ e) $C_9(HO_4)_2$

   Answer: b

3.3 Ionic Compounds: Formulas, Names, and Properties

6. A nitride ion has ________ electrons.
   a) 2 b) 4 c) 7 d) 8 e) 10

   Answer: e

7. An aluminum ion has ________ electrons.
   a) 9 b) 10 c) 12 d) 13 e) 16

   Answer: b

8. Identify the ions present in $KHCO_3$.
   a) $KHCO_3$ is not ionic.
   b) KH^+, and CO_3^-
   c) K^+, H^+, C^{4+}, and O^{2-}
   d) KH^{2+} and CO_3^{2-}
   e) K^+ and HCO_3^-

   Answer: e

9. Identify the ions in $(NH_4)_2Cr_2O_7$.
   a) N^{3-}, H^+, Cr^{3+} and O^{2-}
   b) N^{3-}, H^-, Cr^{3+} and O^{2-}
   c) NH_4^+ and $Cr_2O_7^{2-}$
   d) NH_3 and $H_2Cr_2O_7$
   e) NH_4^+, Cr^{3+} and O^{2-}

   Answer: c

10. Which atom is most likely to form a 1- ion?
    a) I b) S c) Ag d) P e) Na

    Answer: a

11. Which atom is most likely to form a 3+ ion?
    a) Li b) Sr c) Kr d) Fe e) P

    Answer: d

12. What charge is likely on a monatomic silver cation?
    a) 1- b) 1+ c) 2+ d) 3+ e) 4+

    Answer: b

13. For a nonmetal in Group 5A of the periodic table, the most common monatomic ion will have a charge of ________.
    a) 3- b) 2- c) 1- d) 1+ e) 2+

    Answer: a

14. Which of the following formulas is not correct?
a) $Al_2(SO_4)_3$ b) $NaClO_3$ c) Ba_2O_3 d) $Mg(NO_3)_2$ e) KH_2PO_4

Answer: c

15. What is the correct formula for an ionic compound that contains aluminum ions and carbonate ions?
a) $AlCO_3$ b) $Al(CO_3)_2$ c) $Al(CO_3)_3$ d) $Al_2(CO_3)_3$ e) $Al_3(CO_3)_2$

Answer: d

16. What is the correct formula for an ionic compound that contains magnesium ions and fluoride ions?
a) Mg_2F b) MgF c) Mg_2F_2 d) MgF_2 Mg_2F_3

Answer: d

17. Sodium sulfate has the chemical formula Na_2SO_4. Based on this information, the formula for chromium(III) sulfate is ___________.
a) $Cr_3(SO_4)_2$ b) $Cr_2(SO_4)_3$ c) $Cr(SO_4)_2$ d) Cr_2SO_4 e) $CrSO_4$

Answer: b

18. What is the charge on the lead ion in PbS_2?
a) 4- b) 2- c) 0 d) 2+ e) 4+

Answer: e

19. What is the charge on the copper ion in Cu_2O?
a) 2- b) 1- c) 1+ d) 2+ e) 4+

Answer: c

20. What is the correct formula for sodium acetate?
a) Na_2CH_3O b) $NaCH_3CO_2$ c) $NaCH_3O$ d) $NaCH_2O_2$ e) $NaCHO$

Answer: b

21. What is the correct formula for barium nitrate?
a) $Ba(NO_3)_2$ b) BNO_2 c) $Ba(NO_2)_2$ d) BaN e) $BaNO_3$

Answer: a

22. What is the correct formula for potassium dihydrogen phosphate?
    a) $K_2H_2PO_4$ b) KH_2PO_4 c) KH_2PO_3 d) K_2H_2P e) KH_2P

    Answer: b

23. What is the correct formula for cobalt(III) oxide?
    a) CoO b) Co_3O c) Co_3O_2 d) Co_2O_3 e) CoO_3

    Answer: d

24. What is the correct formula for copper(II) chloride?
    a) Cu_2Cl b) CuCl c) $CuCl_2$ d) $CuCl_3$ e) Cu_2Cl_2

    Answer: c

25. What is the correct formula for vanadium(V) sulfide?
    a) V_2S_5 b) V_5S c) VS_5 d) $V_{10}S_5$ e) V_5S_2

    Answer: a

26. All of the following are named correctly EXCEPT
    a) $LiClO_4$; lithium perchlorate.
    b) $CaHPO_4$; calcium hydrogen phosphide.
    c) NaCN; sodium cyanide.
    d) $Mg(OH)_2$; magnesium hydroxide.
    e) $CaSO_3$; calcium sulfite.

    Answer: b

27. All of the following are named correctly EXCEPT
    a) $NiCl_2 \cdot 6H_2O$; nickel(II) chloride hexahydrate.
    b) MnO_2; manganese(IV) oxide.
    c) $TiBr_4$; titanium(IV) bromide.
    d) AuI; gold(I) iodide.
    e) $FeSO_4$; iron(I) sulfate.

    Answer: e

28. What is the correct name for NH_4ClO_4?
    a) ammonia hydrogen chlorate
    b) ammonia hydrogen perchlorate
    c) ammonium perchlorate
    d) ammonium hypochlorite
    e) ammonium hydrochloric acid

    Answer: c

29. What is the correct name for $AlBr_3$?
   a) aluminum tribromium
   b) aluminum tribromide
   c) aluminum(III) tribromide
   d) aluminum bromide
   e) bromine aluminide

   Answer: d

30. What is the correct name for $Cr(NO_3)_2$?
   a) chromium(II) nitrate
   b) chromium(II) dinitrate
   c) chromium dinitrate
   d) dinitrochromate
   e) chromium(II) nitride

   Answer: a

31. What is the formula of sodium bicarbonate, or baking soda?
   a) CH_2ONa b) $NaHCO_3$ c) N_2CO_3 d) Na_2CO_3 e) H_2CO_3

   Answer: b

32. All of the following statements concerning ionic compounds are correct EXCEPT
   a) as the ion charges increase, the attraction between the ions increases.
   b) ionic compounds form extended 3-dimensional networks called crystal lattices.
   c) although not ductile like metals, ionic compounds are often malleable.
   d) positive and negative ions are attracted to each other by electrostatic forces.
   e) ionic crystals tend to be rigid, and they cleave along planes.

   Answer: c

33. Predict which ionic compound has the highest melting point.
   a) NaBr b) $MgBr_2$ c) RbI d) AlN e) KCl

   Answer: d

3.4 Molecular Compounds: Formulas, Names and Properties

34. What is the correct name for SO_2?
 a) sulfur oxide
 b) sulfur dioxide
 c) sulfuric dioxide
 d) sulfuric acid
 e) sulfurous oxide

 Answer: b

35. What is the correct name for PF_5?
 a) phosphorus pentafluoride
 b) phosphorus pentafluorine
 c) phosphorus(V) fluoride
 d) phosphorus fluorate
 e) phosphorus fluorine

 Answer: a

36. What is the common name for N_2H_4?
 a) ammonium ion
 b) hydrazine
 c) nitroglycerin
 d) ammonia
 e) phosphine

 Answer: b

37. What is a common name for N_2O?
 a) nitrous oxide
 b) nitric oxide
 c) nitrogen(I) oxide
 d) hydrazine
 e) ammonia

 Answer: a

3.5 Formulas, Compounds, and the Mole

38. What is the molar mass of cobalt(II) iodide hexahydrate?
 a) 212.8 g/mol b) 293.9 g/mol c) 312.7 g/mol d) 420.8 g/mol e) 465.1 g/mol

 Answer: d

39. What is the molar mass of ammonium sulfate?
a) 110.0 g/mol b) 114.1 g/mol c) 132.1 g/mol d) 144.6 g/mol e) 168.9 g/mol

Answer: c

40. How many moles are in 35.7 g of anhydrous copper(II) sulfate?
a) 0.160 mol b) 0.224 mol c) 0.762 mol d) 4.47 mol e) 5.70×10^3 mol

Answer: b

41. How many moles are in 4.5 g of $AgNO_3$?
a) 1.3×10^{-3} mol b) 2.6×10^{-2} mol c) 0.61 mol d) 38 mol e) 7.6×10^2 mol

Answer: b

42. How many grams are in 0.0751 mol Br_2?
a) 4.70×10^{-4} g b) 0.0833 g c) 2.13 g d) 6.00 g e) 12.0 g

Answer: e

43. How many grams are in 0.50 mol iron(II) sulfide?
a) 0.057 g b) 8.2 g c) 18 g d) 44.0 g e) 1.8×10^2 g

Answer: d

44. What is a correct method to determine the number of oxygen atoms in 8.12 g of carbon dioxide?

a) $8.12\text{ g }CO_2\left(\frac{1\text{ mol }CO_2}{44.01\text{ g }CO_2}\right)\left(\frac{2\text{ mol O atoms}}{1\text{ mol }CO_2}\right)\left(\frac{6.022\times10^{23}\text{ O atoms}}{1\text{ mol O atoms}}\right)=$

b) $8.12\text{ g }CO_2\left(\frac{44.01\text{ g }CO_2}{1\text{ mol }CO_2}\right)\left(\frac{2\text{ mol O atoms}}{1\text{ mol }CO_2}\right)\left(\frac{6.022\times10^{23}\text{ O atoms}}{1\text{ mol O atoms}}\right)=$

c) $8.12\text{ g }CO_2\left(\frac{1\text{ mol }CO_2}{44.01\text{ g }CO_2}\right)\left(\frac{2\text{ mol O atoms}}{1\text{ mol }CO_2}\right)\left(\frac{1\text{ mol O atoms}}{6.022\times10^{23}\text{ O atoms}}\right)=$

d) $8.12\text{ g }CO_2\left(\frac{1\text{ mol }CO_2}{44.01\text{ g }CO_2}\right)\left(\frac{1\text{ mol O atoms}}{2\text{ mol }CO_2}\right)\left(\frac{6.022\times10^{23}\text{ O atoms}}{1\text{ mol O atoms}}\right)=$

e) $8.12\text{ g }CO_2\left(\frac{44.01\text{ g }CO_2}{1\text{ mol }CO_2}\right)\left(\frac{1\text{ mol }CO_2}{2\text{ mol O atoms}}\right)\left(\frac{1\text{ mol O atoms}}{6.022\times10^{23}\text{ O atoms}}\right)=$

Answer: a

45. The density of carbon tetrachloride is 1.59 g/mL. How many chlorine atoms are present in 55 mL of carbon tetrachloride?
a) 3.3×10^{23} atoms
b) 5.4×10^{23} atoms
c) 1.4×10^{24} atoms
d) 8.1×10^{27} atoms
e) 3.2×10^{28} atoms

Answer: c

46. How many oxygen atoms are in 1.50 mol of SO_3?
a) 7.71×10^{21} atoms
b) 1.12×10^{22} atoms
c) 3.01×10^{22} atoms
d) 9.03×10^{23} atoms
e) 2.71×10^{24} atoms

Answer: e

47. What is the mass of one carbon monoxide molecule?
a) 4.65×10^{-23} g
b) 7.31×10^{-23} g
c) 1.66×10^{-22} g
d) 4.22×10^{-22} g
e) 2.81×10^{-21} g

Answer: a

48. If 1.00 g of an unknown molecular compound contains 4.55×10^{21} molecules, what is its molar mass?
a) 44.0 g/mol b) 66.4 g/mol c) 72.1 g/mol d) 98.1 g/mol e) 132 g/mol

Answer: e

49. Which of the following quantities of compounds contains the largest total number of atoms?
a) 1.0 mole of H_3PO_4
b) 2.0 moles of H_2SO_3
c) 3.0 moles of $HClO_4$
d) 4.0 moles of H_2S
e) 5.0 moles of HBr

Answer: c

3.6 Describing Compound Formulas

50. What is the mass percent of chlorine in iron(II) chloride?
a) 14.0% b) 28.0% c) 44.1% d) 55.9% e) 66.7%

Answer: d

51. What is the mass percent of each element in dichloromethane, CH_2Cl_2?
a) 10.06% C, 60.24% H, 29.70% Cl
b) 20.00% C, 20.00% H, 60.00% Cl
c) 24.10% C, 3.11% H, 72.79% Cl
d) 33.87% C, 0.22% H, 65.91% Cl
e) 14.14% C, 2.37% H, 83.48% Cl

Answer: e

52. Nitrogen and oxygen form an extensive series of oxides with the general formula N_xO_y. What is the empirical formula for an oxide that contains 46.68% nitrogen?
a) N_2O b) NO c) NO_2 d) N_2O_3 e) N_2O_5

Answer: b

53. A molecule is found to contain 47.35% C, 10.60% H, and 42.05% O. What is the empirical formula for this molecule?
a) C_2H_6O b) $C_3H_4O_2$ c) $C_3H_8O_2$ d) $C_4H_6O_2$ e) $C_4H_8O_3$

Answer: c

54. Phosphorus, P, is combined with chlorine, Cl_2, to give a gaseous compound with the formula PCl_x. If you start with 2.097 g of P and isolate 9.298 g of PCl_x, what is the value of x?
a) 2 b) 3 c) 4 d) 5 e) 6

Answer: b

55. A ionic compound has the formual MBr_4. The mass of 0.2353 mol of the compound is 103.14 g. What is the identity of the metal?
a) Pb b) Ti c) Sn d) Hg e) Ba

Answer: c

56. In a chemical reaction, 1.000 g of sulfur (S_8) combines with 3.963 g of copper to give a pure compound. What is the empirical formula of this compound?
a) Cu_2S b) CuS c) CuS_2 d) Cu_3S_2 e) Cu_2S_3

Answer: a

3.7 Hydrated Salts

57. A 4.450 g sample of hydrated lithium iodide, $LiI{\cdot}xH_2O$, is dried in an oven. When the anhydrous salt is removed from the oven, its mass is 3.170 g. What is the value of x?
    a) 1   b) 2   c) 3   d) 4   e) 6

    Answer: c

58. A 2.000 g sample of $CoCl_2{\cdot}xH_2O$ is dried in an oven. When the anhydrous salt is removed from the oven, its mass is 1.565 g. What is the value of x?
    a) 1   b) 2   c) 3   d) 4   e) 6

    Answer: b

Short Answer Questions

59. In reactions, metals generally lose electrons to become cations, and nonmetals generally gain electrons to become ________.

    Answer: anions

60. Most cations are metal ions. An exception is ________, a polyatomic cation that contains nitrogen.

    Answer: NH_4^+ (or ammonium ion)

61. The molecular formula of laughing gas is ________.

    Answer: N_2O

62. The formula for potassium permanganate is ________.

    Answer: $KMnO_4$

63. Which group in the periodic table does not form monatomic ions commonly found in ionic compounds?

    Answer: Group 8A

64. A(n) ________ in an instrument that is capable of determining molecular masses.

    Answer: mass spectrometer

65. In which ionic compound, NaBr or KBr, is the force of attraction between anions and cations stronger?

Answer: The force of attraction is stronger for NaBr. The electrostatic attraction between anions and cations decreases as the separation of the ions increases. The potassium ion will be farther from the bromide ion due to its larger ionic radius.

66. The numerical quantity of a mole, 6.022×10^{23}, is defined as the number of atoms in a specific mass of an element. What is the mass and the identity of the element used to define one mole?

Answer: A mole is equal to the number of atoms in 12.00 grams of carbon-12.

Diagnostic Quiz Questions

Goal #1: Interpret, predict, and write formulas for ionic and molecular compounds

1. The formula for acetic acid, CH_3CO_2H, is an example of a(n)
   a) condensed formula.
   b) mathematical formula.
   c) structural formula.
   d) molecular formula.

   Answer: a

2. C_2H_6O is the formula for two possible molecules, ethanol and dimethyl ether. This type of formula is known as a(n)
   a) condensed formula.
   b) mathematical formula.
   c) structural formula.
   d) molecular formula.

   Answer: d

3. Which of the following statements are correct?
   1. Metals generally lose electrons to become cations.
   2. Nonmetals generally gain electrons to become anions.
   3. Group 2A metals form ions with a 2+ charge.

   a) 1 only
   b) 2 only
   c) 3 only
   d) 1,2, and 3

   Answer: d

4. What charge is most commonly observed for a calcium ion?
   a) 2-
   b) 1-
   c) 1+
   d) 2+

   Answer: d

5. What charge is most commonly observed for an oxide ion?
   a) 3-
   b) 2-
   c) 1-
   d) 1+

   Answer: b

6. What charge is most commonly observed for a nickel ion?
   a) 4+
   b) 2+
   c) 1+
   d) 2-

   Answer: b

7. What is the formula for magnesium oxide?
   a) MgO
   b) Mg_2O
   c) MgO_2
   d) Mg_2O_3

   Answer: a

8. What is the formula for aluminum chloride?
   a) $AlCl$
   b) $AlCl_2$
   c) $AlCl_3$
   d) Al_2Cl_3

   Answer: c

9. What is the formula for calcium fluoride?
   a) CaF
   b) CaF_2
   c) Ca_2F
   d) Ca_2F_3

   Answer: b

Goal #2: Name compounds

10. What is the formula for lead(IV) sulfate?
   a) PbS
   b) $Pb(SO_3)_2$
   c) $PbSO_4$
   d) $Pb(SO_4)_2$

   Answer: d

11. What is the name of $Mg(OH)_2$?
   a) magnesium oxide
   b) magnesium dihydroxide
   c) magnesium hydroxide
   d) magnesium(II) hydroxide

   Answer: c

12. What is the name of $KHCO_3$?
   a) hydrogen carbonate potassium ion
   b) potassium hydrogen carbon trioxide
   c) potassium bicarbonate
   d) potassium hydrocarbonate

   Answer: c

13. What is the name of $(NH_4)_2Cr_2O_7$?
   a) ammonium dichromium heptaoxide
   b) dichromate diammine
   c) diammonia dichromate
   d) ammonium dichromate

   Answer: d

14. What is the name of Cu_2S?
    a) copper sulfide
    b) copper(I) sulfide
    c) copper(II) sulfide
    d) dicopper sulfide

    Answer: b

15. What is the name of PCl_5?
    a) phosphorus chloride
    b) phosphorus pentachlorine
    c) phosphorus pentachloride
    d) phosphorus(V) chloride

    Answer: c

16. What is the molecular formula for silicon dioxide?
    a) SiO
    b) Si_2O
    c) SiO_2
    d) $(SiO)_2$

    Answer: c

17. What is the common name for NH_3?
    a) ammonia
    b) nitrogen trihydride
    c) trihydrogen nitride
    d) ammonium

    Answer: a

Goal #3: Understand the properties of ionic compounds

18. Which of the following statements are correct.
    1. The attractive forces between ions of opposite charges increase with increasing separation.
    2. The greater the charges on ions, the greater the attractive forces between ions.
    3. Ions combine to form small molecules, such as NaCl.

    a) 1 only
    b) 2 only
    c) 3 only
    d) 1 and 2

    Answer: b

Goal #4: Calculate and use molar mass

19. Calculate the moles present in 10.5 g $CaCO_3$.
   a) 0.105 mol
   b) 0.410 mol
   c) 9.53 mol
   d) 1.05×10^3 mol

   Answer: a

20. Determine the mass of 0.350 mol Na_3PO_4.
   a) 2.14×10^{-3} g
   b) 0.0174 g
   c) 57.4 g
   d) 163.9 g

   Answer: c

21. How many oxygen atoms are in 0.20 g CO_2?
   a) 2.4×10^{23} oxygen atoms
   b) 2.7×10^{21} oxygen atoms
   c) 5.5×10^{21} oxygen atoms
   d) 1.2×10^{23} oxygen atoms

   Answer: c

Goal #5: Calculate the percent composition for a compound and derive formulas from experimental data.

22. What is the mass percent of oxygen in acetic acid, CH_3CO_2H?
   a) 26.64%
   b) 46.71%
   c) 53.29%
   d) 73.36%

   Answer: c

23. Toluene is composed of 91.25% C and 8.75% H. Determine the empirical formula for toluene.
   a) CH
   b) CH_3
   c) C_4H_5
   d) C_7H_8

   Answer: d

24. Benzene, an organic solvent, has the empirical formula CH. If the molar mass of benzene is 78.11 g/mol, what is the molecular formula of benzene?
    a) C_4H_{30}
    b) C_5H_{18}
    c) C_6H_6
    d) C_7H_8

    Answer: c

25. You want to determine the value of *x* in hydrated iron(II) sulfate, $FeSO_4 \cdot x\ H_2O$. In the laboratory you weigh out 1.983 g of the hydrated salt. After thorough heating, 1.084 g of the anhydrous salt remains. What is the value of *x*?
    a) 2
    b) 3
    c) 6
    d) 7

    Answer: d

Chapter 4: Chemical Equations & Stoichiometry

4.2 Balancing Chemical Equations

1. The combustion of propane produces carbon dioxide and steam.

$$C_3H_8(g) + 5\,O_2(g) \rightarrow 3\,CO_2(g) + 4\,H_2O(g)$$

All of the following statements concerning this reaction are correct EXCEPT
a) three molecule of carbon dioxide are formed per one molecule of propane consumed.
b) five molecules of oxygen are consumed per one molecule of propane consumed.
c) four moles of steam are formed per five moles of oxygen consumed.
d) the combined mass of reactants consumed equals the mass of products formed.
e) three grams of carbon dioxide are formed per five grams of oxygen consumed.

Answer: e

2. Ammonia is prepared by reacting nitrogen and hydrogen gases at high temperature according to the unbalanced chemical equation below.

$$\_\_\ N_2(g) + \_\_\ H_2(g) \rightarrow \_\_\ NH_3(g)$$

What are the respective coefficients when the equation is balanced with the smallest whole numbers?
a) 1, 1, 1 b) 1, 3, 1 c) 1, 3, 2 d) 2, 1, 2 e) 2, 3, 2

Answer: c

3. When methanol undergoes complete combustion, the products are carbon dioxide and water.

$$\_\_\ CH_3OH(\ell) + \_\_\ O_2(g) \rightarrow \_\_\ CO_2(g) + \_\_\ H_2O(g)$$

What are the respective coefficients when the equation is balanced with the smallest whole numbers?
a) 1, 1, 1, 1 b) 1, 2, 1, 2 c) 2, 2, 2, 4 d) 2, 3, 2, 4 e) 2, 4, 6, 4

Answer: d

4. The reaction of calcium with elemental chlorine yields calcium chloride. Write a balanced chemical equation for this reaction.
a) $Ca(s) + Cl_2(g) \rightarrow CaCl(s) + Cl(g)$
b) $Ca(s) + Cl_2(g) \rightarrow CaCl_2(s)$
c) $Ca(s) + 2\,Cl_2(g) \rightarrow CaCl_4(s)$
d) $2\,Ca(s) + Cl_2(g) \rightarrow 2\,CaCl(s)$
e) $4\,Ca(s) + Cl_2(g) \rightarrow 2\,Ca_2Cl(s)$

Answer: b

5. Metals react with oxygen to give oxides with the general formula M_xO_y. Write a balanced chemical equation for the reaction of iron with oxygen to yield iron(III) oxide.
   a) $Fe(s) + O_2(g) \rightarrow FeO_2(s)$
   b) $2\ Fe(s) + 3\ O_2(g) \rightarrow Fe_2O_3(s)$
   c) $3\ Fe(s) + O_2(g) \rightarrow Fe_3O_2(s)$
   d) $4\ Fe(s) + 3\ O_2(g) \rightarrow 2\ Fe_2O_3(s)$
   e) $6\ Fe(s) + O_2(g) \rightarrow 2\ Fe_3O(s)$

   Answer: d

6. The products of the complete combustion of a hydrocarbon are carbon dioxide and water. Write a balanced chemical equation for the combustion of pentane, C_5H_{12}.
   a) $C_5H_{12}(g) + 8\ O_2(g) \rightarrow 5\ CO_2(g) + 6\ H_2O(g)$
   b) $C_5H_{12}(g) \rightarrow 5\ C(s) + 6\ H_2(g)$
   c) $C_5H_{12}(g) + 9\ O_2(g) \rightarrow 4\ CO_2(g) + 5\ H_2O(g)$
   d) $C_5H_{12}(g) + 11\ O_2(g) \rightarrow C_5O_{10}(g) + 6\ H_2O(g)$
   e) $C_5H_{12}(g) + 11\ O_2(g) \rightarrow 5\ CO_2(g) + 12\ H_2O(g)$

   Answer: a

7. The balanced equation for the complete combustion of benzene, C_6H_6, is
   a) $C_6H_6(\ell) \rightarrow 6\ C(s) + 3\ H_2(g)$
   b) $C_6H_6(\ell) + 12\ H_2O(\ell) \rightarrow 6\ CO_2(g) + 15\ H_2(g)$
   c) $C_6H_6(\ell) + O_2(g) \rightarrow CO_2(g) + H_2O(g)$
   d) $2\ C_6H_6(\ell) + 9\ O_2(g) \rightarrow 12\ CO\ (g) + 6\ H_2O(g)$
   e) $2\ C_6H_6(\ell) + 15\ O_2(g) \rightarrow 12\ CO_2(g) + 6\ H_2O(g)$

   Answer: e

4.3 Mass Relationships in Chemical Reactions: Stoichiometry

8. Aluminum reacts with oxygen to produce aluminum oxide.
   $$4\ Al(s) + 3\ O_2(g) \rightarrow 2\ Al_2O_3(s)$$
   If 5.0 moles of Al react with excess O_2, how many moles of Al_2O_3 can be formed?
   a) 1.0 mol b) 2.0 mol c) 2.5 mol d) 5.0 mol e) 10.0 mol

   Answer: c

9. Calcium reacts with fluorine gas to produce calcium fluoride. How many moles of Ca will react with 1.0 moles of F_2?
   a) 0.50 mol b) 1.0 mol c) 1.5 mol d) 2.0 mol e) 4.0 mol

   Answer: b

10. Iron reacts with hydrochloric acid to produce iron(II) chloride and hydrogen gas.

$$Fe(s) + 2\ HCl(aq) \rightarrow FeCl_2(aq) + H_2(g)$$

How many moles of HCl will react with 4.6 moles of Fe?

a) 2.3 mol b) 2.8 mol c) 4.6 mol d) 6.9 mol e) 9.2 mol

Answer: e

11. Copper reacts with nitric acid to produce copper(II) nitrate, nitrogen dioxide gas, and water.

$$Cu(s) + 4\ HNO_3(aq) \rightarrow Cu(NO_3)_2(aq) + 2\ NO_2(g) + 2\ H_2O(\ell)$$

If you have 0.500 moles of Cu,

a) you need at least 0.125 moles of HNO_3 to produce 0.500 moles of $Cu(NO_3)_2$.
b) you need at least 0.250 moles of HNO_3 to produce 0.500 moles of $Cu(NO_3)_2$.
c) you need at least 2.00 moles of HNO_3 to produce 0.500 moles of $Cu(NO_3)_2$.
d) you need at least 2.00 moles of HNO_3 to produce 1.00 moles of $Cu(NO_3)_2$.
e) you need at least 2.00 moles of HNO_3 to produce 2.00 moles of $Cu(NO_3)_2$.

Answer: c

12. The complete combustion of 0.40 moles of butane gas, C_4H_{10}, will

a) consume 0.40 mol O_2 and produce 0.20 mol CO_2.
b) consume 0.80 mol O_2 and produce 0.40 mol CO_2.
c) consume 1.3 mol O_2 and produce 0.80 mol CO_2.
d) consume 2.6 mol O_2 and produce 1.6 mol CO_2.
e) consume 5.2 mol O_2 and produce 3.2 mol CO_2.

Answer: d

13. Nitroglycerine decomposes violently according to the unbalanced chemical equation below. How many total moles of gases are produced from the decomposition of 1.00 mol $C_3H_5(NO_3)_3$?

$$C_3H_5(NO_3)_3(\ell) \rightarrow CO_2(g) + N_2(g) + H_2O(g) + O_2(g)$$

a) 4.00 mol b) 6.50 mol c) 7.25 mol d) 16.5 mol e) 29.0 mol

Answer: c

14. Dinitrogen trioxide, a blue solid, dissociates to form nitrogen monoxide and nitrogen dioxide gases. What mass of nitrogen dioxide is formed from the decomposition of 13.1 g of N_2O_3?

a) 5.17 g b) 7.93 g c) 6.55 g d) 12.8 g e) 21.6 g

Answer: b

15. The reaction of water and coal at a high temperature produces a mixture of hydrogen and carbon monoxide gases. This mixture is known as synthesis gas (or syngas). What mass of hydrogen gas can be formed from the reaction of 51.3 g of carbon with excess water?

$$C(s) + H_2O(g) \rightarrow H_2(g) + CO(g)$$

a) 4.31 g b) 8.61 g c) 17.2 g d) 1.20×10^2 g e) 306 g

Answer: b

16. What mass of carbon reacts completely with 17.8 grams of SiO_2 according to the following equation?

$$SiO_2(s) + 3\ C(s) \rightarrow SiC(s) + 2\ CO(g)$$

a) 1.19 g b) 3.56 g c) 7.12 g d) 10.7 g e) 53.4 g

Answer: d

17. Sodium carbonate reacts with hydrochloric acid as shown below in an **unbalanced** chemical equation. What mass of CO_2 is produced from the reaction of 2.94 g Na_2CO_3 with excess HCl?

$$Na_2CO_3(s) + HCl(aq) \rightarrow NaCl(aq) + CO_2(g) + H_2O(\ell)$$

a) 1.22 g b) 2.44 g c) 2.94 g d) 5.88 g e) 7.08 g

Answer: a

18. What mass of carbon dioxide is produced by the complete combustion of 5.50 grams of hexane, C_6H_{14}?

a) 0.468 g b) 2.81 g c) 3.06 g d) 16.9 g e) 33.7 g

Answer: d

19. The compound P_4S_3 is used in matches. It reacts with oxygen to produce P_4O_{10} and SO_2. The unbalanced chemical equation is shown below.

$$P_4S_3(s) + O_2(g) \rightarrow P_4O_{10}(s) + SO_2(g)$$

What mass of O_2 will react with 0.641 grams of P_4S_3?

a) 0.0466 g b) 0.0932 g c) 0.186 g d) 0.372 g e) 0.746 g

Answer: e

20. What mass of oxygen is consumed by the complete combustion of 23.0 grams of ethylene?

$$C_2H_4(g) + 3\ O_2(g) \rightarrow 2\ CO_2(g) + 2\ H_2O(g)$$

a) 8.75 g b) 26.2 g c) 60.5 g d) 69.0 g e) 78.7 g

Answer: e

21. What is a correct method for determining how many grams of oxygen react with 1.00 gram of pentane?

$$C_5H_{12}(g) + 8\,O_2(g) \rightarrow 5\,CO_2(g) + 6\,H_2O(g)$$

a) $1.00 \text{ g } C_5H_{12}\left(\frac{1 \text{ mol } C_5H_{12}}{72.15 \text{ g}}\right)\left(\frac{8 \text{ mol } O_2}{1 \text{ mol } C_5H_{12}}\right)\left(\frac{32.00 \text{ g } O_2}{1 \text{ mol } O_2}\right) =$

b) $1.00 \text{ g } C_5H_{12}\left(\frac{1 \text{ mol } C_5H_{12}}{72.15 \text{ g}}\right)\left(\frac{1 \text{ mol } O_2}{8 \text{ mol } C_5H_{12}}\right)\left(\frac{32.00 \text{ g } O_2}{1 \text{ mol } O_2}\right) =$

c) $1.00 \text{ g } C_5H_{12}\left(\frac{72.15 \text{ mol } C_5H_{12}}{1 \text{ g}}\right)\left(\frac{8 \text{ mol } O_2}{1 \text{ mol } C_5H_{12}}\right)\left(\frac{16.00 \text{ g } O_2}{1 \text{ mol } O_2}\right) =$

d) $1.00 \text{ g } C_5H_{12}\left(\frac{72.15 \text{ g}}{1 \text{ mol } C_5H_{12}}\right)\left(\frac{1 \text{ mol } C_5H_{12}}{8 \text{ mol } O_2}\right)\left(\frac{32.00 \text{ g } O_2}{1 \text{ mol } O_2}\right) =$

e) $1.00 \text{ g } C_5H_{12}\left(\frac{72.15 \text{ g}}{1 \text{ mol } C_5H_{12}}\right)\left(\frac{1 \text{ mol } C_5H_{12}}{8 \text{ mol } O_2}\right)\left(\frac{1 \text{ mol } O_2}{32.00 \text{ g } O_2}\right) =$

Answer: a

22. What is a correct method for determining how many moles of calcium oxide are produced in the reaction of 10.0 g calcium with excess oxygen?

$$2\,Ca(s) + O_2(g) \rightarrow 2\,CaO(s)$$

a) $10.0 \text{ g Ca}\left(\frac{40.08 \text{ mol Ca}}{1.00 \text{ g}}\right)\left(\frac{2 \text{ mol CaO}}{2 \text{ mol Ca}}\right)\left(\frac{1 \text{ g CaO}}{56.08 \text{ mol CaO}}\right) =$

b) $10.0 \text{ g Ca}\left(\frac{40.08 \text{ mol Ca}}{1.00 \text{ g}}\right)\left(\frac{1 \text{ mol CaO}}{2 \text{ mol Ca}}\right)\left(\frac{1 \text{ g CaO}}{56.08 \text{ mol CaO}}\right) =$

c) $10.0 \text{ g Ca}\left(\frac{1 \text{ g Ca}}{40.08 \text{ mol}}\right)\left(\frac{2 \text{ mol CaO}}{1 \text{ mol Ca}}\right)\left(\frac{56.08 \text{ g CaO}}{1 \text{ mol CaO}}\right) =$

d) $10.0 \text{ g Ca}\left(\frac{1 \text{ mol Ca}}{40.08 \text{ g}}\right)\left(\frac{1 \text{ mol CaO}}{2 \text{ mol Ca}}\right)\left(\frac{56.08 \text{ g CaO}}{1 \text{ mol CaO}}\right) =$

e) $10.0 \text{ g Ca}\left(\frac{1 \text{ mol Ca}}{40.08 \text{ g}}\right)\left(\frac{2 \text{ mol CaO}}{2 \text{ mol Ca}}\right)\left(\frac{56.08 \text{ g CaO}}{1 \text{ mol CaO}}\right) =$

Answer: e

4.4 Reactions in which One Reactant is Present in Limited Supply

23. If 0.250 moles of bromine and 0.600 moles of ammonia react according to the equation below, what is the maximum amount of ammonium bromide (in moles) produced?

$$3\ Br_2(\ell) + 8\ NH_3(g) \rightarrow 6\ NH_4Br(s) + N_2(g)$$

a) 0.250 mol b) 0.450 mol c) 0.500 mol d) 0.600 mol e) 0.800 mol

Answer: b

24. How many moles of potassium bromide can be produced from the reaction of 2.92 moles of potassium with 1.78 moles of bromine gas?

$$2\ K(s) + Br_2(g) \rightarrow 2\ KBr(s)$$

a) 1.46 mol b) 1.78 mol c) 2.92 mol d) 3.56 mol e) 5.84 mol

Answer: c

25. How many moles of $Fe_2O_3(s)$ can be produced from the reaction of 0.45 mol Fe(s) with 0.30 mol $O_2(g)$?

a) 0.15 mol b) 0.20 mol c) 0.23 mol d) 0.45 mol e) 0.75 mol

Answer: b

26. Nitric oxide is made from the oxidation of ammonia. How many moles of nitric oxide can be made from the reaction of 3.80 mol NH_3 with 5.15 mol O_2?

$$4\ NH_3(g) + 5\ O_2(g) \rightarrow 4\ NO(g) + 6\ H_2O(g)$$

a) 3.80 mol b) 4.12 mol c) 5.15 mol d) 6.44 mol e) 8.95 mol

Answer: a

27. Sulfur trioxide is made from the reaction of sulfur dioxide and oxygen. How many grams of SO_3 can be made from the reaction of 3.00 g SO_2 with 2.02 g O_2?

$$2\ SO_2(g) + O_2(g) \rightarrow 2\ SO_3(g)$$

a) 2.53 g b) 3.75 g c) 5.02 g d) 5.05 g e) 10.1 g

Answer: b

28. Magnesium reacts with iodine gas at high temperatures to form magnesium iodide. What mass of MgI_2 can be produced from the reaction of 4.44 g Mg and 13.4 g I_2?

a) 2.76 g b) 14.7 g c) 16.0 g d) 17.8 g e) 50.8 g

Answer: b

29. Aspirin is produced by the reaction of salicylic acid (M = 138.1 g/mol) and acetic anhydride (M = 102.1 g/mol).

$$C_7H_6O_3(s) + C_4H_6O_3(\ell) \rightarrow C_9H_8O_4(s) + C_2H_4O_2(\ell)$$

If you mix 5.00 grams of each reactant, how many grams of aspirin (M = 180.2 g/mol) can theoretically by obtained?
a) 2.83 g b) 3.83 g c) 6.52 g d) 8.82 g e) 10.0 g

Answer: c

30. The reaction of 10.0 g $H_2(g)$ with 10.0 g $O_2(g)$ yields 8.43 g $H_2O(g)$. What is the percent yield of this reaction?
a) 9.43% b) 27.3% c) 42.2% d) 66.8% e) 74.9%

Answer: e

4.5 Percent Yield

31. Aspirin is produced by the reaction of salicylic acid (M = 138.1 g/mol) and acetic anhydride (M = 102.1 g/mol).

$$C_7H_6O_3(s) + C_4H_6O_3(\ell) \rightarrow C_9H_8O_4(s) + C_2H_4O_2(\ell)$$

If 2.04 g of $C_9H_8O_4$ (M = 180.2 g/mol) is produced from the reaction of 3.00 g $C_7H_6O_3$ and 5.40 g $C_4H_6O_3$, what is the percent yield?
a) 14.2% b) 21.4% c) 24.3% d) 52.1% e) 68.0%

Answer: d

32. Under certain conditions the reaction of ammonia with excess oxgyen will produce a 38.2% yield of NO. How many grams of NH_3 must be reacted with excess oxygen to yield 17.5 g of NO?

$$4\ NH_3(g) + 5\ O_2(g) \rightarrow 4\ NO(g) + 6\ H_2O(g)$$

a) 3.79 g b) 9.93 g c) 26.0 g d) 32.2 g e) 45.8 g

Answer: c

33. Under certain conditions the formation of ammonia from nitrogen and hydrogen has a 7.82% yield. Under these conditions, how many grams of NH_3 will be produced from the reaction 25.0 g N_2 with 2.00 g H_2?

$$N_2(g) + 3\ H_2(g) \rightarrow 2\ NH_3(g)$$

a) 0.881 g b) 2.37 g c) 3.12 g d) 11.3 g e) 30.4 g

Answer: a

34. Disulfur dichloride is produced by reacting chlorine gas with molten sulfur.

$$S_8(\ell) + 4\ Cl_2(g) \rightarrow 4\ S_2Cl_2(g)$$

What is the percent yield if 6.25 g S_2Cl_2 is isolated from the reaction of 10.0 g S_8 and 7.50 g Cl_2?

a) 19.9% b) 29.7% c) 31.4% d) 35.7% e) 43.8%

Answer: e

4.6 Chemical Equations and Chemical Analysis

35. A mass of 2.052 g of a metal carbonate, MCO_3, is heated to give the metal oxide and 0.4576 g CO_2.

$$MCO_3(s) \rightarrow MO(s) + CO_2(g)$$

What is the identity of the metal?

a) Cu b) Mg c) Ca d) Ba e) Co

Answer: d

36. A mixture of $MgCO_3$ and $MgCO_3 \cdot 3H_2O$ has a mass of 3.057 g. After heating to drive off all the water the mass is 2.790 g. What is the mass percent of $MgCO_3 \cdot 3H_2O$ in the mixture?

a) 22.4% b) 43.2% c) 67.0% d) 77.6% e) 91.3%

Answer: a

37. When 1.500 g of a metal oxide is heated, the compound decomposes in the reaction shown below.

$$2\ M_2O(s) \rightarrow 4\ M(s) + O_2(g)$$

If 0.1035 g O_2 is produced, what is the identity of the metal?

a) Ag b) Hg c) In d) K e) Na

Answer: a

38. A 2.000 g mixture of $CaCO_3$ and CaO is heated. The $CaCO_3$ decomposes to CaO and CO_2. What was the mass percentage of $CaCO_3$ in the mixture if the mass after heating is 1.809 g?

a) 17.8% b) 21.7% c) 43.4% d) 78.3% e) 95.5%

Answer: b

39. A 4.65 g sample of an oxide of copper, when heated in a stream of hydrogen, produces Cu(s) and 0.585 g H_2O. What is formula for the compound?

a) Cu_2O b) CuO c) CuO_2 d) Cu_2O_3 e) Cu_3O_4

Answer: a

40. The reaction of 1.250 g K with O_2 forms 2.273 g of an oxide of potassium that is used in self-contained breathing devices. Determine the formula for this oxide of potassium.
a) K_2O b) KO c) K_2O_3 d) KO_2 e) KO_4

Answer: d

41. A 4.236 g sample of a hydrocarbon is combusted to give 3.810 g of H_2O and 13.96 g of CO_2. What is the empirical formula of the compound?
a) C_2H_3 b) C_3H_2 c) C_3H_4 d) C_4H_3 e) C_5H_{12}

Answer: c

42. The combustion of 0.1703 mole of a hydrocarbon produces 12.27 g H_2O and 22.48 g CO_2. What is the molar mass of the hydrocarbon?
a) 16.04 g/mol b) 30.07 g/mol c) 44.08 g/mol d) 72.15 g/mol e) 92.13 g/mol

Answer: c

43. Polyethylene is a polymer consisting of only carbon and hydrogen. If 3.000 g of the polymer is burned in oxygen it produces 3.855 g H_2O and 9.414 g CO_2. What is the empirical formula for polyethylene?
a) CH_2 b) C_2H c) C_2H_3 d) C_5H_8 e) C_7H_8

Answer: a

44. Soft drink bottles are made of polyethylene terephthalate (PET), a polymer composed of carbon, hydrogen, and oxygen. If 4.375 g PET is burned in oxygen it produces 1.515 g H_2O and 9.251 g CO_2. What is the empirical formula for PET?
a) CHO b) CH_7O_5 c) C_5H_7O d) $C_8H_{10}O$ e) $C_{10}H_8O_5$

Answer: e

45. Malic acid is a compound that contains only carbon, hydrogen, and oxygen. Combustion of 2.000 g malic acid produces 0.8062 g H_2O and 2.626 g CO_2. What is the empirical formula of malic acid?
a) CHO b) C_2H_3O c) $C_3H_4O_3$ d) $C_4H_6O_5$ e) C_5H_5O

Answer: d

46. Cyclohexene, a hydrocarbon, has a molar mass of 82.1 g/mole. If the combustion of 0.8300 g cyclohexene produces 0.9102 g H_2O and 2.668 g CO_2, what is the molecular formula of this compound?
a) C_3H_5 b) C_5H_{22} c) C_6H_{10} d) C_6H_{14} e) C_7H_{10}

Answer: c

Short Answer Questions

47. The numbers preceding the formulas in chemical equations are referred to as the ________ coefficients.

Answer: stoichiometric

48. The combustion of propane involves the reaction of C_3H_8 with ________.

Answer: oxygen

49. The percent yield of a chemical reaction is calculated by dividing the ________ yield by the theoretical yield and multiplying this ratio by 100%.

Answer: actual

50. A French scientist named __________ introduced the law of conservation of matter.

Answer: Antoine Lavoisier

51. In the reaction below, how many grams of PF_5 can be produced from the reaction of 1.00 g P_4 with 1.00 g F_2?

$$P_4(s) + 10\ F_2(g) \rightarrow 4\ PF_5(g)$$

Answer: 1.33 g

52. The reaction of 2.30 g N_2 with excess H_2 produces 2.00 g NH_3. The percent yield of this reaction is ________.

Answer: 71.5%

Diagnostic Quiz Questions

Goal #1: Balance equations for simple chemical reactions

1. Balance the following chemical equation:

$$Na_2S(s) + HCl(aq) \rightarrow H_2S(g) + NaCl(aq)$$

a) $Na_2S(s) + HCl(aq) \rightarrow H_2S(g) + NaCl(aq)$
b) $Na_2S(s) + 2\ HCl(aq) \rightarrow H_2S(g) + 2\ NaCl(aq)$
c) $NaS(s) + HCl(aq) \rightarrow HS(g) + NaCl(aq)$
d) $Na_2S(s) + H_2Cl(aq) \rightarrow H_2S(g) + Na_2Cl(aq)$

Answer: b

2. Balance the following chemical equation:

$$CH_3OH(\ell) + O_2(g) \rightarrow CO_2(g) + H_2O(g)$$

a) $CH_3OH(\ell) + O_2(g) \rightarrow CO_2(g) + H_2O(g)$

b) $CH_3OH(\ell) + 2\ O_2(g) \rightarrow CO_2(g) + 2\ H_2O(g)$

c) $2\ CH_3OH(\ell) + 3\ O_2(g) \rightarrow 2\ CO_2(g) + 4\ H_2O(g)$

d) $CH_3OH(\ell) + O_3(g) \rightarrow CO_2(g) + 2\ H_2O(g)$

Answer: c

3. Potassium metal reacts with water to form aqueous potassium hydroxide and hydrogen gas. Write a balanced chemical equation for this reaction.

a) $K(s) + H_2O(\ell) \rightarrow KOH(aq) + H_2(g)$

b) $2\ K(s) + 2\ H_2O(\ell) \rightarrow 2\ KOH(aq) + H_2(g)$

c) $2\ K(s) + H_2O(\ell) \rightarrow 2\ KOH(aq) + H_2(g)$

d) $4\ K(s) + 2\ H_2O(\ell) \rightarrow 4\ KOH(aq) + H_2(g)$

Answer: b

Goal #2: Perform stoichiometric calculations using balanced chemical equations

4. What amount of carbon dioxide (in moles) is produced from the reaction of 2.24 moles of ethanol with excess oxygen?

$$C_2H_5OH(aq) + 3\ O_2(g) \rightarrow 2\ CO_2(g) + 3\ H_2O(g)$$

a) 1.12 mol
b) 2.24 mol
c) 4.48 mol
d) 6.72 mol

Answer: c

5. What amount of bromine (in moles) reacts with 4.0 mol of aluminum to produce $AlBr_3$?

$$2\ Al(s) + 3\ Br_2(g) \rightarrow 2\ AlBr_3(s)$$

a) 2.0 mol
b) 4.0 mol
c) 6.0 mol
d) 12 mol

Answer: c

6. What amount of sodium (in moles) reacts with 1.5 moles of fluorine gas? The unbalanced chemical equation is given below.

   $Na(s) + F_2(g) \rightarrow NaF(s)$

   a) 0.75 mol
   b) 1.5 mol
   c) 3.0 mol
   d) 6.0 mol

   Answer: c

7. What mass of magnesium iodide is produced from the reaction of 0.555 g Mg with excess iodine?

   $Mg(s) + I_2(s) \rightarrow MgI_2(s)$

   a) 0.0228 g
   b) 0.0485 g
   c) 0.555 g
   d) 6.35 g

   Answer: d

8. What mass of sodium will react with 1.23 g of chlorine gas to produce sodium chloride?

   $2\ Na(s) + Cl_2(g) \rightarrow 2\ NaCl(s)$

   a) 0.199 g
   b) 0.399 g
   c) 0.798 g
   d) 2.46 g

   Answer: c

9. The unbalanced equation for the process of reducing iron ore to the metal is given below. What mass of iron metal is produced from the reduction of 10.0 g Fe_2O_3?

   $Fe_2O_3(s) + CO(g) \rightarrow Fe(s) + CO_2(g)$

   a) 3.50 g
   b) 6.99 g
   c) 24.6 g
   d) 57.2 g

   Answer: b

Goal #3: Understand the meaning of a limiting chemical reactant

10. What amount of ammonia (in moles) is produced by the reaction of 4.00 mol H_2 with 3.00 mol N_2?

$$3\ H_2(g) + N_2(g) \rightarrow 2\ NH_3(g)$$

a) 2.67 mol
b) 2.00 mol
c) 6.00 mol
d) 7.00 mol

Answer: a

11. What mass of H_2S is produced by the reaction of 1.55 g Na_2S with 2.11 g HCl?

$$Na_2S(s) + 2\ HCl(g) \rightarrow 2\ NaCl(s) + H_2S(g)$$

a) 0.677 g
b) 0.986 g
c) 1.97 g
d) 3.66 g

Answer: a

12. The combustion of methane produces carbon dioxide and water. If 5.00 g CH_4 reacts with 25.0 g O_2, what mass of water is produced?

a) 2.82 g
b) 5.62 g
c) 11.2 g
d) 14.1 g

Answer: c

Goal #4: Calculate the theoretical and percent yield of a chemical reaction

13. The decomposition of 3.32 g $CaCO_3$ yields 1.24 g CaO. What is the percent yield of this reaction?

$$CaCO_3(s) \rightarrow CaO(s) + CO_2(g)$$

a) 1.86%
b) 37.3%
c) 62.7%
d) 66.7%

Answer: d

14. The reaction of 23.1 g NH_3 and 18.3 g O_2 produces 4.10 g NO. What is the percent yield of this reaction?

$$4\ NH_3(g)\ +\ 5\ O_2(g)\ \rightarrow\ 4\ NO(g)\ +\ 6\ H_2O(g)$$

a) 9.90%
b) 10.1%
c) 22.4%
d) 29.9%

Answer: d

Goal #5: Use stoichiometry to analyze a mixture of compounds or to determine the formula of a compound

15. A mixture of $CaCO_3$ and CaO has a mass of 2.500 g. After heating, the mass of the sample is reduced to 2.016 g. What is the mass percent of $CaCO_3$ in the mixture?

$$CaCO_3(s)\ \rightarrow\ CaO(s)\ +\ CO_2(g)$$

a) 19.4%
b) 44.0%
c) 54.6%
d) 80.6%

Answer: b

16. Cyclooctene is a hydrocarbon containing only C and H atoms. When burned in oxygen, 1.000 g of cyclooctene produces 3.195 g CO_2 and 1.144 g H_2O. What is the empirical formula of cyclooctene?
a) CH_2
b) C_2H_5
c) C_3H_7
d) C_4H_7

Answer: d

17. Benzoic acid contains C, H, and O atoms. When 1.500 g benzoic acid is burned in oxygen, 3.784 g CO_2 and 0.6639 g H_2O are produced. What is the empirical formula of benzoic acid?
a) $C_3H_5O_2$
b) C_4H_6O
c) $C_5H_6O_2$
d) $C_7H_6O_2$

Answer: d

CHAPTER 5: Reactions in Aqueous Solution

5.1. Properties of Compounds in Aqueous Solution

1. Which one of the following compounds is a nonelectrolyte when dissolved in water?
   a) CH_3CH_2OH b) $ZnBr_2$ c) LiCl d) $Ca(NO_3)_2$ e) KOH

   Answer: a

2. Which one of the following compounds is a weak electrolyte when dissolved in water?
   a) sodium hydroxide (NaOH)
   b) sodium acetate ($NaCH_3CO_2$)
   c) acetic acid (CH_3CO_2H)
   d) glucose ($C_6H_{12}O_6$)
   e) calcium chloride ($CaCl_2$)

   Answer: c

3. If 0.10 mole of each of the following compounds is dissolved to a volume of 1.0 L in water, which one will have the highest concentration of dissolved ions?
   a) HF b) NaOH c) $MgSO_4$ d) $Al(NO_3)_3$ e) CaI_2

   Answer: d

4. Which of the following statements are correct?
   1. Salts of phosphate ion are insoluble in water, with the exception of those with NH_4^+ and alkali metal cations.
   2. All salts of nitrate ion are soluble in water.
   3. Most salts of ammonium ion are soluble in water.

   a) 1 only b) 2 only c) 3 only d) 1 and 2 e) 1, 2, and 3

   Answer: e

5. Which of the following statements are correct?
   1. All salts of carbonate ion are soluble in water.
   2. Most salts of chloride ion are soluble in water.
   3. All salts of hydroxide ion are soluble in water.

   a) 1 only b) 2 only c) 3 only d) 1 and 2 e) 1, 2, and 3

   Answer: b

6. Which of the following compounds will be soluble in water: $LiOH$, $Mg(OH)_2$, $Cu(OH)_2$, and $Fe(OH)_3$?
   a) $LiOH$ only
   b) $LiOH$ and $Mg(OH)_2$
   c) $Cu(OH)_2$ and $Fe(OH)_3$
   d) $Mg(OH)_2$ and $Cu(OH)_2$
   e) $LiOH$, $Mg(OH)_2$, and $Fe(OH)_3$

   Answer: a

7. Which of the following compounds will be soluble in water: KNO_3, $Ca_3(PO_4)_2$, $CuCl_2$, and Fe_2S_3?
   a) KNO_3 and $Ca_3(PO_4)_2$
   b) KNO_3 and $CuCl_2$
   c) $Ca_3(PO_4)_2$ and Fe_2S_3
   d) $CuCl_2$ and Fe_2S_3
   e) KNO_3, $Ca_3(PO_4)_2$, and Fe_2S_3

   Answer: b

5.2 Precipitation Reactions

8. A precipitate will form when aqueous $AgNO_3$ is added to an aqueous solution of ________.
   a) KNO_3 b) $Cu(NO_3)_2$ c) KI d) $Ca(ClO_4)_2$ e) $NaNO_3$

   Answer: c

9. A precipitate will form when aqueous iron(II) bromide is added to ________.
   a) Na_2SO_4 b) $Cu(NO_3)_2$ c) KI d) $Ba(OH)_2$ e) $NaCl$

   Answer: d

10. A white solid is either NaI or $NaNO_3$. If an aqueous solution of the solid is prepared, the addition of which reagent will allow you to distinguish between the two compounds?
    a) $NaOH$ b) HCl c) K_3PO_4 d) NH_4Br e) $Pb(NO_3)_2$

    Answer: e

11. A white solid is either $Pb(NO_3)_2$ or $Zn(NO_3)_2$. If an aqueous solution is prepared, which reagent will allow you to distinguish between the two compounds?
    a) KBr b) HNO_3 c) CH_3CO_2H d) NH_4ClO_4 e) $LiNO_3$

    Answer: a

12. What is the net ionic equation for the reaction of aqueous lead(II) nitrate with aqueous sodium bromide?
a) $Pb(NO_3)_2(aq) + 2\ NaBr(aq) \rightarrow PbBr_2(aq) + 2\ NaNO_3(s)$
b) $Na^+(aq) + NO_3^-(aq) \rightarrow NaNO_3(s)$
c) $Pb^{2+}(aq) + 2\ Br^-(aq) \rightarrow PbBr_2(s)$
d) $Pb^{2+}(aq) + 2\ Na^+(aq) \rightarrow PbNa_2(s)$
e) $Pb(NO_3)_2(aq) + 2\ NaBr(aq) \rightarrow PbBr_2(s) + 2\ NaNO_3(aq)$

Answer: c

13. What is the net ionic equation for the reaction of aqueous sodium carbonate with aqueous iron(III) chloride?
a) $3\ Na^+(aq) + Fe^{3+}(aq) \rightarrow Na_3Fe(s)$
b) $Na^+(aq) + Cl^-(aq) \rightarrow NaCl(s)$
c) $3\ Na_2CO_3(aq) + 2\ FeCl_3(aq) \rightarrow 6\ NaCl(aq) + Fe_2(CO_3)_3(aq)$
d) $2\ Fe^{3+}(aq) + 3\ CO_3^{2-}(aq) \rightarrow Fe_2(CO_3)_3(s)$
e) $CO_3^{2-}(aq) + Cl^-(aq) \rightarrow CO_2(g) + Cl_2O(s)$

Answer: d

5.3 Acids and Bases

14. Which of the following compounds is a weak acid?
a) H_3PO_4 b) HI c) HBr d) $HClO_4$ e) H_2SO_4

Answer: a

15. Which of the following compounds is a weak base?
a) LiOH b) KCl c) CH_3CO_2H d) $Sr(OH)_2$ e) NH_3

Answer: e

16. Which of the following species are strong acids: HNO_3, HBr, H_3PO_4,CH_3CO_2H, and HSO_4^-?
a) HNO_3 and HBr
b) HNO_3 and H_3PO_4
c) HBr and HSO_4^-
d) H_3PO_4, CH_3CO_2H, and HSO_4^-
e) HNO_3, H_3PO_4, and HSO_4^-

Answer: a

17. Sulfuric acid is the product of the reaction of ________ and H_2O.
a) H_2S b) SO_2 c) SO_3^{2-} d) SO_4^{2-} e) SO_3

Answer: e

18. Metal oxides react with water to produce ________.
   a) hydrogen gas
   b) bases
   c) oxygen gas
   d) acids
   e) reduced metals

   Answer: b

5.4 Reactions of Acids and Bases

19. What is the net ionic equation for the reaction of aqueous lithium hydroxide and aqueous nitric acid?
   a) $H^+(aq) + LiOH(aq) \rightarrow H_2O(\ell) + Li^+(aq)$
   b) $H^+(aq) + OH^-(aq) \rightarrow H_2O(\ell)$
   c) $HNO_3(aq) + LiOH(aq) \rightarrow H_2O(\ell) + LiNO_3(aq)$
   d) $Li^+(aq) + NO_3^-(aq) \rightarrow LiNO_3(aq)$
   e) $LiOH(aq) + H_2O(\ell) \rightarrow H^+(aq) + Li(OH)_2(s)$

   Answer: b

20. What are the spectator ions in the reaction between aqueous perchloric acid and aqueous potassium hydroxide?
   a) H^+, ClO_4^-, K^+, and OH^-
   b) H^+ and OH^-
   c) K^+ and ClO_4^-
   d) H^+ and ClO_4^-
   e) K^+ and OH^-

   Answer: c

21. Write a net ionic equation for the reaction of aqueous acetic acid and aqueous potassium hydroxide.
   a) $CH_3CO_2H(aq) + KOH(aq) \rightarrow K^+(aq) + CH_3CO_2^-(aq) + H_2O(\ell)$
   b) $CH_3CO_2H(aq) + KOH(aq) \rightarrow KCH_3CO_2(aq) + H_2O(\ell)$
   c) $H^+(aq) + OH^-(aq) \rightarrow H_2O(\ell)$
   d) $CH_3CO_2H(aq) + OH^-(aq) \rightarrow CH_3CO_3H_2^-(aq)$
   e) $CH_3CO_2H(aq) + OH^-(aq) \rightarrow CH_3CO_2^-(aq) + H_2O(\ell)$

   Answer: e

22. One use of calcium oxide (lime) is in pollution control at coal-burning power plants. Which reaction below occurs in the smokestack of a power plant?
a) $CaO(s) + C(s) \rightarrow Ca(s) + CO(g)$
b) $CaO(s) + SO_3(g) \rightarrow CaSO_4(s)$
c) $CaO(s) + SO_3(g) \rightarrow CaS(s) + 2\,O_2(g)$
d) $2\,CaO(s) + C(s) \rightarrow Ca_2C(s) + O_2(g)$
e) $CaO(s) + SO_3(g) \rightarrow Ca(s) + SO_2(g) + O_2(g)$

Answer: b

5.5 Gas-Forming Reactions

23. Write a net ionic equation for the reaction of limestone and a strong acid.
a) $Ca(s) + 2\,H^+(aq) \rightarrow Ca^{2+}(aq) + H_2(g)$
b) $CaCO_3(s) + 2\,H^+(aq) \rightarrow Ca^{2+}(aq) + H_2CO_3(s)$
c) $CaCO_3(s) + 2\,HCl(aq) \rightarrow CaCl_2(aq) + H_2CO_3(aq)$
d) $CaCO_3(s) + 2\,H^+(aq) \rightarrow Ca^{2+}(aq) + CO_2(g) + H_2O(\ell)$
e) $CaO(s) + 2\,H^+(aq) \rightarrow Ca^{2+}(aq) + H_2O(\ell)$

Answer: d

24. Write a balanced equation for the reaction of aqueous solutions of baking soda and acetic acid.
a) $NaHCO_3(aq) + CH_3CO_2H(aq) \rightarrow NaCH_3CO_2(aq) + H_2O(\ell) + CO_2(g)$
b) $2\,NaHCO_3(aq) + CH_3CO_2H(aq) \rightarrow 2\,Na_2CO_3(aq) + CH_4(aq) + 2H_2O(\ell) + CO_2(g)$
c) $NaHCO_3(aq) + CH_3CO_2H(aq) \rightarrow H_2CO_3(s) + NaCH_3CO_2(aq)$
d) $NaHCO_3(aq) + CH_3CO_2H(aq) \rightarrow NaOH(aq) + CH_3CO_2H(aq) + CO_2(g)$
e) $NaHCO_3(aq) + CH_3CO_2H(aq) \rightarrow CH_4(aq) + 2\,CO_2(g) + O_2(g) + NaOH(aq)$

Answer: a

25. Write a balanced equation for the reaction of aqueous solutions of potassium sulfide and hydrochloric acid.
a) $K_2S(aq) + 2\,HCl(aq) \rightarrow 2\,KH(aq) + SCl_2(g)$
b) $K_2S(aq) + HCl(aq) \rightarrow HS(g) + KCl(aq)$
c) $K_2S(aq) + 2\,HCl(aq) \rightarrow H_2S(g) + 2\,KCl(aq)$
d) $K_2S(aq) + 2\,HCl(aq) \rightarrow 2\,K(s) + H_2(g) + SCl_2(g)$
e) $K_2S(aq) + 2\,HCl(aq) \rightarrow S(s) + H_2(g) + 2\,KCl(aq)$

Answer: c

5.6 Classifying Reactions in Aqueous Solution

26. Which of the following chemical equations is an acid-base reaction?
    a) $2\ HCl(aq) + Zn(s) \rightarrow H_2(g) + ZnCl_2(aq)$
    b) $HCl(aq) + NH_3(aq) \rightarrow NH_4Cl(aq)$
    c) $HCl(aq) + AgNO_3(aq) \rightarrow AgCl(s) + HNO_3(aq)$
    d) $Ba(OH)_2(aq) + Na_2SO_4(aq) \rightarrow BaSO_4(s) + 2\ NaOH(aq)$
    e) $2\ NaOH(aq) + CuCl_2(aq) \rightarrow Cu(OH)_2(s) + 2\ NaCl(aq)$

    Answer: b

5.7 Oxidation-Reduction Reactions

27. Which molecule in the reaction below undergoes oxidation?
    $$2\ C_2H_2(g) + 5\ O_2(g) \rightarrow 4\ CO_2(g) + 2\ H_2O(g)$$
    a) C_2H_2 b) O_2 c) H_2O d) CO_2 e) C_2H_2 and O_2

    Answer: a

28. Which species in the reaction below undergoes reduction?
    $$Sn(s) + 2\ H^+(aq) \rightarrow Sn^{2+}(aq) + H_2(g)$$
    a) Sn
    b) H^+
    c) Sn^{2+}
    d) H_2
    e) No compound is reduced.

    Answer: b

29. Which compound is the oxidizing agent in the reaction below?
    $$I^-(aq) + ClO^-(aq) \rightarrow IO^-(aq) + Cl^-(aq)$$
    a) I^- b) H_2O c) Cl^- d) IO^- e) ClO^-

    Answer: e

30. What is the oxidation number of iodine in IO_3^-?
    a) -1 b) 0 c) +3 d) +5 e) +7

    Answer: d

31. What is the oxidation number of phosphorus in P_2O_5?
    a) -3 b) 0 c) +3 d) +5 e) +7

    Answer: d

32. What is the oxidation number of each atom in sodium perbromate, $NaBrO_4$?
   a) Na = +1, Br = -1, O = -2
   b) Na = +1, Br = +1, O = -2
   c) Na = +1, Br = +7, O = -2
   d) Na = -1, Br = -1, O = +2
   e) Na = +1, Br = -1, O = 0

   Answer: c

33. What is the oxidation number of each atom in potassium nitrate, KNO_3?
   a) K = +1, N = -3, O = -2
   b) K = +1, N = +5, O = -2
   c) K = +1, N = -3, O = +2
   d) K = -1, N = +3, O = -2
   e) K = 0, N = 0, O = 0

   Answer: b

34. All of the following are oxidation-reduction reactions EXCEPT
   a) $CaO(s) + H_2O(\ell) \rightarrow Ca(OH)_2(aq)$
   b) $H_2(g) + Br_2(g) \rightarrow 2\ HBr(g)$
   c) $Ca(s) + 2\ HCl(aq) \rightarrow CaCl_2(aq) + H_2(g)$
   d) $2\ NaBr(aq) + F_2(g) \rightarrow 2\ NaF(aq) + Br_2(g)$
   e) $2\ H_2O(\ell) \rightarrow 2\ H_2(g) + O_2(g)$

   Answer: a

35. Which of the following elements generally acts as a reducing agent?
   a) O_2 b) H^+ c) F^- d) F_2 e) Na

   Answer: e

36. Which of the following are oxidation-reduction reactions?
   1. $Zn(s) + CuSO_4(aq) \rightarrow ZnSO_4(aq) + Cu(s)$
   2. $Pb(ClO_4)_2(aq) + 2\ KI(aq) \rightarrow PbI_2(s) + 2\ NaClO_4(aq)$
   3. $CaCO_3(s) \rightarrow CO_2(g) + CaO(s)$

   a) 1 only b) 2 only c) 1 and 2 d) 1 and 3 e) 2 and 3

   Answer: a

5.8 Measuring Concentrations of Compounds in Solution

37. If 2.891 g $MgCl_2$ is dissolved in enough water to make 500.0 mL of solution, what is the molarity of the magnesium chloride solution?
a) 5.782×10^{-3} M
b) 1.518×10^{-2} M
c) 6.073×10^{-2} M
d) 0.5505 M
e) 5.782 M

Answer: c

38. How many liters of 0.1107 M KCl(aq) contain 15.00 g of KCl?
a) 0.02227 L b) 0.5502 L c) 1.661 L d) 1.818 L e) 123.8 L

Answer: d

39. If 5.15 g $FeCl_3$ is dissolved in enough water to make exactly 150.0 mL of solution, what is the molar concentration of chloride ion?
a) 0.103 M b) 0.212 M c) 0.578 M d) 0.635 M e) 16.7 M

Answer: d

40. What is the mass of solute in 225 mL of 5.91×10^{-2} M KIO_3?
a) 0.0133 g b) 0.0562 g c) 0.263 g d) 1.51 g e) 2.85 g

Answer: e

41. If 5.00 mL of 0.314 M KOH is diluted to exactly 125 mL with water, what is the concentration of the resulting solution?
a) 5.02×10^{-4} M b) 1.26×10^{-2} M c) 0.127 M d) 0.281 M e) 7.85 M

Answer: b

5.9 pH, A Concentration Scale for Acids and Bases

42. A aqueous hydrochloric acid solution has a pH of 4.15. What mass of HCl is present in 1.0 L of this solution?
a) 7.2×10^{-5} g b) 2.6×10^{-3} g c) 0.58 g d) 23 g e) 150 g

Answer: b

43. What is the pH of 0.27 M HNO_3?
a) 0.57 b) 1.31 c) 1.86 d) 2.70 e) 13.43

Answer: a

44. The pH of an aqueous KOH solution is 12.38. What is the hydrogen ion concentration of this solution?
a) 8.1×10^{-14} M b) 4.2×10^{-13} M c) 8.1×10^{-6} M d) 0.21 M e) 1.1 M

Answer: b

5.10 Stoichiometry of Reactions in Aqueous Solution

45. A 25.00 mL sample of NaOH is titrated with 17.13 mL of 0.3150 M HCl. What is the concentration of the NaOH solution?
a) 0.001360 M b) 0.1233 M c) 0.2158 M d) 0.4597 M e) 0.7356 M

Answer: c

46. If 0.3000 g of impure soda ash (Na_2CO_3) is titrated with 17.66 mL of 0.1187 M HCl, what is the percent purity of the soda ash?

$$Na_2CO_3(aq) + 2\ HCl(aq) \rightarrow 2\ NaCl(aq) + H_2O(\ell) + CO_2(g)$$

a) 11.11% b) 22.22% c) 57.91% d) 37.03% e) 74.06%

Answer: d

47. What volume of 0.300 M NaCl is required to precipitate all the Pb^{2+} ion from 25.0 mL of aqueous 0.440 M $Pb(NO_3)_2$?

$$Pb(NO_3)_2(aq) + 2\ NaCl(aq) \rightarrow PbCl_2(s) + 2\ NaNO_3(aq)$$

a) 17.0 mL b) 36.7 mL c) 73.3 mL d) 84.4 mL e) 147 mL

Answer: c

Short Answer Questions

48. A(n) ________ agent gains electrons in an oxidation-reduction reaction.

Answer: oxidizing

49. The pH of 1.0×10^{-5} M HNO_3 is ____ .

Answer: 5.00

50. ________ acid is produced in a larger quantity than any other chemical in the United States. This chemical is used in the production of fertilizers, pigments, alcohol, paper and detergents.

Answer: Sulfuric

51. The net ionic equation for the reaction of iron(II) chloride and potassium oxalate is shown below.

$$Fe^{2+}(aq) + C_2O_4^{2-}(aq) \rightarrow FeC_2O_4(s)$$

Chloride and potassium ions are referred to as ________ ions because they are not involved in the reaction.

Answer: spectator

52. A base is a substance that, when dissolved in water, increases the ______ concentration of the solution

Answer: hydroxide ion

53. Give the name of a basic oxide and write a balanced chemical equation for the reaction of the oxide with water.

Answer: Calcium oxide. $CaO(aq) + H_2O(\ell) \rightarrow Ca(OH)_2(aq)$. Other basic oxides include the group 1A and 2A metal oxides.

54. You mix 50.0 mL of 0.185 M $AgNO_3$ with 70.0 mL of 0.142 M KBr. What is the maximum mass of precipitate formed?

Answer: 1.74 g

55. The pH of an aqueous sodium hydroxide solution gradually decreases if the solution is left in contact with air. In fact, the process can be hastened if a person exhales over a sodium hydroxide solution. Write a balanced chemical equation that describes the process by which the hydroxide ion concentration decreases.

Answer: $OH^-(aq) + CO_2(aq) \rightarrow HCO_3^-(aq)$

Diagnostic Quiz Questions

Goal #1: Understand the nature of ionic substances dissolved in water

1. Which of the following compounds is a nonelectrolyte?
   a) calcium chloride, $CaCl_2$
   b) acetic acid, CH_3CO_2H
   c) potassium iodide, KI
   d) ethanol, CH_3CH_2OH

   Answer: d

2. Which of the following statements concerning electrolytes are correct?
   1. All ionic compounds are strong electrolytes.
   2. Weak acids, such as acetic acid, are weak electrolytes.
   3. Molecular species are weak electrolytes, provided they dissolve in water.

   a) 1 only
   b) 2 only
   c) 3 only
   d) 1, 2, and 3

   Answer: b

3. Which compound is insoluble in water?
   a) $BaSO_4$
   b) K_2HPO_4
   c) NaOH
   d) $Ni(NO_3)_2$

   Answer: a

4. Which compound is soluble in water?
   a) AgI
   b) $Cu_3(PO_4)_2$
   c) $Zn(OH)_2$
   d) $(NH_4)_2CO_3$

   Answer: d

5. What ions are formed when $NaHCO_3$ dissolves in water?
   a) Na^+, H^+, and CO_3^-
   b) NaH^+, and CO_3^{2-}
   c) Na^+ and HCO_3^-
   d) Na^-, H^+, and CO_3^{2-}

   Answer: c

Goal #2: Recognize common acids and bases and understand their behavior in aqueous solution.

6. Which of the following acids is a weak acid?
   a) HF
   b) HCl
   c) HNO_3
   d) H_2SO_4

   Answer: a

7. What ions are produced from the dissolution of sodium hydroxide in water?
   a) Na^+ and H^+
   b) Na^- and H^+
   c) Na^+ and OH^-
   d) Na^+ and O^{2-}

   Answer: c

8. Metal oxides are often referred to as basic oxides. Write a balanced chemical equation for the reaction of CaO and water.
   a) $CaO(s) + H_2O(\ell) \rightarrow CaO_2(s) + H_2(g)$
   b) $CaO(s) + H_2O(\ell) \rightarrow Ca(OH)_2(aq)$
   c) $CaO(s) + H_2O(\ell) \rightarrow CaOH(aq) + OH^-(aq)$
   d) $CaO(s) + H_2O(\ell) \rightarrow H_2CO_2(aq)$

   Answer: b

9. Nonmetal oxides are often referred to as acidic oxides. Write a balanced chemical equation for the reaction of SO_3 and water.
   a) $SO_3(g) + H_2O(\ell) \rightarrow H_2SO_4(aq)$
   b) $SO_3(g) + H_2O(\ell) \rightarrow SO_2(g) + 2\ OH^-(aq)$
   c) $SO_3(g) + H_2O(\ell) \rightarrow SO_2(g) + H_2O_2(aq)$
   d) $SO_3(g) + H_2O(\ell) \rightarrow H_2SO_3(aq) + O^{2-}(aq)$

   Answer: a

Goal #3: Recognize and write equations for the common types of reactions in aqueous solutions

10. Write a balanced net ionic equation for the reaction of NaOH with $FeCl_3$.
    a) $Na^+(aq) + Cl^-(aq) \rightarrow NaCl(s)$
    b) $Fe^{3+}(aq) + 3\ Na^+(aq) \rightarrow Fe(Na)_3(s)$
    c) $Fe^{3+}(aq) + 3\ OH^-(aq) \rightarrow Fe(OH)_3(s)$
    d) $Cl^-(aq) + OH^-(aq) \rightarrow HOCl(s)$

    Answer: c

11. Write a balanced equation for the reaction of aqueous solutions of silver nitrate with calcium bromide.
   a) $2\ AgNO_3(aq) + CaBr_2(aq) \rightarrow CaAg_2(s) + 2\ BrNO_3(aq)$
   b) $Ag(NO_3)_2(aq) + CaBr_2(aq) \rightarrow AgBr_2(aq) + Ca(NO_3)_2(s)$
   c) $AgNO_3(aq) + CaBr_2(aq) \rightarrow AgBr_2(s) + CaNO_3(aq)$
   d) $2\ AgNO_3(aq) + CaBr_2(aq) \rightarrow 2\ AgBr(s) + Ca(NO_3)_2(aq)$

   Answer: d

12. Write a balanced equation for the reaction of acetic acid with potassium hydroxide.
   a) $CH_3CO_2H(aq) + KOH(aq) \rightarrow KCH_3CO_2(aq) + H_2O(\ell)$
   b) $CH_3CO_2H(aq) + KOH(aq) \rightarrow KCH_3O(aq) + OH^-(aq) + CO_2(g)$
   c) $CH_3CO_2H(aq) + KOH(aq) \rightarrow KCH_3(aq) + H_2O(\ell) + CO_2(g)$
   d) $CH_3CO_2H(aq) + KOH(aq) \rightarrow KCH_3CO_3H(aq)$

   Answer: a

13. Write a net ionic equation for the reaction of hydrobromic acid with sodium hydroxide.
   a) $Na^+(aq)$ and $Br^-(aq) \rightarrow NaBr(aq)$
   b) $HBr(aq) + NaOH(aq) \rightarrow H_2O(\ell) + NaBr(aq)$
   c) $H^+(aq) + OH^-(aq) \rightarrow H_2O(\ell)$
   d) $Na^+(aq) + Br^-(aq) \rightarrow NaBr(s)$

   Answer: c

14. Write a balanced chemical equation for the reaction of zinc(II) carbonate with nitric acid.
   a) $ZnCO_3(s) + 2\ HNO_3(aq) \rightarrow Zn(NO_3)_2(aq) + H_2O(\ell) + CO_2(g)$
   b) $ZnCO_3(s) + HNO_3(aq) \rightarrow ZnO(s) + CO_2(g) + HNO_3(aq)$
   c) $ZnCO_3(s) + 2\ HNO_3(aq) \rightarrow Zn(OH)_2(s) + N_2O_5(s) + CO_2(g)$
   d) $ZnCO_3(s) + 2\ HNO_3(aq) \rightarrow Zn(NO_3)_2(aq) + H_2CO_3(aq)$

   Answer: a

15. The chemical equation below,

$$2\ Na(s) + 2\ H_2O(\ell) \rightarrow 2\ NaOH(aq) + H_2(g)$$

   is an example of a(n) __________ reaction.
   a) precipitation
   b) oxidation-reduction
   c) acid-base
   d) both oxidation-reduction and acid-base

   Answer: b

16. The chemical equation below,

$$AlCl_3(aq) + 3\ NaOH(aq) \rightarrow Al(OH)_3(s) + 3\ NaCl(aq)$$

is an example of a(n) __________ reaction.

a) precipitation
b) oxidation-reduction
c) strong acid-strong base
d) gas-forming

Answer: a

Goal #4: Recognize common oxidizing and reducing agents and identify oxidation-reduction reactions.

17. Determine the oxidation number of sulfur in SO_4^{2-}.

a) +2
b) +4
c) +6
d) +8

Answer: c

18. Determine the oxidation number of each atom in CH_2Cl_2.

a) C = 0, H = +1, Cl = -1
b) C = +2, H =0, Cl = -1
c) C = 0, H = 0, Cl = 0
d) C = -2, H = +1, Cl = 0

Answer: a

19. Which of the following lists contains only common oxidizing agents?

a) K, Ca, and F_2
b) Li, Na, and K
c) F_2, MnO_4^-, and $IINO_3$
d) F^-, Cl^-, and Br^-

Answer: c

Goal #5: Define and use solution conentrations in reaction stoichiometry

20. If 8.33 g $Ca(NO_3)_2$ is dissolved in enough water to make 0.250 L of solution, what is the molar concentration of $Ca(NO_3)_2$?
   a) 0.0127 M
   b) 0.203 M
   c) 3.33 M
   d) 4.92 M

   Answer: b

21. What volume of 0.250 M KOH contains 25.0 g KOH?
   a) 0.111 L
   b) 0.781 L
   c) 0.625 L
   d) 1.78 L

   Answer: d

22. If 45.0 mL of 0.244 M $NiCl_2$ is diluted to 275 mL with pure water, what is the molar concentration of $NiCl_2$ in the diluted solution?
   a) 0.0399 M
   b) 0.105 M
   c) 0.671 M
   d) 1.49 M

   Answer: a

23. If 0.40 M of acetic acid solution has a pH of 2.57, what is the H^+ concentration of the solution?
   a) 2.7×10^{-3} M
   b) 0.039 M
   c) 0.40 M
   d) 3.7×10^{2} M

   Answer: a

24. What is the pH of a 0.34 M HNO_3 solution?
   a) -0.47
   b) 0.34
   c) 0.47
   d) 2.13

   Answer: c

25. If you combine 35.0 mL of 0.100 M $AgNO_3$ with 45.0 mL of 0.0800 M NaBr, what mass of AgBr is produced?
   a) 0.537 g
   b) 0.657 g
   c) 0.676 g
   d) 0.818 g

   Answer: b

26. A 25.00 mL sample of sulfuric acid, H_2SO_4, requires 42.13 mL of 0.1533 M NaOH for titration to the equivalence point. What is the concentration of the sulfuric acid?

$$H_2SO_4(aq) + 2\ NaOH \rightarrow Na_2SO_4(aq) + 2\ H_2O(\ell)$$

   a) 0.04558 M
   b) 0.1292 M
   c) 0.2583 M
   d) 0.5167 M

   Answer: b

CHAPTER 6: Principles of Reactivity: Energy and Chemical Reactions

6.1. Energy: Some Basic Principles

1. A battery-operated power tool, such as a cordless drill, converts
   a) electrostatic energy to chemical potential energy.
   b) mechanical energy to thermal energy.
   c) thermal energy to mechanical energy.
   d) thermal energy to chemical potential energy.
   e) chemical potential energy to mechanical energy.

   Answer: e

2. Which of the following types of energy is a form of potential energy?
   a) sound energy
   b) electrical energy
   c) thermal energy
   d) gravitational energy
   e) mechanical energy

   Answer: d

3. Of the following types of energy, which are classified as kinetic energy?
   1. thermal energy
   2. mechanical energy
   3. electrical energy

   a) 1 only b) 2 only c) 3 only d) 2 and 3 e) 1, 2, and 3

   Answer: e

4. Which one of the following statements is INCORRECT?
   a) The heat required to raise the temperature of 1.00 g $H_2O(\ell)$ by 1.00 °C is 1.00 J.
   b) 1.000 calories is equal to 4.184 J.
   c) 1.00 J is equal to 1.00 $kg{\cdot}m^2{\cdot}s^{-2}$.
   d) A dietary Calorie is equal to 1000 calories.
   e) 1000 calories is equal to 1 kcal.

   Answer: a

5. Which one of the following statements is INCORRECT?
   a) Energy is neither created nor destroyed in chemical reactions.
   b) Kinetic energy is the energy associated with motion.
   c) Exothermic processes transfer heat from the surrounding into the system.
   d) Increasing the thermal energy of a gas increases the motion of its atoms.
   e) Energy is the capacity to do work.

   Answer: c

6. How many joules are equivalent to 235 calories?
   a) 0.0178 J b) 56.2 J c) 983 J d) 5.62×10^3 J e) 1.67×10^4 J

   Answer: c

7. How many nutritional calories are equivalent to 875 kJ?
   a) 0.875 Cal b) 209 Cal c) 3.66×10^3 Cal d) 2.09×10^5 Cal e) 3.66×10^6 Cal

   Answer: b

6.2 Specific Heat Capacity and Heat Transfer

8. Specific heat capacity is
   a) the quantity of heat needed to change 1.00 g of a substance by 1.00 K.
   b) the mass of a substance that 1.00 J of energy will heat by 1.00 K.
   c) the mass of a substance that 1.00 cal of energy will heat by 1.00 K.
   d) the temperature change undergone when 1.00 g of a substance absorbs 4.184 J.
   e) the quantity of heat required to melt 4.184 g of a substance.

   Answer: a

9. Heat capacity is defined as
   a) the amount of heat required to raise the temperature of 1 gram of substance by 1 K.
   b) the amount of heat required to raise the temperature of a substance by 1 K.
   c) the amount of heat required to vaporize a solid or liquid.
   d) the amount of heat required to melt a solid.
   e) 4.18 cal/g·K.

   Answer: b

10. Which one of the following statements is INCORRECT?
    a) In an exothermic process heat is transferred from the system to the surroundings.
    b) The greater the heat capacity of an object, the more thermal energy it can store.
    c) The SI unit of specific heat capacity is joules per gram per kelvin.
    d) The specific heat capacity has a positive value for liquids and a negative value for gases.
    e) When heat is transferred from the system to the surroundings, q is negative.

    Answer: d

11. If 136 J is required to change the temperature of 8.75 g of nickel by 35.0 K, what is the specific heat capacity of mercury?
    a) 0.0294 J/g·K b) 0.311 J/g·K c) 0.417 J/g·K d) 0.444 J/g·K e) 2.25 J/g·K

    Answer: d

12. If the same amount of heat is added to 5.00 g samples of each of the metals below, which metal will experience the smallest temperature change?

Metal	Specific Heat Capicity (J/g·K)
Al	0.897
Au	0.129
Cu	0.385
Fe	0.449
K	0.753

a) Al b) Au c) Cu d) Fe e) K

Answer: a

13. If 1.00 mole of ethanol, CH_3CH_2OH, at 22.0 °C absorbs 1.45 kJ of heat, what is the final temperature of the ethanol? The specific heat capacity of ethanol is 2.44 J/g·K.
a) 9.1 °C b) 22.0 °C c) 34.9 °C d) 47.0 °C e) 616 °C

Answer: c

14. How much energy is required to change the temperature of 21.5 g Cu from 27.0 °C to 88.1 °C? The specific heat capacity of copper is 0.385 J/g·K.
a) 223 J b) 506 J c) 641 J d) 729 J e) 3.41×10^3 J

Answer: b

15. If 50.0 g H_2O at 13.6 °C is combined with 85.0 g H_2O at 93.7 °C, what is the final temperature of the mixture? The specific heat capacity of water is 4.184 J/g·K.
a) 26.1 °C b) 40.0 °C c) 56.1 °C d) 64.0 °C e) 80.1 °C

Answer: d

16. If 46.1 g Zn at 18.0 °C is placed in 80.0 g H_2O at 75.0 °C, what is the final temperature of the mixture? The specific heat capacities of zinc and water are 0.388 J/g·K and 4.184 J/g·K, respectively.
a) 20.9 °C b) 27.3 °C c) 64.5 °C d) 65.7 °C e) 72.1 °C

Answer: e

17. When 27.0 g of an unknown metal at 88.4 °C is placed in 115 g H_2O at 21.0 °C, the final temperature of the water is 23.7 °C. What is the specific heat capacity of the metal? The specific heat capacity of water is 4.184 J/g·K.
a) 0.34 J/g·K b) 0.51 J/g·K c) 0.74 J/g·K d) 0.94J/g·K e) 1.4 J/g·K

Answer: c

6.3 Energy and Changes of State

18. Calculate the amount of heat required to change 50.0 g ice at –20.0 °C to steam at 135 °C. (Heat of fusion = 333 J/g; heat of vaporization = 2260 J/g; specific heat capacities: ice = 2.09 J/g·K, liquid water = 4.18 J/g·K, steam = 1.84 J/g·K)
 a) 4.18 kJ b) 32.4 kJ c) 78.8 kJ d) 135 kJ e) 156 kJ

 Answer: e

19. 45.0 g of ice at -12.0 °C is mixed with 256 g of water at 31.4 °C. What is the final temperature of the mixture? (Heat of fusion = 333 J/g; specific heat capacities: ice = 2.09 J/g·K, liquid water = 4.184 J/g·K)
 a) –1.6 °C b) 0.0 °C c) 13.9 °C d) 21.0 °C e) 25.8 °C

 Answer: c

20. A mass of 95.0 g of ice at 0.00 °C is combined with 105 g water at 25.0 °C. The final temperature of the mixture is 0.00 °C. What mass of ice melts? (Heat of fusion = 333 J/g; specific heat capacities: ice = 2.09 J/g·K, liquid water = 4.184 J/g·K)
 a) 33.0 g b) 36.6 g c) 41.8 g d) 47.9 g e) 95.0 g

 Answer: a

21. The heat of vaporization of benzene, C_6H_6, is 30.8 kJ/mol at its boiling point of 80.1 °C. How much heat is required to vaporize 128 g benzene at its boiling point?
 a) 4.04 kJ b) 18.8 kJ c) 19.3 kJ d) 50.5 kJ e) 4.04×10^3 kJ

 Answer: d

6.4 The First Law of Thermodynamics

22. Calculate ΔE for the system in which 16 J of work is done on a gas by the surroundings and the gas absorbs 51 J of heat?
 a) –67 J b) –35 J c) +35 J d) +51 J e) +67 J

 Answer: e

23. Calculate ΔE for the system in which a gas absorbs 31 J of heat and does 18 J of work on the surroundings?
 a) –49 J b) –13 J c) +13 J d) +31 J e) +49 J

 Answer: c

24. Which of the following statements is CORRECT?
   a) If a reaction occurs at constant pressure, $w = \Delta E$.
   b) If a reaction occurs at constant pressure, $q = \Delta H$.
   c) If a reaction occurs at constant pressure, $q = \Delta E$.
   d) If a reaction occurs at constant volume, $\Delta E > \Delta H$.
   e) If a reaction occurs at constant volume, $w = \Delta E$.

   Answer: b

25. One statement of the first law of thermodynamics is that
   a) the amount of work done on a system is dependent of pathway.
   b) the total work done on a system must equal the heat absorbed by the system.
   c) the heat flow in or out of a system is independent of pathway.
   d) the total energy flow in or out of a system is equal to the sum of the heat transferred to or from the system and the work done by or on the system.
   e) in any chemical process the heat flow must equal the change in enthalpy.

   Answer: d

26. Which of the following thermodynamic quantities are state functions: heat (q), work (w), enthalpy change (ΔH), and internal energy change (ΔE)?
   a) ΔE only   b) q only   c) ΔE and w d) ΔH and q e) ΔH and ΔE

   Answer: e

6.5 Enthalpy Changes for Chemical Reactions

27. The thermochemical equation for the combustion of hexane is shown below.
   $$C_6H_{14}(g) + 19/2\ O_2(g) \rightarrow 6\ CO_2(g) + 7\ H_2O(g) \qquad \Delta H° = -4163 \text{ kJ}$$
   What is the enthalpy change for the combustion of 2.50 g C_6H_{14}?
   a) -121 kJ   b) -1.66×10^3 kJ   c) -1.04×10^4 kJ   d) -1.43×10^5 kJ   e) -3.59×10^5 kJ

   Answer: a

28. The thermochemical equation for the combustion of butane is shown below.
   $$C_4H_{10}(g) + 13/2\ O_2(g) \rightarrow 4\ CO_2(g) + 5\ H_2O(\ell) \qquad \Delta H° = -2877 \text{ kJ}$$
   What is the enthalpy change for the following reaction?
   $$16\ CO_2(g) + 20\ H_2O(\ell) \rightarrow 4\ C_4H_{10}(g) + 26\ O_2(g)$$
   a) -5754 kJ   b) -719.2 kJ   c) +719.2 kJ   d) +5754 kJ   e) $+1.151 \times 10^4$ kJ

   Answer: e

29. Hydrazine, N_2H_4, is a liquid used as a rocket fuel. It reacts with oxygen to yield nitrogen gas and water.

$$N_2H_4(\ell) + O_2(g) \rightarrow N_2(g) + 2\,H_2O(\ell)$$

The reaction of 3.80 g N_2H_4 evolves 73.7 kJ of heat. Calculate the enthalpy change per mole of hydrazine combusted.
a) -8.74 kJ/mol
b) -19.4 kJ/mol
c) -2.80×10^2 kJ/mol
d) -622 kJ/mol
e) -8.98×10^3 kJ/mol

Answer: d

6.5 Calorimetry

30. If 1.86 g MgO is combined with 100.0 mL of 1.00 M HCl (density 100.0 g/mL) in a coffee cup calorimeter, the temperature of the resulting solution increases from 21.3 °C to 35.7 °C. Calculate the enthalpy change for the reaction per mole of MgO. Assume that the specific heat capacity of the solution is 4.18 J/g·K.
a) -6.13 kJ b) -28.3 kJ c) –133 kJ d) –329 kJ e) -613 kJ

Answer: c

31. Commercial cold packs consist of solid ammonium nitrate and water. NH_4NO_3 absorbs 3.30×10^2 J of heat per gram dissolved in water. In a coffee-cup calorimeter, 4.40 g NH_4NO_3 is dissolved in 100.0 g of water at 21.0 °C. What is the final temperature of the solution? Assume that the solution has a specific heat capacity of 4.18 J/g·K.
a) 3.3 °C b) 17.7 °C c) 20.2 °C d) 20.9 °C e) 24.3 °C

Answer: b

32. When 14.2 g KBr is dissolved in 100.0 g of water in a coffee-cup calorimeter, the temperature drops from 28.88 °C to 23.38 °C. What is the enthalpy change per gram of KBr dissolved in the water? Assume that the solution has a specific heat capacity of 4.18 J/g·K.
a) 28.6 J/g b) 185 J/g c) 263 J/g d) 326 J/g e) 2.63×10^3 J/g

Answer: b

33. A chemical reaction in a bomb calorimeter evolves 5.17 kJ of heat. If the heat capacity of the calorimeter is 1.08 kJ/°C, what is the temperature change of the calorimeter?
a) 0.209 °C b) 4.09 °C c) 4.79 °C d) 5.58 °C e) 6.25 °C

Answer: c

34. A 2.885 g sample of methanol, CH_3OH, is combusted in a bomb calorimeter. The temperature of the calorimeter increases by 11.38 K. If the heat capacity of the bomb is 727.1 J/K and it contains 1.200 kg of water, what is the heat evolved per mole of methanol combusted? The specific heat capacity of water is 4.184 J/g·K and the molar mass of methanol is 32.04 g/mol.
a) -65.41 kJ/mol
b) -91.89 kJ/mol
c) -634.5 kJ/mol
d) -726.5 kJ/mol
e) -1.019×10^6 kJ/mol

Answer: d

35. The molar enthalpy of combustion for glucose, $C_6H_{12}O_6$, is -2803 kJ. A mass of 1.150 g glucose is combusted in a bomb calorimeter. If the calorimeter contains 925 g H_2O and the bomb has a heat capacity of 622 J/K, what is the temperature increase of the bomb calorimeter? The specific heat capacity of water is 4.184 J/g·K and the molar mass of glucose is 180.2 g/mol.
a) 0.72 K   b) 2.76 K   c) 3.98 K   d) 5.12 K   e) 7.20 K

Answer: c

6.7 Hess's Law

36. Determine the heat of vaporization of titanium(IV) chloride given the enthalpies of reaction below.

$$Ti(s) + 2\,Cl_2(g) \rightarrow TiCl_4(\ell) \qquad \Delta H = -804.2 \text{ kJ}$$
$$Ti(s) + 2\,Cl_2(g) \rightarrow TiCl_4(g) \qquad \Delta H = -763.2 \text{ kJ}$$

a) –1567.4 kJ   b) -41.0 kJ   c) +1.054 kJ   d) +41.0 kJ   e) +1567.4 kJ

Answer: d

37. Determine the heat of reaction for the oxidation of iron,

$$4\,Fe(s) + 3\,O_2(g) \rightarrow 2\,Fe_2O_3(s)$$

given the thermochemical equations below.

$$2\,Fe(s) + 6\,H_2O(\ell) \rightarrow 2\,Fe(OH)_3(s) + 3\,H_2(g) \qquad \Delta H = +321.8 \text{ kJ}$$
$$2\,H_2O(\ell) \rightarrow 2\,H_2(g) + O_2(g) \qquad \Delta H = +571.7 \text{ kJ}$$
$$Fe_2O_3(s) + 3\,H_2O(\ell) \rightarrow 2\,Fe(OH)_3(s) \qquad \Delta H = +288.6 \text{ kJ}$$

a) –1648.7 kJ   b) –1182.1 kJ   c) –505.3 kJ   d) +360.5 kJ   e) +1447.1 kJ

Answer: a

38. Determine ΔH for the following reaction,

$N_2(g) + 3\ H_2(g) \rightarrow 2\ NH_3(g)$

given the thermochemical equations below.

$N_2(g) + O_2(g) \rightarrow 2\ NO(g)$ $\Delta H = +180.8$ kJ

$4\ NH_3(g) + 5\ O_2(g) \rightarrow 4\ NO(g) + 6\ H_2O(g)$ $\Delta H = -906.2$ kJ

$2\ H_2(g) + O_2(g) \rightarrow 2\ H_2O(g)$ $\Delta H = -483.6$ kJ

a) -1209.0 kJ b) -1189.0 kJ c) -756.5 kJ d) -241.8 kJ e) -91.5 kJ

Answer: e

39. Calculate ΔH for the following reaction,

$CaO(s) + CO_2(g) \rightarrow CaCO_3(s)$

given the thermochemical equations below.

$2\ Ca(s) + O_2(g) \rightarrow 2\ CaO(s)$ $\Delta H = -1270.2$ kJ

$C(s) + O_2(g) \rightarrow CO_2(g)$ $\Delta H = -393.5$ kJ

$2\ Ca(s) + 2\ C(s) + 3\ O_2(g) \rightarrow 2\ CaCO_3(s)$ $\Delta H = -2413.8$ kJ

a) -4077.3 kJ b) -750.1 kJ c) -178.3 kJ d) +350.2 kJ e) +2870.6 kJ

Answer: c

6.8 Standard Enthalpies of Formation

40. Calculate ΔH_f° for sulfur dioxide,

$S(s) + O_2(g) \rightarrow SO_2(g)$

given the thermochemical equations below.

$2\ S(s) + 3\ O_2(g) \rightarrow 2\ SO_3(g)$ $\Delta H^\circ = -791.5$ kJ

$2\ SO_2(g) + O_2(g) \rightarrow 2\ SO_3(g)$ $\Delta H^\circ = -197.9$ kJ

a) -296.8 kJ/mol b) -395.7 kJ/mol c) -494.7 kJ/mol d) -593.6 kJ/mol e) -989.4 kJ/mol

Answer: a

41. Which of the following chemical equations corresponds to the standard molar enthalpy of formation of N_2O?

a) $NO(g) + 1/2\ N_2(g) \rightarrow N_2O(g)$
b) $N_2(g) + 1/2\ O_2(g) \rightarrow N_2O(g)$
c) $2N(g) + O(g) \rightarrow N_2O(g)$
d) $N_2(g) + O(g) \rightarrow N_2O(g)$
e) $2\ N_2(g) + O_2(g) \rightarrow 2\ N_2O(g)$

Answer: b

42. Which of the following chemical equations does not correspond to a standard molar enthalpy of formation?
a) $Ca(s) + C(s) + 3/2\ O_2(g) \rightarrow CaCO_3(s)$
b) $C(s) + 1/2\ O_2(g) \rightarrow CO(g)$
c) $H_2(g) + 1/2\ O_2(g) \rightarrow H_2O(\ell)$
d) $N_2(g) + 2\ O_2(g) \rightarrow N_2O_4(g)$
e) $SO_2(g) + 1/2\ O_2(g) \rightarrow SO_3(g)$

Answer: e

43. Calculate ΔH°_{rxn} for the combustion of gaseous ethanol,

$$C_2H_5OH(g) + 3\ O_2(g) \rightarrow 2\ CO_2(g) + 3\ H_2O(g)$$

using standard molar enthalpies of formation.

molecule	ΔH_f° (kJ/mol)
$C_2H_5OH(g)$	-235.3
$CO_2(g)$	-393.5
$H_2O(g)$	-241.8

a) -1747.7 kJ b) -1277.1 kJ c) -793.5 kJ d) -400.0 kJ e) -83.6 kJ

Answer: b

44. The standard enthalpy change for the combustion of 1 mole of benzene is -3267.4 kJ.

$$C_6H_6(\ell) + 15/2\ O_2(g) \rightarrow 6\ CO_2(g) + 3\ H_2O(\ell)$$

Calculate ΔH_f° for benzene based on the following standard molar enthalpies of formation.

molecule	ΔH_f° (kJ/mol)
$CO_2(g)$	-393.5
$H_2O(\ell)$	-285.8

a) -6485.8 kJ b) -3946.7 kJ c) -3218.4 kJ d) +49.0 kJ e) +2588.1 kJ

Answer: d

45. The standard molar enthalpy of formation of $NH_3(g)$ is -45.9 kJ/mol. What is the enthalpy change if 5.38 g $N_2(s)$ and 3.32 g $H_2(g)$ react to produce $NH_3(g)$?
a) -75.6 kJ b) -50.4 kJ c) -17.6 kJ d) -8.81 kJ e) -1.20 kJ

Answer: c

Short Answer Questions

46. ______ energy is often referred to as the energy of motion. Examples of this type of energy include thermal, mechanical, and electrical energies.

Answer: kinetic

47. A form of potential energy that is associated with an object's height is known as ________ potential energy.

Answer: gravitational

48. A system is defined as the object, or collection of objects being studied. The ________ include everything outside the system.

Answer: surroundings

49. The heat required to convert a solid at its melting point to a liquid is called the heat of ________ .

Answer: fusion

50. The enthalpy change is the heat absorbed or evolved in a reaction that occurs at constant ________.

Answer: pressure

51. Why are you at greater risk from being burned by steam at 100 °C than from liquid water at the same temperature?

Answer: At 100 °C, steam has approximately 40 kJ/mole more potential energy than liquid water at the same temperature.

52. Enthalpy change is a state function. What is meant by this statement?

Answer: The difference between the initial and final enthalpies of a system is independent of the path between the two states.

Diagnostic Quiz Questions

Goal #1: Assess heat transfer associated with changes in temperature and changes in state

1. The combustion of gasoline to propel a car is an example of
   a) the conversion of chemical potential energy to mechanical energy.
   b) the conversion of mechanical energy to thermal energy.
   c) the conversion of thermal energy to electrical energy.
   d) the conversion of electrical energy to chemical potential energy.

   Answer: a

2. A nutritional Calorie equals 1.0×10^3 calories. What is the energy in joules in a 1500 nutritional Calorie diet?
   a) 2.8×10^{-6} J
   b) 3.6×10^2 J
   c) 3.6×10^5 J
   d) 6.3×10^6 J

   Answer: d

3. The specific heat capacity of aluminum is 0.897 J/g·K. What quantity of heat (in joules) is required to heat 15.0 g Al from 22.0 °C to 75.0 °C?
   a) 1.13×10^{-3} J
   b) 3.17 J
   c) 713 J
   d) 886 J

   Answer: c

4. The quantity of heat required to heat 5.01 g of ice from -25.0 °C to -10.0 °C is 155 J. What is the specific heat capacity of ice?
   a) 0.485 J/g·K
   b) 2.06 J/g·K
   c) 51.8 J/g·K
   d) 1.16×10^4 J/g·K

   Answer: b

5. The specific heat of water is 4.184 J/g·K. A 25.0 g sample of water at 37.3 °C is mixed with 45.0 g of water at 72.9 °C. What is the final temperature of the water after thermal equilibrium is reached?
   a) 33.5 °C
   b) 47.1 °C
   c) 52.8 °C
   d) 60.2 °C

   Answer: d

6. The heat of vaporization of bromine is 29.6 kJ/mol at its boiling point of 59 °C. What quantity of heat is released when 10.0 g Br_2 is condensed at 59 °C?
   a) -1.85 kJ
   b) -2.96 kJ
   c) -109 kJ
   d) -296 kJ

   Answer: a

7. What quantity of heat (in joules) is evolved when 25.0 g $H_2O(g)$ at 125 °C is converted to $H_2O(\ell)$ at 35.0 °C? The specific heats of steam and liquid water are 2.03 J/g·K and 4.184 J/g·K, respectively. The heat of vaporization of water is 2256 J/g.
   a) -9.41×10^3 J
   b) -8.07×10^3 J
   c) -5.64×10^4 J
   d) -6.45×10^4 J

   Answer: d

Goal #2: Apply the first law of thermodynamics
Goal #3: Define and understand the state functions enthalpy and internal energy

8. According to the first law of thermodynamics as applied to chemical reactions,
   a) energy is conserved in any reaction.
   b) the enthalpy change in a reaction equals the heat change.
   c) the heat evolved in a chemical reaction is equal and opposite in sign to the work done on the system.
   d) the work done on a system is independent of the pathway.

   Answer: a

Goal #4: Calculate the energy changes occurring in chemical reactions and learn how these changes are measured

9. Which one of the following processes is endothermic?
   a) Steam condenses to liquid water.
   b) Propane gas burns in a gas grill.
   c) Water evaporates from your skin after stepping out of the shower.
   d) Water freezes to form ice.

   Answer: c

10. You mix 45 mL of 0.20 *M* HCl with 55 mL of 0.20 *M* KOH in a perfectly insulated solution calorimeter. The following reaction occurs.

$$HCl(aq) + KOH(aq) \rightarrow KCl(aq) + H_2O(\ell)$$

   The temperature of both solutions before mixing is 21.5 °C. The temperature of the solution rises to 22.7 °C after mixing. What is the enthalpy change per mole of water produced? The specific heat capacity of water and the solution is 4.2 J/g·K.
   a) -0.50 kJ
   b) -2.5 kJ
   c) -48 kJ
   d) -56 kJ

   Answer: d

11. A 3.90 g sample of methanol, CH_3OH, is combusted in a constant volume calorimeter with excess $O_2(g)$. The temperature of the calorimeter increases by 16.02 °C. If the heat capacity of the calorimeter is 788 J/K and it contains 1.20 kg of water, what is the heat evolved per mole of methanol combusted? The specific heat capacity of water is 4.184 J/g·K and the molar mass of methanol is 32.04 g/mol.
   a) -93.1 kJ/mol
   b) -764 kJ/mol
   c) -987 kJ/mol
   d) -1.04×10^3 kJ/mol

   Answer: b

12. Ethylene, C_2H_4, reacts with excess oxygen to produce carbon dioxide, water, and heat. The standard molar enthalpy of combustion is –1411 kJ.

$$C_2H_4(g) + 3\ O_2(g) \rightarrow 2\ CO_2(g) + 2\ H_2O(\ell)$$

   What is the enthalpy change for the following reaction?

$$4\ CO_2(g) + 4\ H_2O(\ell) \rightarrow 2\ C_2H_4(g) + 6\ O_2(g)$$

   a) -2822 kJ
   b) -705 kJ
   c) +1411 kJ
   d) +2822 kJ

   Answer: d

13. Hydrazine, N_2H_4, is a liquid used as a rocket fuel. It reacts with oxygen to yield nitrogen and water.

$$N_2H_4(\ell) + O_2(g) \rightarrow N_2(g) + 2\ H_2O(\ell)$$

   The reaction of 2.50 g N_2H_4 evolves 48.5 kJ of heat. Calculate the enthalpy change per mole of hydrazine combusted.
   a) -3.90×10^3 kJ/mol
   b) -831 kJ/mol
   c) -622 kJ/mol
   d) -121 kJ/mol

   Answer: c

14. Calculate the enthalpy for the formation of calcium carbonate from calcium oxide and carbon dioxide,

$$CaO(s) + CO_2(g) \rightarrow CaCO_3(s)$$

given the enthalpies of the reactions below.

$$2\ Ca(s) + O_2(g) \rightarrow 2\ CaO(s) \qquad \Delta H = -1270.2\ kJ$$
$$C(s) + O_2(g) \rightarrow CO_2(g) \qquad \Delta H = -393.5\ kJ$$
$$2\ Ca(s) + 2\ C(s) + 3\ O_2(g) \rightarrow 2\ CaCO_3(s) \qquad \Delta H = -2413.8\ kJ$$

a) -178.3 kJ
b) -965.3 kJ
c) -2235.5 kJ
d) -4077.5 kJ

Answer: a

15. Which one of the following reactions is not a standard enthalpy of formation reaction?
a) $CaO(s) + CO_2(g) \rightarrow CaCO_3(s)$
b) $S(s) + O_2(g) \rightarrow SO_2(g)$
c) $S(s) + 3/2\ O_2(g) \rightarrow SO_3(g)$
d) $C(s) + \frac{1}{2}\ O_2(g) \rightarrow CO(g)$

Answer: a

16. Determine the standard enthalpy of reaction for the following reaction,

$$2\ CH_3OH(\ell) + 3\ O_2(g) \rightarrow 2\ CO_2(g) + 4\ H_2O(\ell)$$

given the following standard enthalpies of formation.

Species	ΔH_f° (kJ/mol)
$CH_3OH(\ell)$	-238.7
$CO_2(g)$	-393.5
$H_2O(g)$	-241.8
$H_2O(\ell)$	-285.8

a) -440.6 kJ
b) -1276.8 kJ
c) -1452.8 kJ
d) -2407.6 kJ

Answer: c

CHAPTER 7: Atomic Structure

7.1 Electromagnetic Radiation

1. Which of the following regions of the electromagnetic spectrum have longer wavelengths than visible light?
   1. FM radio waves
   2. x-rays
   3. microwaves

   a) 1 only b) 2 only c) 3 only d) 1 and 2 e) 1 and 3

   Answer: e

2. Which of the following regions of the electromagnetic spectrum has the longest wavelengths?
   a) microwave b) infrared c) x-ray d) gamma ray e) visible

   Answer: a

3. Which of the following colors of visible light has the longest wavelength?
   a) orange b) blue c) red d) green e) yellow

   Answer: c

4. Which of the following regions of the electromagnetic spectrum has the lowest frequency?
   a) ultraviolet b) infrared c) visible d) x-ray e) gamma ray

   Answer: b

5. An argon ion laser emits light at 488 nm. What is the frequency of this radiation?
   a) 4.07×10^{-19} s^{-1} b) 1.63×10^{-15} s^{-1} c) 1.46×10^{2} s^{-1} d) 2.05×10^{6} s^{-1} e) 6.14×10^{14} s^{-1}

   Answer: e

6. If an AM radio station broadcasts at 995 kHz, what is the wavelength of this radiation?
   a) 6.59×10^{-28} m b) 1.01×10^{-6} m c) 3.32×10^{-3} m d) 301 m e) 2.98×10^{14} m

   Answer: d

7. A microwave oven emits radiation at a wavelength of 0.500 cm. What is the frequency of this radiation?
   a) 1.67×10^{-11} s^{-1} b) 6.67×10^{-7} s^{-1} c) 2.00 s^{-1} d) 1.50×10^{6} s^{-1} e) 6.00×10^{10} s^{-1}

   Answer: e

8. An radium-266 atom may emit a photon of light with a frequency of 4.50×10^{19} s^{-1} when it undergoes nuclear fission. What is the wavelength (in pm) of this photon?
a) 0.0298 pm b) 6.66 pm c) 15.0 pm d) 29.8 pm e) 135 pm

Answer: b

7.2 Planck, Einstein, Energy, and Photons

9. The ________ of a photon of light is ________ proportional to its frequency and ________ proportional to its wavelength.
a) amplitude, directly, inversely
b) energy, inversely, directly
c) energy, directly, inversely
d) velocity, inversely, directly
e) velocity, directly, inversely

Answer: c

10. According to experiments concerned with the photoelectric effect, which of the following will increase the kinetic energy of an electron ejected from a metal surface?
   1. increasing the wavelength of the light striking the surface
   2. increasing the frequency of the light striking the surface
   3. increasing the number of photons of light striking the surface

a) 1 only b) 2 only c) 3 only d) 1 and 2 e) 2 and 3

Answer: b

11. Planck suggested that all energy gained or lost by an atom must be some integral multiple of a minimum amount of energy called a ________.
a) Bohr b) frequency c) photon d) transition e) quantum

Answer: e

12. A common infrared laser operates at 1.06×10^3 nm. What is the energy of a photon with this wavelength?
a) 7.02×10^{-40} J b) 6.25×10^{-28} J c) 3.54×10^{-15} J d) 1.87×10^{-19} J e) 2.83×10^{14} J

Answer: d

13. Excited hydrogen atoms emit light in the ultraviolet at 2.47×10^{15} Hz. What is the energy of a single photon with this frequency?
a) 8.04×10^{-41} J b) 2.68×10^{-19} J c) 1.64×10^{-18} J d) 6.11×10^{-17} J e) 1.21×10^{-7} J

Answer: c

14. A red laser pointer emits light at a wavelength of 635 nm. If the laser emits 7.5×10^{-4} J of energy per second in the form of visible radiation, how many photons per second are emitted from the laser?
a) 4.2×10^{-16} photons/sec
b) 2.4×10^{15} photons/sec
c) 6.3×10^{17} photons/sec
d) 2.7×10^{22} photons/sec
e) 1.8×10^{36} photons/sec

Answer: b

15. What is the energy of a mole of photons of red light with a wavelength of 632 nm?
a) 189 kJ b) 252 kJ c) 314 kJ d) 515 kJ e) 756 kJ

Answer: a

16. If the energy of 1.00 mole of photons is 441 kJ, what is the wavelength of the light?
a) 271 nm b) 451 nm c) 680 nm d) 732 nm e) 891 nm

Answer: a

17. What is the binding energy of an electron in a photosensitive metal (in kJ/mol) if the minimum frequency of light that can eject electrons from the metal is 6.30×10^{14} Hz?
a) 121 kJ/mol b) 212 kJ/mol c) 251 kJ/mol d) 315 kJ/mol e) 476 kJ/mol

Answer: c

18. The energy required to break one mole of chlorine-chlorine bonds in Cl_2 is 242 kJ/mol. What is the longest wavelength of light capable of breaking a single chlorine-chlorine bond?
a) 124 nm b) 242 nm c) 421 nm d) 494 nm e) 821 nm

Answer: d

7.3 Atomic Line Spectra and Niels Bohr

19. For a hydrogen atom, calculate the wavelength of the line in the Lyman series that results from the transition $n = 4$ to $n = 1$. The Rydberg constant equals -2.18×10^{-18} J.
a) 30.4 nm b) 97.2 nm c) 114 nm d) 121 nm e) 182 nm

Answer: b

20. According to the Bohr model for the hydrogen atom, the energy necessary to excite an electron from $n = 2$ to $n = 3$ is ________ the energy necessary to excite an electron from $n = 4$ to $n = 5$.
a) less than
b) greater than
c) equal to
d) either equal to or greater than
e) either less than or equal to

Answer: b

21. Which of the following transitions in a hydrogen atom would **emit** the lowest energy photon?
a) $n = 1$ to $n = 2$ b) $n = 3$ to $n = 2$ c) $n = 5$ to $n = 1$ d) $n = 2$ to $n = 8$ e) $n = 6$ to $n = 5$

Answer: e

22. For a hydrogen atom, calculate the energy of a photon in the Balmer series that results from the transition $n = 3$ to $n = 2$. What is the region of the electromagnetic spectrum? The Rydberg constant is -2.18×10^{-18} J.
a) 1.09×10^{-17} J; ultraviolet region
b) 2.18×10^{-18} J; ultraviolet region
c) 2.18×10^{-18} J; visible region
d) 3.03×10^{-19} J; visible region
e) 3.03×10^{-19} J; ultraviolet region

Answer: d

23. If a hydrogen atom in the excited $n = 3$ state relaxes to the ground state, what is the maximum number of possible emission lines?
a) 0 b) 1 c) 2 d) 3 e) infinite

Answer: d

<u>7.4 The Wave Properties of the Electron</u>

24. For a neutron (mass = 1.675×10^{-27} kg) moving with a velocity of 5.2×10^3 m/s, what is the de Broglie wavelength?
a) 7.6×10^{-11} m b) 4.5×10^{-9} m c) 2.1×10^{-6} m d) 486 m e) 1.3×10^{10} m

Answer: a

25. If the de Broglie wavelength of an electron is 555 nm, what is its velocity? The mass of an electron is 9.1×10^{-31} kg.
a) 4.0×10^{-10} m/s b) 7.6×10^{-4} m/s c) 5.2×10^2 m/s d) 1.3×10^3 m/s e) 2.5×10^9 m/s

Answer: d

26. What is the de Broglie wavelength of a 120 g golf ball traveling at 65 km/hr?
a) 8.5×10^{-38} m b) 3.1×10^{-34} m c) 0.15 m d) 3.3×10^{33} m e) 1.2×10^{37} m

Answer: b

7.5 Quantum Mechanical View of the Atom

27. Which of the following types of experiments demonstrate that an electron has the properties of a particle?
a) nuclear fission
b) electron diffraction
c) light emission from atomic gases
d) mass spectroscopy
e) photoelectric effect

Answer: e

28. The Schrödinger wave equation
a) calculates the position and momentum of an electron at any given time.
b) can be solved to determine the probability of finding an electron in a region of space.
c) proves that energy is equal to mass times the speed of light squared.
d) incorrectly predicts circular orbits of electrons around nuclei.
e) is used to calculate the velocity of an electron.

Answer: b

29. Which of the following statements is INCORRECT?
a) The location and the energy of an electron in an atom can be simultaneously known with little or no uncertainty.
b) The energies of an atom's electrons are quantized.
c) Quantum numbers define the energy states and the orbitals available to an electron.
d) The behavior of an atom's electrons can be described by standing waves.
e) Electrons have both wave and particle properties.

Answer: a

30. What type of orbital is designated $n = 3,\ \ell = 2,\ m_\ell = -1$?
a) 3*s* b) 3*p* c) 3*d* d) 2*f* e) 2*d*

Answer: c

31. What type of orbital is designated $n = 2,\ \ell = 0,\ m_\ell = 0$?
a) 2*s* b) 2*p* c) 2*d* d) 2*f* e) none

Answer: a

32. What type of orbital is designated $n = 4$, $\ell = 3$, $m_\ell = -3$?

a) 4*s* b) 4*p* c) 4*d* d) 4*f* e) none

Answer: d

33. What shell contains a total of 9 orbitals?

a) $n = 1$ b) $n = 2$ c) $n = 3$ d) $n = 4$ e) $n = 5$

Answer: c

34. Which of the following sets of quantum numbers refers to a 4*p* orbital?

a) $n = 1, \ell = 1, m_\ell = -1$

b) $n = 1, \ell = 2, m_\ell = -1$

c) $n = 4, \ell = 1, m_\ell = 0$

d) $n = 4, \ell = 2, m_\ell = -1$

e) $n = 4, \ell = 3, m_\ell = +2$

Answer: c

35. How many orbitals have the following set of quantum numbers: $n = 6$, $\ell = 3$, $m_\ell = -2$?

a) 0 b) 1 c) 3 d) 6 e) 7

Answer: b

36. All of the following sets of quantum numbers are allowed EXCEPT

a) $n = 6, \ell = 0, m_\ell = +1$

b) $n = 5, \ell = 4, m_\ell = 0$

c) $n = 4, \ell = 1, m_\ell = -1$

d) $n = 3, \ell = 2, m_\ell = +2$

e) $n = 1, \ell = 0, m_\ell = 0$

Answer: a

37. All of the following sets of quantum numbers are allowed EXCEPT

a) $n = 5$, $\ell = 3$, $m_\ell = +2$

b) $n = 3$, $\ell = 2$, $m_\ell = -1$

c) $n = 3$, $\ell = 0$, $m_\ell = 0$

d) $n = 4$, $\ell = 4$, $m_\ell = -2$

e) $n = 5$, $\ell = 3$, $m_\ell = +3$

Answer: d

38. What is the total number of orbitals having $n = 5$ and $\ell = 1$?

a) 1 b) 2 c) 3 d) 5 e) 10

Answer: c

39. Which of the following properties is associated with the value of the m_ℓ quantum number?

a) the number of electrons in an orbital
b) the size of an orbital
c) the shape of an orbital
d) the energy of an orbital
e) the orientation in space of an orbital

Answer: e

40. Which of the following properties is associated with the value of the ℓ quantum number?

a) the number of electrons in an orbital
b) the size of an orbital
c) the orientation in space of an orbital
d) the energy of an orbital
e) the shape of an orbital

Answer: e

41. Which of the following orbitals might have $m_\ell = -2$?

a) *s* b) *s* and *p* c) *p* and *d* d) *d* and *f* e) *p*, *d*, and *f*

Answer: d

42. How many nodal planes exist for a 2*p* orbital?

a) 0 b) 1 c) 2 d) 3 e) 4

Answer: b

43. The n = _____ shell is the lowest that may contain *p*-orbitals.
a) 2 b) 3 c) 4 d) 5 e) 6

Answer: a

44. Which of the following diagrams represent *d*-orbitals?

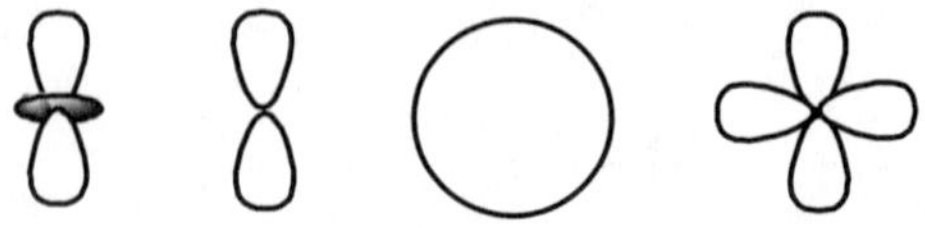

(I) (II) (III) (IV)

a) (I) only b) (II) only c) (III) only d) (IV) only e) (I) and (IV)

Answer: e

45. Which of the following diagrams represent *p*-orbitals?

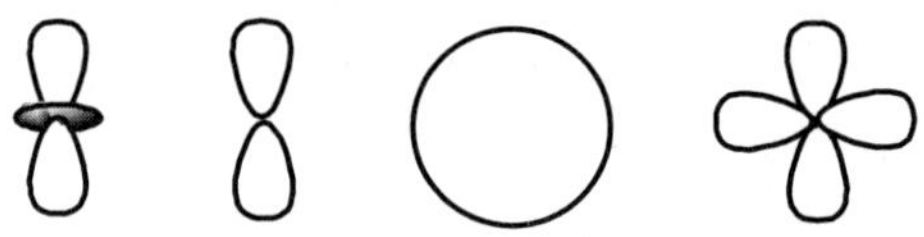

(I) (II) (III) (IV)

a) (I) only b) (II) only c (III) only d) (IV) only e) (I) and (II)

Answer: b

<u>Short Answer Questions</u>

46. According to Heisenberg's uncertainty principle, it is impossible to simultaneously measure the exact ________ and momentum of an electron.

Answer: position (or location)

47. A ________ is a region of an electron probability density graph where the probability of finding an electron is zero.

Answer: nodal surface

48. A(n) ________ is a massless particle of radiation.

Answer: photon

49. The ________ quantum number is given the symbol n. This quantum number is the primary factor in determining the energy and size of an orbital.

Answer: principal

50. The ________ of a photon is equal to the speed of light divided by the wavelength of the photon.

Answer: frequency

51. The Bohr model predicts that the energy of an atom'a electron is ________, meaning that the electron can only occupy orbitals of specific energies.

Answer: quantized

52. A photon wavelength of 1.0×10^{-5} m falsl in the ________ region of the electromagnetic spectrum.

Answer: infrared

53. What evidence does the photoelectric effect provide that photons are not only waves?

Answer: The photoelectric effect provides evidence that an electron is ejected when a single particle of light (as opposed to a wave of light) interacts with a metal surface. When photons of sufficient energy strike a metal surface, electrons are ejected. The number of ejected electrons is proportion to the number of photons striking the surface (which is consistent with both wave and particle theories of light). However, the energy of an ejected electron is independent of the number of photons striking the surface. If the interaction of matter and light was dependent on wave properties, then increasing the intensity of the light would increase the energy striking the surface and thus increase the energy of an ejected electron. Furthermore, there is a minimum wavelength of light required to eject an electron from a metal surface. At longer wavelengths, no photons are ejected from the metal regardless of the intensity of the light.

54. The size of an electron orbital is often chosen to be that where 90% of the electron density is found. Why is the orbital radius not chosen to be the distance within which 100% of the electron density is found?

Answer: The probability of finding an electron drops exponentially as distance from the nucleus increases. However, the probability does not reach zero until the electron is an infinite distance from the nucleus.

Diagnostic Quiz Questions

Goal #1: Describe the properties of electromagnetic radiation

1. Which of the following types of radiation has the highest energy?
   a) microwave
   b) radio-wave
   c) infrared
   d) visible

   Answer: d

2. Which of the following types of radiation has the longest wavelength?
   a) x-ray
   b) ultraviolet
   c) microwave
   d) visible

   Answer: c

3. A typical laser pointer emits lights with a wavelength of 6.50×10^2 nm. What is the frequency of this light?
   a) 2.17×10^{-15} s^{-1}
   b) 5.13×10^{-3} s^{-1}
   c) 1.95×10^{2} s^{-1}
   d) 4.61×10^{14} s^{-1}

   Answer: d

4. What is the wavelength (in nm) of x-ray radiation with a frequency of 6.39×10^{17} s^{-1}?
   a) 0.469 nm
   b) 2.13 nm
   c) 46.9 nm
   d) 192 nm

   Answer: a

5. What is the energy (in Joules) of a 455 nm photon?
   a) 4.86×10^{-36} J
   b) 9.04×10^{-32} J
   c) 4.37×10^{-19} J
   d) 1.46×10^{-18} J

   Answer: c

6. An energy of 353 kJ/mol is required to cause a magnesium atom on a metal surface to lose an electron. Calculate the lowest frequency of light that can ionize a magnesium atom.
   a) $3.38 \times 10^7\ s^{-1}$
   b) $8.85 \times 10^{14}\ s^{-1}$
   c) $1.07 \times 10^{16}\ s^{-1}$
   d) $5.33 \times 10^{35}\ s^{-1}$

   Answer: b

7. What is the wavelength (in nanometers) of a 3.44×10^{-18} J photon?
   a) 57.7 nm
   b) 173 nm
   c) 562 nm
   d) 8.72×10^4 nm

   Answer: a

Goal #2: Understand the origin of light from excited atoms and its relationship to atomic structure.

8. For hydrogen, emission energies can by calculated using the Bohr model.

$$\Delta E = -2.18 \times 10^{-18}\ \text{J/atom}\left(\frac{1}{n^2_{\text{final}}} - \frac{1}{n^2_{\text{initial}}}\right)$$

   The Lyman series corresponds to transitions from excited electronic states to a final state of $n = 1$. If the wavelength of a transition in the Lyman series is 121.5 nm, what is the initial state of the atom?
   a) 2
   b) 3
   c) 4
   d) 5

   Answer: a

9. For hydrogen, which of the following electron transitions requires the least energy?
   a) $n = 1$ to $n = 2$
   b) $n = 1$ to $n = 3$
   c) $n = 2$ to $n = 3$
   d) $n = 3$ to $n = 4$

   Answer: d

Goal #3: Describe the experimental evidence for the wave-particle duality

10. Calculate the wavelength (in nanometers) associated with a proton moving at 1.0×10^3 m/s. The mass of a proton is 1.67×10^{-27} kg.
   a) 0.032 nm
   b) 0.40 nm
   c) 2.5 nm
   d) 14 nm

   Answer: b

Goal #4: Describe the basic ideas of quantum mechanics.

11. Which of the following statements are correct?
   1. The behavior of an electron in an atom can be described by a standing wave.
   2. The energy of an electron in an atom is quantized.
   3. Quantum numbers are used to define the energy states and orbitals available to electrons in an atom.

   a) 1 only
   b) 2 only
   c) 3 only
   d) 1, 2, and 3

   Answer: d

Goal #5: Define the three quantum numbers (n, ℓ, and m_ℓ) and their relation to atomic structure.

12. Which of the following statements are correct?
   1. The principal quantum number, *n*, describes the shape of an atomic orbital.
   2. No more than two electrons in an atom can have the same principal quantum number (*n*).
   3. The magnetic quantum number, m_ℓ, is related to the orientation in space of an orbital within a given subshell.

   a) 1 only
   b) 2 only
   c) 3 only
   d) 2 and 3

   Answer: c

13. Which of the following is a possible set of quantum numbers for an electron in an atom?
    a) $n = 1, \ell = 1, m_\ell = 1$
    b) $n = 2, \ell = 0, m_\ell = -1$
    c) $n = 0, \ell = 0, m_\ell = 0$
    d) $n = 3, \ell = 1, m_\ell = -1$

    Answer: d

14. What is the maximum number of orbitals that can be occupied by electrons in the $n = 2$ shell?
    a) 1
    b) 2
    c) 4
    d) 9

    Answer: c

15. How many subshells occur in an electron shell with the principal quantum number $n = 3$?
    a) 1
    b) 2
    c) 3
    d) 4

    Answer: c

16. What is the maximum number of orbitals that can be identified with the following quantum numbers: $n = 5, \ell = 3$?
    a) 1
    b) 3
    c) 5
    d) 7

    Answer: d

17. What is the maximum number of orbitals that can be identified with the following quantum numbers: $n = 3, \ell = 1, m_\ell = 0$?
    a) 0
    b) 1
    c) 3
    d) 5

    Answer: b

18. If a subshell is labeled *p*, what is the maximum number of orbitals that can occupy the subshell?
    a) 1
    b) 3
    c) 5
    d) 7

    Answer: b

19. A particular orbital has $n = 4$, $\ell = 2$. Identify the orbital.
    a) 2*f*
    b) 2*g*
    c) 4*p*
    d) 4*d*

    Answer: d

20. How many nodal surfaces are associated with a 3*s* orbital?
    a) 0
    b) 1
    c) 2
    d) 3

    Answer: a

CHAPTER 8: Atomic Electron Configurations and Chemical Periodicity

8.1 Electron Spin

1. Which of the following statements are CORRECT?
   1. A paramagnetic substance is attracted to a magnetic field.
   2. Atoms with one or more unpaired electrons are paramagnetic.
   3. Electron spin is quantized.

   a) 1 only b) 2 only c) 3 only d) 1 and 2 e) 1, 2, and 3

   Answer: e

2. Which of the following statements are CORRECT?
   1. Substances that retain their magnetism after they are withdrawn from a magnetic field are called ferromagnetic.
   2. A diamagnetic substance is slightly repelled by a magnetic field.
   3. An atom with an even number of electrons is diamagnetic.

   a) 1 only b) 2 only c) 3 only d) 1 and 2 e) 1, 2, and 3

   Answer: d

8.2 The Pauli Exclusion Principle

3. The Pauli exclusion principle states that
   a) no two electrons in an atom can have the same four quantum numbers.
   b) electrons can have either ±½ spins.
   c) electrons with opposing spins are attracted to each other.
   d) no two electrons in an atom can have the same spin.
   e) atoms with no unpaired electrons are diamagnetic.

   Answer: a

4. How many electrons can be described by the following quantum numbers: $n = 4$, $\ell = 3$, $m_\ell = -2$, $m_s = +1/2$?

   a) 0 b) 1 c) 2 d) 3 e) 6

   Answer: b

5. How many electrons can be described by the quantum numbers $n = 5$, $\ell = 3$, $m_s = -1/2$?

   a) 1 b) 3 c) 5 d) 7 e) 9

   Answer: d

6. How many electrons can be described by the quantum numbers $n = 7$, $\ell = 1$?

   a) 0 b) 2 c) 6 d) 7 e) 14

   Answer: c

7. What is the maximum number of electrons that can occupy the $n = 5$ shell?
a) 5 b) 10 c) 14 d) 25 e) 50

Answer: e

8. What is the maximum number of electrons that can occupy the $n = 2$ shell?
a) 4 b) 8 c) 18 d) 32 e) 50

Answer: b

9. Which one of the following sets of quantum numbers is NOT allowed?

a) $n = 6, \ell = 0, m_\ell = 0, m_s = +1/2$

b) $n = 5, \ell = 3, m_\ell = 1, m_s = +1/2$

c) $n = 4, \ell = 2, m_\ell = -3, m_s = -1/2$

d) $n = 3, \ell = 1, m_\ell = -1, m_s = +1/2$

e) $n = 2, \ell = 0, m_\ell = 0, m_s = -1/2$

Answer: c

10. Which of the following sets of quantum numbers is allowed?

a) $n = 2, \ell = 1, m_\ell = 1, m_s = -1/2$

b) $n = 3, \ell = 2, m_\ell = 1, m_s = +1$

c) $n = 3, \ell = 1, m_\ell = -3, m_s = -1/2$

d) $n = 4, \ell = 4, m_\ell = -1, m_s = +1/2$

e) $n = 5, \ell = 2, m_\ell = +2, m_s = -1$

Answer: a

8.3 Atomic Subshell Energies and Electron Assignments

11. Which of the following statements regarding subshell filling order are CORRECT?
    1. Electrons are assigned to the 4*s* subshell before they are assigned to the 3*d* subshell.
    2. Electrons are assigned to the 4*f* subshell before they are assigned to the 6*s* subshell.
    3. Electrons are assigned to the 5*d* subshell before they are assigned to the 6*p* subshell.

a) 1 only b) 2 only c) 3 only d) 1 and 3 e) 1, 2, and 3

Answer: d

12. Which of the following statements concerning ground state electron configurations are CORRECT?
    1. In a hydrogen atom with one electron, the 2*s* and 2*p* orbitals have different energy.
    2. For a lithium atom with three electrons, the 2*s* and 2*p* orbitals have different energies.
    3. The effective nuclear charge felt by an electron in a 2*p* orbital is larger for carbon than for fluorine.

    a) 1 only b) 2 only c) 3 only d) 1 and 3 e) 1, 2, and 3

    Answer: b

8.4 Atomic Electron Configurations

13. Which of the following elements is an *s*-block element?
    a) C b) Cd c) Cs d) Cl e) Cf

    Answer: c

14. Which of the following elements is an *f*-block element?
    a) C b) Cd c) Cs d) Cl e) Cf

    Answer: e

15. Which of the following atoms is paramagnetic?
    a) Se b) Cd c) Ar d) He e) Ca

    Answer: a

16. Which of the following atoms is diamagnetic?
    a) Rb b) C c) F d) Cu e) Sr

    Answer: e

17. Which element has the electron configuration $1s^2 2s^2 2p^6 3s^2 3p^4$?
    a) Mg b) S c) Si d) Se e) Ga

    Answer: b

18. Which element has the electron configuration $[Ar]3d^{10}4s^1$?
    a) Cu b) Zn c) Ga d) Ag e) Co

    Answer: a

19. What is the electron configuration for a lead atom?
   a) $[Xe]6s^2 6p^2$
   b) $[Kr]5d^{10} 6s^2 6p^2$
   c) $[Xe]4f^{14} 5d^{10} 6s^2 6p^2$
   d) $[Xe]5d^{10} 6s^2 6p^2$
   e) $[Kr]4d^{10} 4f^{14} 5d^2$

   Answer: c

20. Hund's rule states that the most stable arrangement of electrons (for a ground state electron configuration)
   a) has a filled valence shell of electrons.
   b) has two electrons per orbital, each with identical spins.
   c) has m_ℓ values greater than or equal to zero.
   d) has the maximum number of unpaired electrons, all with the same spin.
   e) has two electrons per orbital, each with opposing spins.

   Answer: d

21. Which element has the following electron configuration?

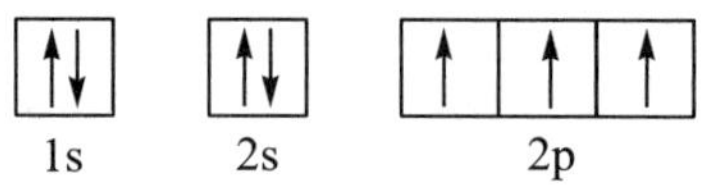

a) P b) S c) O d) N e) F

   Answer: d

22. Which element has the following electron configuration?

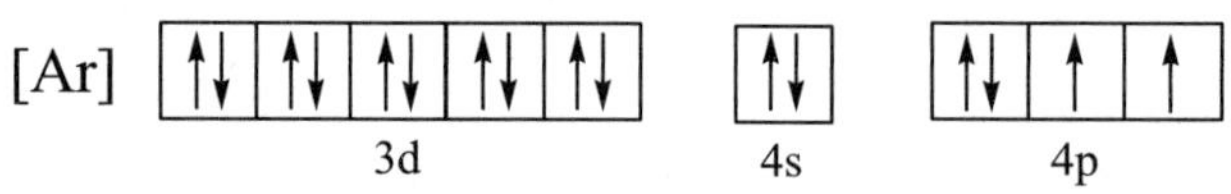

a) Br b) Se c) S d) Cl e) Ge

   Answer: b

23. What is a possible set of quantum numbers for the unpaired electron in the orbital box diagram below?

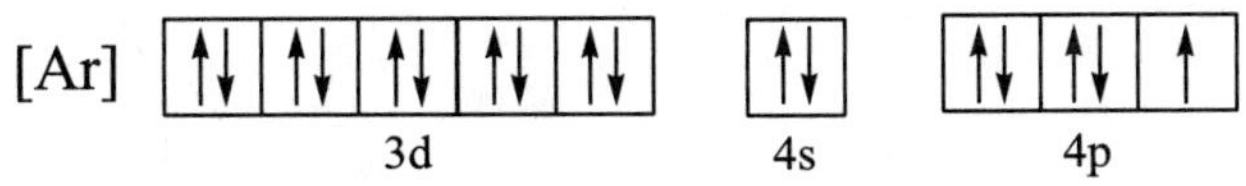

a) $n = 3$, $\ell = 1$, $m_\ell = -1$, $m_s = +1/2$

b) $n = 3$, $\ell = 2$, $m_\ell = -1$, $m_s = -1/2$

c) $n = 3$, $\ell = 2$, $m_\ell = -2$, $m_s = +1/2$

d) $n = 4$, $\ell = 0$, $m_\ell = 0$, $m_s = +1/2$

e) $n = 4$, $\ell = 1$, $m_\ell = -1$, $m_s = +1/2$

Answer: e

8.5 Electron Configurations of Ions

24. For which of the following atoms is the common ion paramagnetic?
a) Al b) Mg c) Ni d) F e) H

Answer: c

25. Which 3+ ion has the ground state electron configuration $[Kr]4d^5$?
a) Ru b) Fe c) Ag d) Tc e) Nb

Answer: a

26. If the ground state electron configuration of an element is $[Ar]3d^{10}4s^2 4p^3$, what is the typical charge on the monatomic anion of the element?
a) 2+ b) 1+ c) 1- d) 2- e) 3-

Answer: e

27. What is the ground state electron configuration for Cr^{3+}?
a) [Ar] b) $[Ar]3d^3 4s^2$ c) $[Ar]3d^4 4s^1$ d) $[Ar]3d^3$ e) $[Ar]3d^7 4s^2$

Answer: d

28. What is the ground state electron configuration for Sn^{2+}?
a) $[Kr]4d^{10}5s^2$
b) $[Kr]4d^{10}5p^2$
c) $[Kr]5s^2$
d) $[Kr]4f^{14}5d^{10}$
e) $[Kr]4f^{14}5d^{10}5s^2$

Answer: a

29. Which of the following ions have the same ground state electron configuration: Sn^{4+}, Pb^{4+}, Sr^{2+}, and Br^-?
a) Sn^{4+} and Pb^{4+}
b) Pb^{4+} and Sr^{2+}
c) Sr^{2+} and Br^-
d) Pb^{4+} and Br^-
e) Sn^{4+}, Sr^{2+}, and Br^-

Answer: c

30. What 3- ion has the following ground state electron configuration?

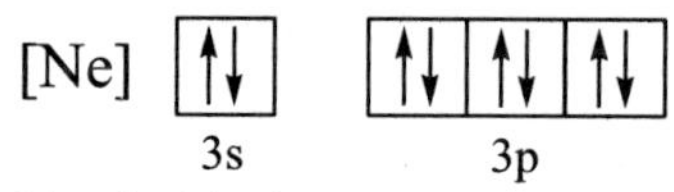

a) oxide ion b) nitride ion c) sodium ion d) aluminum ion e) phosphide ion

Answer: e

31. What 2+ ion has the following ground state electron configuration?

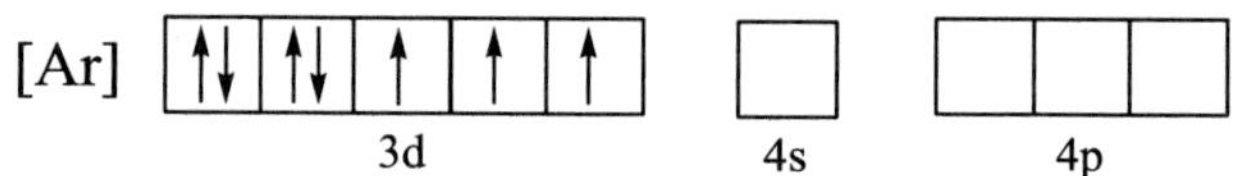

a) Ni^{2+} b) Co^{2+} c) Ca^{2+} d) Cu^{2+} e) Mn^{2+}

Answer: b

32. What 1+ ion has the following ground state electron configuration?

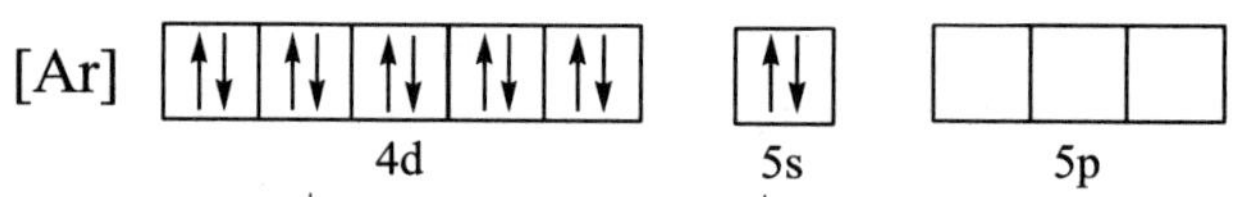

a) K^+ b) Rb^+ c) In^+ d) Tl^+ e) Y^+

Answer: c

33. Which of the following cations has four unpaired electrons in the ground state?
a) Ni^{2+} b) Fe^{2+} c) Pb^{2+} d) Mn^{2+} e) Cu^{2+}

Answer: b

8.6 Atomic Properties and Periodic Trends

34. In general, atomic radii
    a) decrease down a group and remain constant across a period.
    b) decrease down a group and increase across a period.
    c) increase down a group and increase across a period.
    d) increase down a group and remain constant across a period.
    e) increase down a group and decrease across a period.

    Answer: e

35. Place the following atoms in order of increasing atomic radii: Ca, Mg, P, and Cl.
    a) Cl < P < Mg < Ca
    b) Mg < P < Cl < Ca
    c) Ca < Mg < P < Cl
    d) P < Cl < Mg < Ca
    e) Ca < Cl < P < Mg

    Answer: a

36. Place the following atoms in order of increasing atomic radii: Se, O, S, and As?
    a) O < S < Se < As
    b) O < S < As > Se
    c) As < Se < S < O
    d) Se < As < S < O
    e) S < As < O < Se

    Answer: a

37. Which one of the following statements is INCORRECT?
    a) Ionization energy is always a positive value.
    b) Ionization energy is the energy required to remove an electron from a gaseous atom.
    c) For any element, the second ionization energy is larger than the first ionization energy.
    d) Ionization energy decreases across a periodic of the periodic table.
    e) Ionization energy decreases down a group of the periodic table

    Answer: d

38. Which of the following chemical equations refers to the second ionization of Mg?
    a) $Mg(s) + 2e^- \rightarrow Mg^{2-}(s)$
    b) $Mg(s) \rightarrow Mg^+(s) + e^-$
    c) $Mg(g) \rightarrow Mg^{2+}(g) + 2e^-$
    d) $Mg^+(g) \rightarrow Mg^{2+}(g) + e^-$
    e) $Mg^{2+}(g) + e^- \rightarrow Mg^+(g)$

    Answer: d

39. Which of the following elements would have the greatest difference between the first and the second ionization energies?
   a) Mg b) Al c) K d) Ca e) Sr

   Answer: c

40. Rank K, Li, and Cs in order of increasing first ionization energy.
   a) K < Cs < Li
   b) Cs < K < Li
   c) Cs < Li < K
   d) Li < Cs < K
   e) Li < K < Cs

   Answer: b

41. The change in energy for the following reaction is referred to as the ________ for fluorine.
   $$F(g) + e^- \rightarrow F^-(g)$$
   a) oxidation energy
   b) electron affinity
   c) electronegativity energy
   d) first ionization energy
   e) second ionization energy

   Answer: b

42. Which of the following elements has the most negative electron affinity?
   a) Li b) B c) C d) N e) F

   Answer: e

43. Which elements have no affinity for electrons?
   a) transition metals
   b) main group metals
   c) main group nonmetals
   d) semiconductors
   e) noble gases

   Answer: e

44. Which group of the periodic table of elements forms only 2+ ions?
   a) group 1A b) group 2A c) group 7B d) group 7A e) group 8A

   Answer: b

45. Place the following ions in order from smallest to largest ionic radii: P^{3-}, K^+, Na^+, and Cl^-.
   a) $Na^+ < K^+ < Cl^- < P^{3-}$
   b) $K^+ < Na^+ < Cl^- < P^{3-}$
   c) $Cl^- < P^{3-} < Na^+ < K^+$
   d) $Cl^- < P^{3-} < K^+ < Na^+$
   e) $P^{3-} < Cl^- < K^+ < Na^+$

   Answer: *a*

Short Answer Questions

46. The ________ principle leads to the conclusion that no more than two electrons can occupy the same orbital and that the two electrons must have opposite spins.

   Answer: Pauli exclusion

47. The element ________ has the following electron configuration: $[Rn]5f^7 7s^2$

   Answer: americium

48. As one moves horizontally from left to right across a period, the effective ________ charge increases, resulting in decreasing atomic radii.

   Answer: nuclear

49. The f-block elements are also referred to as the lanthanides and ________.

   Answer: actinides

50. The maximum number of electrons that can be placed in an $n = 4$ shell is ________.

   Answer: 32

51. Explain why the first ionization energy for boron is lower than that for beryllium.

   Answer: Beryllium has a $1s^2 2s^2$ electron configuration and boron has a $1s^2 2s^2 2p^1$ electron configuration. The $2s$ electrons partially shield the $2p$ electron from the nuclear charge, which results in the $2p$ electron being slightly higher in energy and easier to ionize than a $2s$ electron.

52. Explain the difference between paramagnetic and ferromagnetic.

Answer: Elements with unpaired electrons are paramagnetic. In the absence of a magnetic field, the unpaired electrons in paramagnetic elements are randomly oriented. In the presence of an external magnetic field, the unpaired electrons align with the magnetic field. For some paramagnetic elements (notably iron, cobalt, and nickel), clusters of unpaired electrons will remain aligned even in the absence of an external magnetic field. These elements are called ferromagnetic.

Diagnostic Quiz Questions

Goal #1: Understand the role magnetism plays in determining and revealing atomic structure

1. What is the maximum number of electrons that can be identified with the following set of quantum numbers: $n = 4$, $\ell = 1$?
   a) 2
   b) 4
   c) 6
   d) 10

   Answer: c

2. What is the maximum number of electrons that can occupy an $n = 4$ shell?
   a) 8
   b) 18
   c) 32
   d) 50

   Answer: c

3. Which one of the following sets of quantum numbers is possible?
   a) $n = 1, \ell = 1, m_\ell = +1, m_s = +1/2$
   b) $n = 1, \ell = 0, m_\ell = 0, m_s = 0$
   c) $n = 2, \ell = 1, m_\ell = -1, m_s = -1/2$
   d) $n = 3, \ell = 0, m_\ell = -1, m_s = +1/2$

   Answer: c

4. What is a possible set of quantum numbers for the outermost electron in the ground state electron configuration of Na?
    a) $n = 3, \ell = 0, m_\ell = 0, m_s = +1/2$
    b) $n = 3, \ell = 1, m_\ell = +1, m_s = -1/2$
    c) $n = 2, \ell = 0, m_\ell = 0, m_s = +1/2$
    d) $n = 3, \ell = 1, m_\ell = 0, m_s = -1/2$

    Answer: a

5. Which of the following statements are correct?
    1. The Pauli exclusion principle states that no two electrons can have the same set of four quantum numbers.
    2. An atom is diamagnetic if it has an even number of unpaired electrons in atomic orbitals.
    3. An atom is paramagnetic if it has no unpaired electrons in its atomic orbitals.

    a) 1 only
    b) 2 only
    c) 3 only
    d) 1 and 3

    Answer: a

6. Which of the following atoms is paramagnetic in its ground electronic state?
    a) He
    b) Mg
    c) Li
    d) Ar

    Answer: c

7. Which of the following sets of quantum numbers represents a 4*d* electron?
    a) $n = 4, \ell = 0, m_\ell = 0, m_s = -1/2$
    b) $n = 4, \ell = 1, m_\ell = +1, m_s = 0$
    c) $n = 4, \ell = 2, m_\ell = +1, m_s = -1$
    d) $n = 4, \ell = 2, m_\ell = -1, m_s = -1/2$

    Answer: d

Goal #2: Understand effective nuclear charge and its role in determining atomic properties.

8. Which of the following statements concerning effective nuclear charge is correct?
    1. In a hydrogen atom with only 1 electron, the $2s$ and $2p$ orbitals have the same energy.
    2. For carbon, the $2p$ electrons experience a greater effective nuclear charge than the $2s$ electrons.
    3. In lithium, all three electrons experience an identical effective nuclear charge.

   a) 1 only
   b) 2 only
   c) 3 only
   d) 1 and 2

   Answer: a

Goal #3: Write the electron configurations for elements and monatomic ions.

9. Which element has the ground state electron configuration $1s^2 2s^2 2p^6 3s^2 3p^2$?
   a) Mg
   b) Al
   c) Si
   d) P

   Answer: c

10. Which 3+ ion has the ground state electron configuration $[Ar]3d^5$?
    a) Sc^{3+}
    b) V^{3+}
    c) Cr^{3+}
    d) Fe^{3+}

    Answer: d

11. What is the ground state electron configuration for a bromine atom?
    a) $[Ar]3s^2 3p^5$
    b) $[Ar]4s^2 4p^5$
    c) $[Ar]3d^{10} 4s^2 4p^3$
    d) $[Ar]3d^{10} 4s^2 4p^5$

    Answer: d

12. What is the ground state electron configuration of a calcium ion?
    a) $1s^22s^22p^63s^23p^6$
    b) $1s^22s^22p^63s^23p^64s^1$
    c) $1s^22s^22p^63s^23p^64s^2$
    d) $1s^22s^22p^63s^23p^64s^24p^2$

Answer: a

13. What element has the following ground state electron configuration: $[Rn]5f^67s^2$?
    a) Th
    b) U
    c) Pu
    d) Bk

Answer: c

14. What element has the following ground state electron configuration?

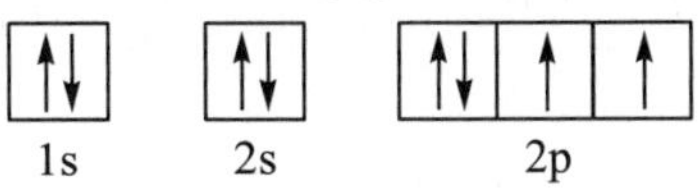

    a) N
    b) O
    c) F
    d) S

Answer: b

15. What 2+ ion has the following ground state electron configuration?

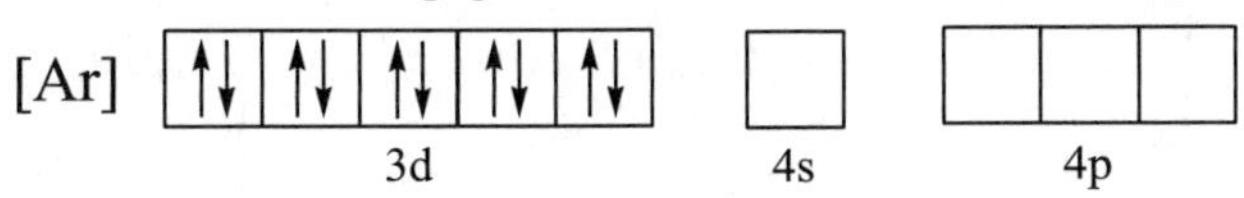

    a) Ca^{2+}
    b) Ni^{2+}
    c) Zn^{2+}
    d) Ge^{2+}

Answer: c

Goal #4: Understand the fundamental physical properties of the elements and their periodic trends

16. Which of the following ions have the same ground state electron configuration: Rb^+, K^+, P^{3-}, and Al^{3+}?
    a) Rb^+ and K^+
    b) K^+ and P^{3-}
    c) P^{3-} and Al^{3+}
    d) K^+ and Al^{3+}

    Answer: b

17. Place the following atoms in order from smallest to largest atomic radii: F, S, and Cl.
    a) F < Cl < S
    b) F < S < Cl
    c) S < Cl < F
    d) S < F < Cl

    Answer: a

18. Place the following ions in order from smallest to largest ionic radii: N^{3-}, F^-, and Mg^{2+}.
    a) $N^{3-} < F^- < Mg^{2+}$
    b) $F^- < N^{3-} < Mg^{2+}$
    c) $N^{3-} < F^- < Mg^{2+}$
    d) $Mg^{2+} < F^- < N^{3-}$

    Answer: d

19. Which of the following groups of elements is arranged correctly in order of increasing first ionization energy?
    a) O < S < Se
    b) Na < K < Rb
    c) O < F < Ne
    d) Cl < Ar < K

    Answer: c

20. Which of the following statements are correct?
    1. Electron affinity generally decreases from top to bottom in a group.
    2. Nonmetals generally have greater (i.e., more negative) electron affinities than metals.
    3. The greater the affinity an atom has for an electron, the more negative the electron affinity will be.

a) 1 only
b) 2 only
c) 3 only
d) 1, 2, and 3

Answer: d

CHAPTER 9: Bonding and Molecular Structure: Fundamental Concepts

9.1 Valence Electrons

1. For an atom of a main group element, the number of valence electrons is equal to
   a) its number of core electrons.
   b) the principle quantum number of its outer shell.
   c) its period number.
   d) its group number.
   e) eight minus its group number.

   Answer: d

2. What is the expected number of valence electrons for an atom of a group 6A element?
   a) 0 b) 1 c) 2 d) 4 e) 6

   Answer: e

3. A silicon atom has _____ valence electrons.
   a) 0 b) 2 c) 4 d) 6 e) 10

   Answer: c

9.2 Chemical Bond Formation

4. Which combination of atoms is most likely to produce a compound with ionic bonds?
   a) Al and F b) P and H c) C and O d) Si and O e) S and Br

   Answer: a

5. Which of the following statements are CORRECT?
   1. Ionic bonds form when one or more valence electrons are transferred from one atom to another.
   2. Covalent bonds involve sharing of electrons between atoms.
   3. Ionic bond formation is always exothermic; covalent bond formation is always endothermic.

   a) 1 only b) 2 only c) 3 only d) 1 and 2 e) 1, 2, and 3

   Answer: d

9.3 Bonding in Ionic Compounds

6. Which of the following compounds is expected to have the strongest ionic bonds?
   a) RbF b) NaF c) NaI d) CsBr e) CsI

   Answer: b

7. Which of the following compounds is expected to have the strongest ionic bonds?
a) MgO b) KBr c) NaI d) SrO e) CaS

Answer: a

8. The lattice energy of LiBr is -815 kJ/mol. This energy corresponds to which reaction below?
a) $Li(s) + \frac{1}{2} Br_2(g) \rightarrow LiBr(s)$
b) $Li(g) + Br(g) \rightarrow LiBr(g)$
c) $Li(g) + Br(g) \rightarrow LiBr(s)$
d) $Li^+(aq) + Br^-(aq) \rightarrow LiBr(s)$
e) $Li^+(g) + Br^-(g) \rightarrow LiBr(s)$

Answer: e

9. Calculate the lattice energy of NaBr(s), given the following thermochemical equations.

$Na(s) \rightarrow Na(g)$ $\Delta H_f^\circ = 107$ kJ
$Na(g) \rightarrow Na^+(g) + e^-$ $\Delta IE = 496$ kJ
$\frac{1}{2} Br_2(g) \rightarrow Br(g)$ $\Delta H_f^\circ = 112$ kJ
$Br(g) + e^- \rightarrow Br^-(g)$ $\Delta EA = -325$ kJ
$Na(s) + \frac{1}{2} Br_2(g) \rightarrow NaBr(s)$ $\Delta H_f^\circ = -361$ kJ

a) -963 kJ b) -751 kJ c) –290 kJ d) +290 kJ e) +1403 kJ

Answer: b

9.4 Covalent Bonding and Lewis Structures

10. What is the total number of valence electrons in a carbon dioxide molecule?
a) 4 b) 8 c) 16 d) 20 e) 22

Answer: c

11. What is the total number of valence electrons in a carbonate ion?
a) 20 b) 22 c) 24 d) 26 e) 30

Answer: c

12. Which of the following molecules or ions are isoelectronic: SO_2, CO_2, NO_2^-, ClO_2^-?
a) SO_2 and CO_2
b) SO_2 and NO_2^-
c) CO_2 and ClO_2^-
d) NO_2^- and ClO_2^-
e) SO_2, NO_2^-, and ClO_2^-

Answer: b

13. If two or more species have the same number of electrons resulting in similar Lewis structures, they are said to be ________.
 a) neutral
 b) resonant structures
 c) symmetric
 d) isoelectronic
 e) covalent

Answer: d

14. How many lone-pair electrons are on the carbon atom in CO_2?
 a) 0 b) 1 c) 2 d) 3 e) 4

Answer: a

15. Which of the following species will have a Lewis structure most like that of carbon disulfide, CS_2?
 a) NO_2^- b) SO_2 c) H_2S d) SF_2 e) SCN^-

Answer: e

16. Which of the following species will have a Lewis structure most like that of the hydronium ion, H_3O^+?
 a) NO_3^- b) NH_3 c) SO_3 d) CO_3^{2-} e) H_2CO

Answer: b

17. How many lone-pair electrons surround the each nitrogen atom in hydrazine, N_2H_4?
 a) 0 b) 1/2 c) 1 d) 3 e) 4

Answer: c

18. Which of the following is a correct Lewis structure for ozone, O_3?

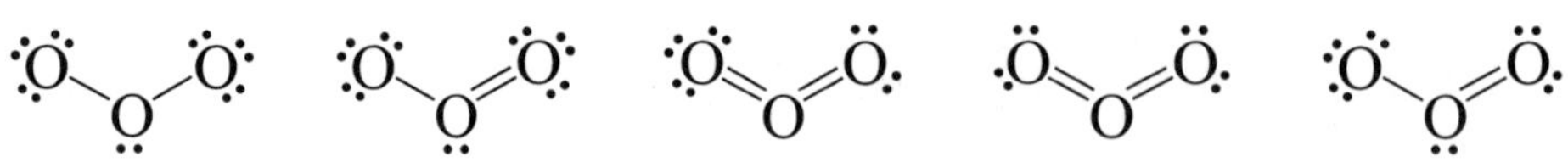

(1) (2) (3) (4) (5)

 a) 1 b) 2 c) 3 d) 4 e) 5

Answer: e

19. Which of the following is a correct Lewis structure for N_2O?

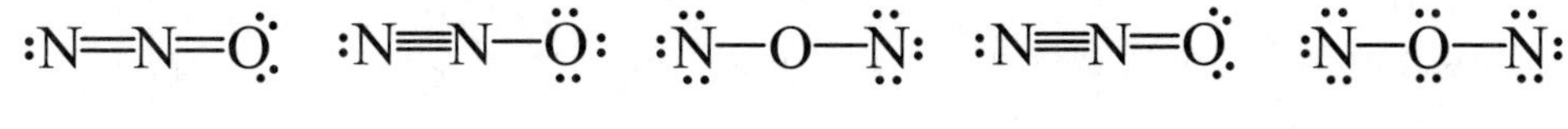

(1) (2) (3) (4) (5)

a) 1 b) 2 c) 3 d) 4 e) 5

Answer: b

20. Which of the following is a correct Lewis structure for the phosphate ion?

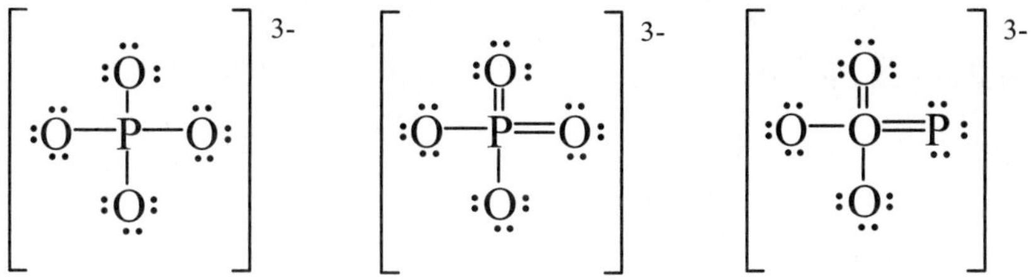

(1) (2) (3)

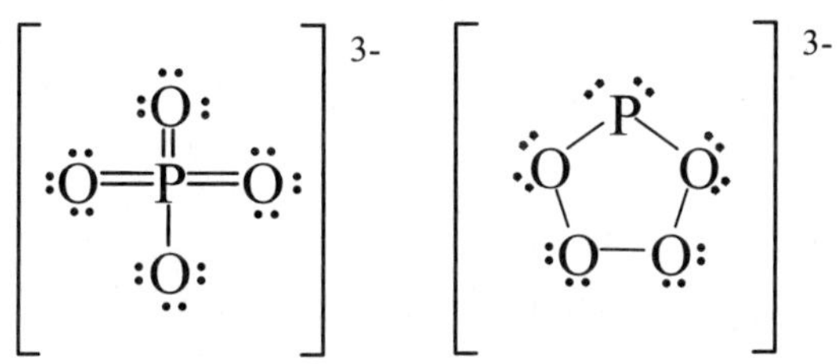

(4) (5)

a) 1 b) 2 c) 3 d) 4 e) 5

Answer: a

21. Which of the following are possible Lewis structures for C_2H_6O?

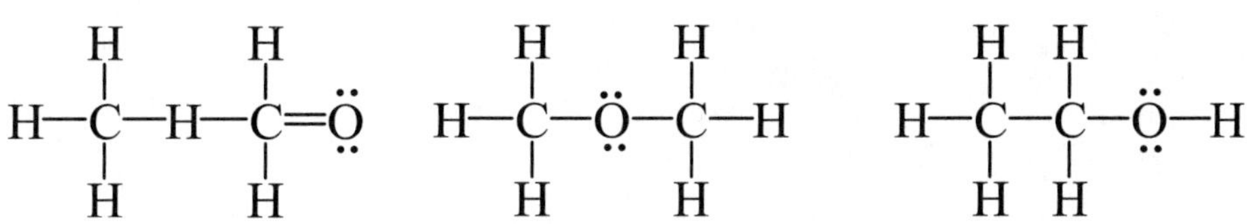

(1) (2) (3)

a) 1 b) 2 c) 3 d) 2 and 3 e) 1, 2, and 3

Answer: d

9.5 Resonance

22. Which of the following are resonance structures for sulfur dioxide, SO_2?

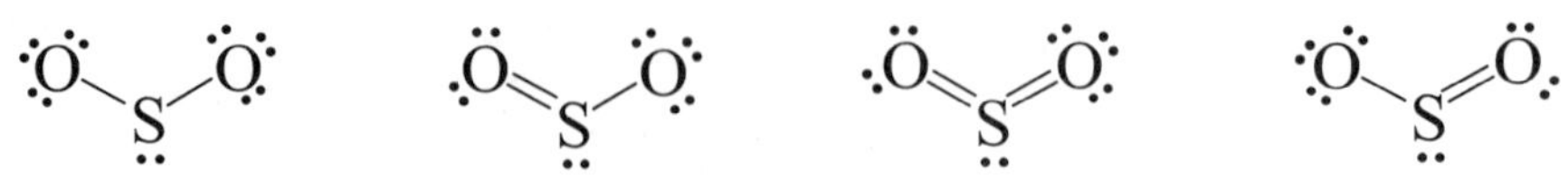

(1) (2) (3) (4)

a) 1 and 2 b) 2 and 4 c) 3 and 4 d) 1, 2, and 3 e) 2, 3, and 4

Answer: b

23. Which of the following are resonance structures for the formate ion, HCO_2^-?

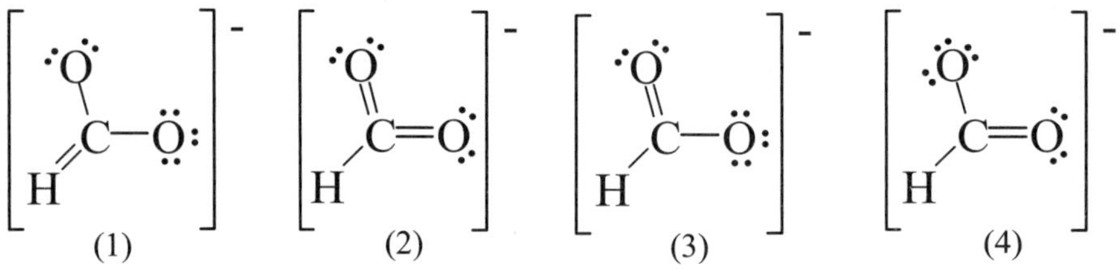

(1) (2) (3) (4)

a) 1 and 2 b) 2 and 3 c) 3 and 4 d) 1, 3, and 4 e) 2, 3, and 4

Answer: c

24. How many resonance structures can be drawn for the thiocyanate ion, SCN^-? The carbon atom is in the center of this ion.
a) 1 b) 2 c) 3 d) 4 e) 5

Answer: c

9.6 Exceptions to the Octet Rule

25. Which of the following elements is most likely to form a molecular structure that disobeys the octet rule?
a) Ne b) C c) B d) F e) Cl

Answer: e

26. Which of the following elements is most likely to form a molecular structure that disobeys the octet rule?
a) B b) C c) N d) O e) F

Answer: a

27. When both of the electrons in a molecular bond originate from the same atom, the bond is called a(n)
   a) double bond.
   b) coordinate covalent bond.
   c) pi bond.
   d) sigma bond.
   e) ionic bond.

   Answer: b

28. What is the correct Lewis structure for ICl_4^-?

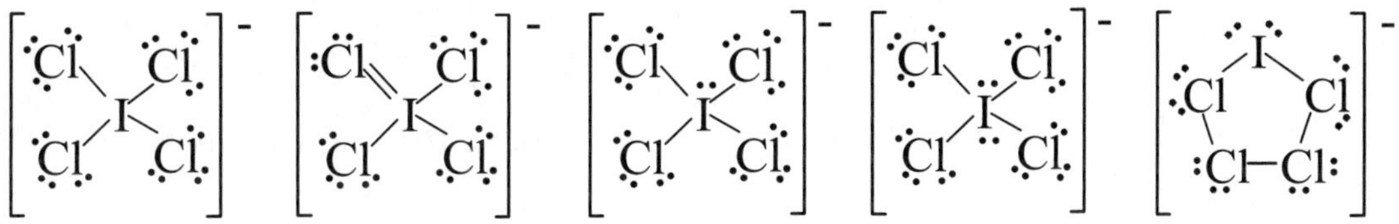

(1) (2) (3) (4) (5)

   a) 1 b) 2 c) 3 d) 4 e) 5

   Answer: d

29. What is the correct Lewis structure for IF_3?

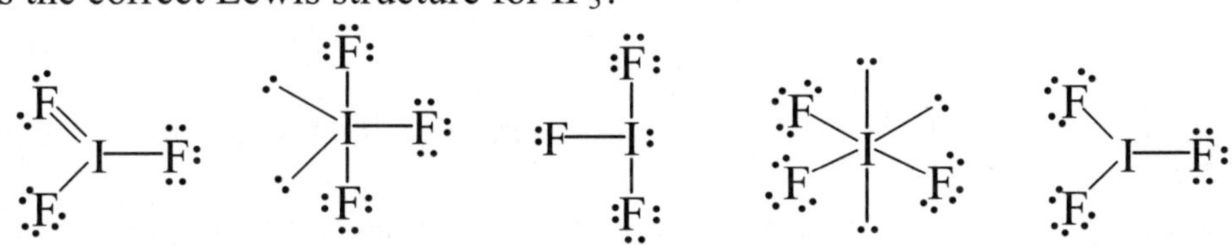

(1) (2) (3) (4) (5)

   a) 1 b) 2 c) 3 d) 4 e) 5

   Answer: b

30. The central atom in XeF_4 is surrounded by
   a) 3 single bonds, 1 double bond, and no lone pairs of electrons.
   b) 2 single bonds, 2 double bonds, and no lone pairs of electrons.
   c) 3 single bonds, 1 double bond, and 1 lone pair of electrons.
   d) 4 single bonds, no double bonds, and no lone pairs of electrons.
   e) 4 single bonds, no double bonds, and 2 lone pairs of electrons.

   Answer: e

31. The central atom in IF_2^- is surrounded by
 a) no single bonds, 2 double bonds, and no lone pairs of electrons.
 b) no single bonds, 2 double bonds, and 2 lone pairs of electrons.
 c) 1 single bond, 1 double bond, and 1 lone pair of electrons.
 d) 2 single bonds, no double bonds, and 2 lone pairs of electrons.
 e) 2 single bonds, no double bonds, and 3 lone pairs of electrons.

Answer: e

32. Which of the following species will have a Lewis structure most like that of SF_4?
 a) XeF_4 b) SO_4^{2-} c) ICl_4^+ d) PF_4^+ e) IO_4^-

Answer: c

33. Which of the following molecules or ions are likely to be free radicals: N_2O, NO, OCl^-, and ClO_2?
 a) N_2O only
 b) N_2O and NO
 c) NO and OCl^-
 d) NO and ClO_2
 e) OCl^- and ClO_2

Answer: d

9.7 Molecular Shapes

34. Use VSEPR theory to predict the electron-pair geometry and the molecular geometry of the nitrite ion, NO_2^-.
 a) The electron-pair geometry is linear, the molecular geometry is linear.
 b) The electron-pair geometry is trigonal-planar, the molecular geometry is bent.
 c) The electron-pair geometry is trigonal-planar, the molecular geometry is linear.
 d) The electron-pair geometry is tetrahedral, the molecular geometry is bent.
 e) The electron-pair geometry is tetrahedral, the molecular geometry is linear.

Answer: b

35. Use VSEPR theory to predict the electron-pair geometry and the molecular geometry of iodine trichloride, ICl_3.
 a) The e^--pair geometry is trigonal-planar, the molecular geometry is trigonal-planar.
 b) The e^--pair geometry is tetrahedral, the molecular geometry is trigonal-pyramidal.
 c) The e^--pair geometry is tetrahedral, the molecular geometry is trigonal-planar.
 d) The e^--pair geometry is trigonal-bipyramidal, the molecular geometry is T-shaped.
 e) The e^--pair geometry is trigonal-bipyramidal, the molecular geometry is trigonal-planar.

Answer: d

36. Use VSEPR theory to predict the molecular geometry of H_3O^+.
   a) trigonal-pyramidal
   b) trigonal-planar
   c) bent
   d) T-shaped
   e) linear

   Answer: a

37. Use VSEPR theory to predict the molecular geometry of CS_2.
   a) linear b) bent c) trigonal-planar d) tetrahedral e) octahedral

   Answer: a

38. Use VSEPR theory to predict the molecular geometry of IF_5.
   a) tetrahedral
   b) see-saw
   c) trigonal-bipyramidal
   d) square-pyramidal
   e) octahedral

   Answer: d

39. Use VSEPR theory to predict the molecular geometry of CO_3^{2-}.
   a) bent
   b) tetrahedral
   c) trigonal-pyramidal
   d) T-shaped
   e) trigonal-planar

   Answer: e

40. What are the Cl-P-Cl bond angles in PCl_3?
   a) 109.5° b) 120° c) 109.5° and 120° d) 90° and 120° e) 90° and 180°

   Answer: a

41. What is the bond angle in N_2O?
   a) 90° b) 107° c) 109.5° d) 120° e) 180°

   Answer: e

42. What are the bond angles in ClO_4^-?
   a) 90° b) 109.5° c) 120° d) 180° e) 90° and 180°

   Answer: b

43. Place the following molecules in order from smallest to largest bond angles: NH_4^+, NH_3, and NH_2^-.
   a) $NH_4^+ < NH_3 < NH_2^-$
   b) $NH_4^+ < NH_2^- < NH_3$
   c) $NH_2^- < NH_3 < NH_4^+$
   d) $NH_2^- < NH_4^+ < NH_3$
   e) $NH_3 < NH_2^- < NH_4^+$

   Answer: c

9.8 Charge Distribution in Covalent Bonds and Molecules

44. What is the formal charge on each atom in a hypobromite ion, OBr^-?
   a) O = -2, Br = -1
   b) O = -2, Br = +1
   c) O = -1, Br = +1
   d) O = -1, Br = 0
   e) O = 0, Br = -1

   Answer: d

45. One resonance structure for thiocyanate ion is drawn below. What is the formal charge on each atom?

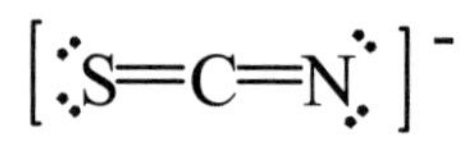

   a) S = 0, C = -1, N = 0
   b) S = 0, C = 0, N = -1
   c) S = -1, C = 0, N = 0
   d) S = 0, C = +2, N = -3
   e) S = -2, C = +4, N = -3

   Answer: b

46. Predict which of the following compounds will have the most polar covalent bond(s).
   a) HF b) CBr_4 c) H_2S d) NCl_3 e) HI

   Answer: a

9.9 Molecular Polarity

47. Which one of the following molecules has a dipole moment?
   a) CI_4 b) PF_5 c) NCl_3 d) SO_3 e) O_2

   Answer: c

48. Which one of the following molecules or ions has a dipole moment?
a) IO_4^- b) ICl_2^- c) SF_4 d) XeF_4 e) CO_2

Answer: c

9.10 Bond Properties: Order, Length, and Energy

49. When bonding electrons are unequally shared, we say that the bond is polar. Linus Pauling noticed that the energy of a polar bond is often greater than expected. He attributed the greater bond energy to
a) a coulombic attraction between atoms with partially positive and negative charges.
b) the greater bond lengths of the heteronuclear bonds.
c) one of the many unexplainable phenomena that scientists encounter.
d) the ability of heteronuclear species to form double and triple bonds.
e) the higher number of valence electrons found in heteronuclear molecules.

Answer: a

50. In molecules, as bond order increases,
a) both bond length and bond energy increase.
b) both bond length and bond energy decrease.
c) bond length increases and bond energy is unchanged.
d) bond length is unchanged and bond energy increases.
e) bond length decreases and bond energy increases.

Answer: e

51. Use Lewis structures to predict the bond order for a nitrogen-oxygen bond in the nitrite ion, NO_2^-.
a) 1/2 b) 1 c) 4/3 d) 3/2 e) 3

Answer: d

52. The standard molar enthalpy of formation of ClF_3 is -405 kJ.

$$1/2\ Cl_2(g) + 3/2\ F_2(g) \rightarrow ClF_3(g)$$

The bond energies of Cl_2 and F_2 are 242 kJ and 155 kJ/mol, respectively. Calculate the energy of an Cl-F bond.
a) 8 kJ/mol b) 87 kJ/mol c) 207 kJ/mol d) 253 kJ/mol e) 802 kJ/mol

Answer: d

53. When heated, azomethane decomposes into nitrogen gas and ethane gas.

$$CH_3N{=}NCH_3(g) \rightarrow N_2(g) + C_2H_6(g)$$

Bond	Bond Energy (kJ/mol)	Bond	Bond Energy (kJ/mol)
C-H	413	N-N	163
C-N	305	N=N	418
C-C	346	N≡N	945

Using average bond energies, calculate the enthalpy of reaction.

a) -611 kJ b) -527 kJ c) -429 kJ d) -313 kJ e) -263 kJ

Answer: e

Short Answer Questions

54. The molecular geometry of a molecule whose central atom has three single bonds and two lone pairs of electrons is ________.

Answer: T-shaped

55. If a molecule has a positive and a negative end, the molecule is said to have a ________ moment.

Answer: dipole

56. In PCl_5, the Cl-P-Cl bond angle between an axial and an equatorial chlorine is ____ degrees.

Answer: 90

57. The ________ energy is the energy of formation of one mole of a solid crystalline ionic compound from gas phase ions.

Answer: lattice

58. Three equivalent Lewis structures can be drawn for carbonate ion. Each structure consists of two carbon-oxygen single bonds and one carbon-oxygen double bond. These three equivalent structures are referred to as ________ structures.

Answer: resonance

59. One might expect H_2O to have a linear molecular geometry since its central atom is surrounded by only two outer atoms. However, water has a bent geometry. Explain.

Answer: The shape of a molecule is dependent upon the number of electron-pairs surrounding the central atom. The oxygen atom in water has four electron-pairs which are distributed around the oxygen atom in a tetrahedral geometry. Thus, the bond angles are approximately 109°, not 180°.

60. The second ionization energy of sodium is much greater than that of magnesium. Explain.

Answer: Valence shell electrons are more easily removed than inner shell electrons because they are partially shielded from the nuclear charge by inner shell electrons. Only one electron occupies sodium's valence shell. The second ionization of sodium requires removal of an electron from an inner shell. For magnesium, both the first and second ionizations remove electrons from the valence shell.

Diagnostic Quiz Questions

Goal #1: Understand the difference between ionic and covalent bonds

1. What is the expected number of valence electrons for a group 3A element?
   a) 1
   b) 2
   c) 3
   d) 5

   Answer: c

2. Which one of the following compounds is ionic?
   a) $SnCl_2$
   b) $SiCl_4$
   c) PCl_3
   d) SCl_2

   Answer: a

3. The lattice energy of LiF is –1037 kJ/mol. This energy corresponds to which reaction below?
   a) $Li(s) + \frac{1}{2} F_2(g) \rightarrow LiF(s)$
   b) $Li^+(g) + F^-(g) \rightarrow LiF(s)$
   c) $Li(g) + F(g) \rightarrow LiF(s)$
   d) $Li(g) + F(g) \rightarrow LiF(g)$

   Answer: b

4. Calculate the lattice energy, $\Delta E_{Lattice}$, of LiBr(s),

$Li^+(g) + Br^-(g) \rightarrow LiBr(s)$ $\quad \Delta E_{Lattice} = ?$

given the following thermochemical equations.

$Li(s) \rightarrow Li(g)$ $\quad \Delta H_f^\circ = +159$ kJ

$Li(g) \rightarrow Li^+(g) + e^-$ $\quad \Delta IE = +513$ kJ

$\frac{1}{2} Br_2(\ell) \rightarrow Br(g)$ $\quad \Delta H_f^\circ = +112$ kJ

$Br(g) + e^- \rightarrow Br^-(g)$ $\quad \Delta EA = -325$ kJ

$Li(s) + \frac{1}{2} Br_2(\ell) \rightarrow LiBr(s)$ $\quad \Delta H = -356$ kJ

a) +103 kJ
b) -103 kJ
c) -815 kJ
d) -1465 kJ

Answer: c

Goal #2: Draw Lewis electron structures for small molecules and ions.

5. Which of the following are correct Lewis structures for SO_2?

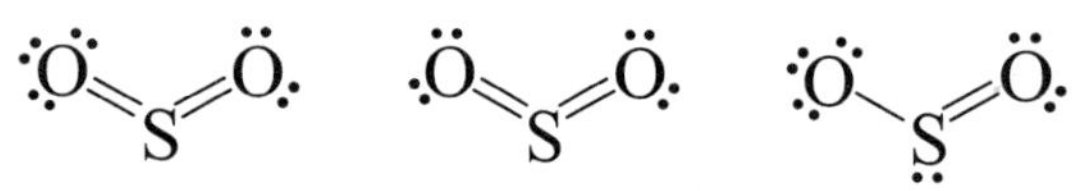

(1) (2) (3)

a) 1 only
b) 2 only
c) 3 only
d) 2 and 3

Answer: c

6. The central atom in carbon dioxide is surrounded by
a) 2 single bonds and 2 lone pair of electrons.
b) 1 double bond, 1 single bond, and 1 lone pair of electrons.
c) 1 double bond, 1 single bond, and 2 lone pairs of electrons.
d) 2 double bonds and no lone pairs of electrons.

Answer: d

7. The central atom in SF_4 is surrounded by
a) 4 single bonds, no double bonds, and no lone pairs of electrons.
b) 4 single bonds, no double bonds, and 1 lone pair of electrons.
c) 3 single bonds, 1 double bond, and 1 lone pair of electrons.
d) 2 single bonds, 2 double bonds, and no lone pairs of electrons.

Answer: b

8. Which of the following species will have a Lewis structure most like that of a sulfate ion, SO_4^{2-}? Assume that the Lewis structure has no double bonds.
   a) NH_3
   b) CBr_4
   c) SO_3
   d) H_2CO

   Answer: b

9. Which of the following are resonance structures for a nitrite ion, NO_2^-?

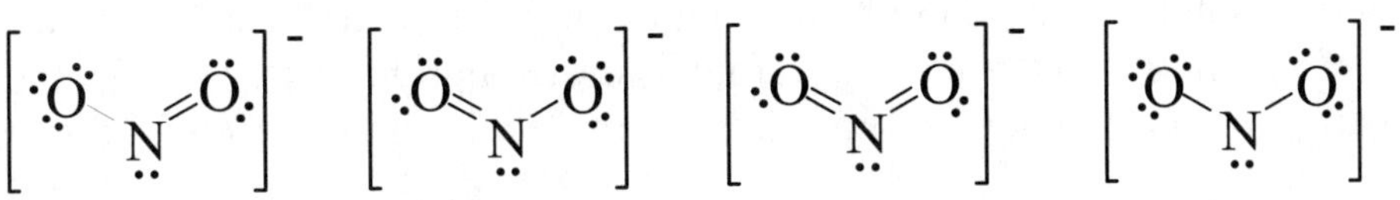

(1) (2) (3) (4)

   a) 1 and 2
   b) 1, 2, and 3
   c) 2 and 3
   d) 3 and 4

   Answer: a

Goal #3: Use the valence shell electron pair repulsion theory (VSEPR) to predict the shapes of simple molecules and ions and to understand the structures of more complex molecules.

10. Use VSEPR theory to predict the molecular geometry of SCl_2.
    a) bent
    b) trigonal planar
    c) trigonal pyramidal
    d) linear

    Answer: a

11. Use VSEPR theory to predict the bond angles of CCl_4.
    a) 90°
    b) 109.5°
    c) 120°
    d) 180°

    Answer: b

12. Use VSEPR theory to predict the molecular geometry of HCN.
    a) bent
    b) linear
    c) trigonal planar
    d) tetrahedral

    Answer: b

13. Use VSEPR theory to predict the electron-pair geometry and the molecular geometry of ICl_3.
    a) electron-pair geometry = trigonal planar; molecular geometry = trigonal planar.
    b) electron-pair geometry = tetrahedral; molecular geometry = trigonal planar.
    c) electron-pair geometry = trigonal bipyramidal; molecular geometry = trigonal planar.
    d) electron-pair geometry = trigonal bipyramidal; molecular geometry = T-shaped

    Answer: d

14. Place the following molecules in order from smallest to largest bond angles: CH_4, NH_3, and NH_2^-.
    a) $NH_2^- < NH_3 < CH_4$
    b) $CH_4 < NH_2^- < NH_3$
    c) $CH_4 < NH_3 < NH_2^-$
    d) $NH_3 < NH_2^- < CH_4$

    Answer: a

15. Use VSEPR theory to predict the electron-pair geometry and the molecular geometry of PF_6^-.
    a) electron-pair geometry = tetrahedral; molecular geometry = tetrahedral
    b) electron-pair geometry = trigonal bipyramidal; molecular geometry = seesaw
    c) electron-pair geometry = octahedral; molecular geometry = square planar
    d) electron-pair geometry = octahedral; molecular geometry = octahedral

    Answer: d

Goal #4: Use electronegativity to predict the charge distribution in molecules and ions and to define the polarity of bonds.

16. What is the formal charge on each atom in OCl^-?
    a) O = 0, Cl = -1
    b) O = -1, Cl = 0
    c) O = -1, Cl = +1
    d) O = -2, Cl = +1

    Answer: b

17. What is the formal charge on each atom in CO?
   a) C = -1, O = +1
   b) C = 0, O = 0
   c) C = +1, O = -1
   d) C = +2, O = -2

Answer: a

18. Electronegativity is a measure of
   a) the charge on an electron.
   b) a molecule's polarity.
   c) the charge on an atom.
   d) an atom's ability to attract electrons to itself when bonded to another atom.

Answer: d

19. Predict which of the following molecules has the most polar bonds.
   a) H_2O
   b) H_2S
   c) H_2Se
   d) H_2Te

Answer: a

Goal #5: Predict the polarity of molecules

20. Which one of the following molecules is nonpolar?
   a) CH_3Cl
   b) SO_2
   c) BF_3
   d) H_2O

Answer: c

21. Which one of the following molecules is polar?
   a) SF_4
   b) XeF_4
   c) F_2
   d) CF_4

Answer: a

Goal #6: Understand the properties of covalent bonds and their influence on molecular structure.

22. What is the bond order for a nitrogen-oxygen bond in a nitrate ion, NO_3^-?
    a) 1
    b) 4/3
    c) 3/2
    d) 2

    Answer: b

23. Based on bond order, predict which molecule has the shortest carbon-oxygen bond length.
    a) CO
    b) CO_2
    c) CH_3OH
    d) H_2CO

    Answer: a

24. When heated, azomethane decomposes into nitrogen gas and methane gas.

$$CH_3N{=}NCH_3(g) \rightarrow N_2(g) + C_2H_6(g)$$

Bond	Bond Energy (kJ/mol)	Bond	Bond Energy (kJ/mol)
C-H	413	N-N	163
C-N	305	N=N	418
C-C	346	N≡N	?

The enthalpy change for this reaction is -263 kJ/mol. Determine the bond enthalpy of a nitrogen-nitrogen triple bond.
    a) 419 kJ
    b) 618 kJ
    c) 725 kJ
    d) 945 kJ

    Answer: d

CHAPTER 10: Bonding and Molecular Structure: Orbital Hybridization and Molecular Orbitals

10.1 Orbitals and Bonding Theories

1. All of the following statements concerning valence bond (VB) and molecular orbital (MO) bond theories are correct EXCEPT
   a) MO theory predicts that electrons are delocalized over the molecule.
   b) in VB theory, bonding electrons are localized between pairs of atoms.
   c) VB theory describes a molecular bond as the overlap between two atomic orbitals.
   d) MO theory can describe molecular bonding in excited states.
   e) VB theory is used to predict the colors of compounds.

   Answer: e

10.2 Valence Bond Theory

2. How many sigma (σ) bonds and pi (π) bonds are in the following molecule?

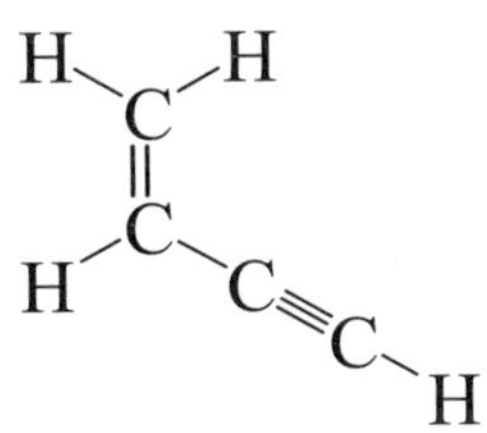

   a) five σ and two π
   b) five σ and three π
   c) five σ and five π
   d) seven σ and two π
   e) seven σ and three π

   Answer: e

3. How many sigma (σ) bonds and pi (π) bonds are in acetic acid?

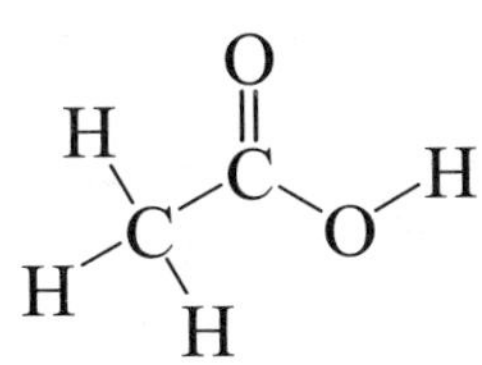

   a) six σ and one π
   b) six σ and two π
   c) seven σ and one π
   d) eight σ and zero π
   e) eight σ and one π

   Answer: c

4. How many sigma (σ) bonds and pi (π) bonds are in acetylene, C_2H_2?
   a) one σ, one π b) two σ, two π c) three σ, one π d) three σ, two π e) four σ, one π

   Answer: d

5. How many sigma (σ) bonds and pi (π) bonds are in formaldehyde, H_2CO?
   a) one σ, one π b) two σ, two π c) three σ, one π d) three σ, two π e) four σ, one π

   Answer: c

6. To form a set of sp^3d^2 hybrid atomic orbitals, what set of pure atomic orbitals must be mixed?
   a) one *s*, one *p*, and one *d*
   b) one *s*, three *p*, and one *d*
   c) one *s*, three *p*, and two *d*
   d) two *s*, six *p*, and two *d*
   e) two *s*, six *p*, and four *d*

   Answer: c

7. What is the maximum number of hybridized orbitals formed by an oxygen atom?
   a) 2 b) 3 c) 4 d) 5 e) 6

   Answer: c

8. What is the hybridization of the carbon atom is carbon disulfide?
   a) sp b) sp^2 c) sp^3 d) sp^3d e) sp^3d^2

   Answer: a

9. What is the hybridization of the sulfur atom in SO_2?
   a) sp b) sp^2 c) sp^3 d) sp^3d e) sp^3d^2

   Answer: b

10. What is the hybridization of the sulfur atom in SF_4?
    a) sp b) sp^2 c) sp^3 d) sp^3d e) sp^3d^2

    Answer: d

11. What is the hybridization of the central oxygen atom in ozone, O_3?
    a) sp b) sp^2 c) sp^3 d) sp^3d e) sp^3d^2

    Answer: b

12. What is the hybridization of the iodine atom in periodate ion, IO_4^-?
a) *sp* b) sp^2 c) sp^3 d) sp^3d e) sp^3d^2

Answer: c

13. What is the hybridization of the central atom in a molecule with a square-planar molecular geometry?
a) *sp* b) sp^2 c) sp^3 d) sp^3d e) sp^3d^2

Answer: e

14. What is the hybridization of the central atom in a molecule with a trigonal-planar geometry?
a) *sp* b) sp^2 c) sp^3 d) sp^3d e) sp^3d^2

Answer: b

15. In which of the following molecules and ions does the central carbon atom have *sp* hybridization: Cl_2CO, CH_2Br_2, CO_2, and OCN^-?
a) Cl_2CO only
b) Cl_2CO and CH_2Br_2
c) CH_2Br_2 and CO_2
d) CH_2Br_2 and OCN^-
e) CO_2 and OCN^-

Answer: e

16. In which of the following molecules and ions does the central sulfur atom have sp^3 hybridization: SO_4^{2-}, SF_4, SO_2, and SO_3?
a) SO_4^{2-} only
b) SO_4^{2-} and SF_4
c) SF_4 and SO_2
d) SO_2 and SO_3
e) SO_4^{2-}, SF_4, and SO_3

Answer: a

17. In which of the following molecules and ions does the central atom have sp^2 hybridization: BF_3, NO_2^-, NH_2^-, H_2O?
a) BF_3 and NO_2^-
b) NO_2^- and NH_2^-
c) NH_2^- and H_2O
d) BF_3 and NH_2^-
e) NO_2^- and H_2O

Answer: a

18. What is the molecular geometry around a central atom that is sp^3 hybridized and has one lone electron-pair?
a) bent
b) linear
c) trigonal-planar
d) trigonal-pyramidal
e) trigonal-bipyramidal

Answer: d

19. What is the molecular geometry around a central atom that is sp^3d^2 hybridized and has two lone electron-pairs?
a) tetrahedral
b) trigonal-bipyramidal
c) square-planar
d) square-pyramidal
e) see-saw

Answer: c

20. What is the molecular geometry around a central atom that is sp^2 hybridized, has three sigma bonds, and has one pi bond?
a) trigonal-planar
b) trigonal-pyramidal
c) trigonal-bipyramidal
d) linear
e) tetrahedral

Answer: a

21. What is the molecular geometry around a central atom that is sp^3d hybridized and has three sigma bonds and two lone electron-pairs?
a) linear
b) trigonal-pyramidal
c) trigonal-planar
d) tetrahedral
e) T-shaped

Answer: e

22. What is the molecular geometry around a central atom that has four sigma bonds and has no lone electron-pairs?
a) square-pyramidal
b) tetrahedral
c) trigonal-planar
d) square-planar
e) see-saw

Answer: b

23. Carbon dioxide reacts with an aqueous solution of sodium hydroxide to form carbonate ion. What change in the hybridization of carbon occurs in this reaction?
a) sp to sp^2 b) sp^2 to sp^3 c) sp^3 to sp^3d d) sp^3 to sp^3d^2 e) no change

Answer: a

24. One product of the combustion of ethylene, C_2H_4, is carbon dioxide. What change in hybridization of the carbon occurs in this reaction?
a) sp^3 to sp^2 b) sp^3 to sp c) sp^3 to sp^3d d) sp^2 to sp^3d^2 e) sp^2 to sp

Answer: e

25. Sulfur dioxide reacts with oxygen to form sulfur trioxide. What change in hybridization of the sulfur occurs in this reaction?
a) sp^2 to sp^3 b) sp^2 to sp c) sp^3 to sp^3d d) sp^2 to sp^3d^2 e) no change

Answer: e

26. Which of the following hybridized atoms is not possible?
a) an sp hybridized carbon atom
b) an sp^2 hybridized nitrogen atom
c) an sp^3 hybridized oxygen atom
d) an sp^3d hybridized fluorine atom
e) an sp^3d^2 hybridized sulfur atom

Answer: d

27. Which of the following characteristics apply to PCl_3?
1. nonpolar molecule
2. polar bonds
3. trigonal-pyramidal molecular geometry
4. sp^2 hybridized

a) 1 and 2 b) 2 and 3 c) 3 and 4 d) 1, 2, and 3 e) 1, 2, 3, and 4

Answer: b

28. Sulfur trioxide is known to be planar with all the oxygen atoms equidistant from the central sulfur atom. On the basis of these facts, which of the following statements are CORRECT?
    1. It can be represented by three equivalent resonance structures.
    2. The sulfur atom is sp^2 hybridized.
    3. The molecule is nonpolar.

    a) 1 only b) 2 only c) 3 only d) 1 and 2 e) 1, 2, and 3

    Answer: e

29. For which of the following compounds is it possible for *cis* and *trans* isomers to exist?

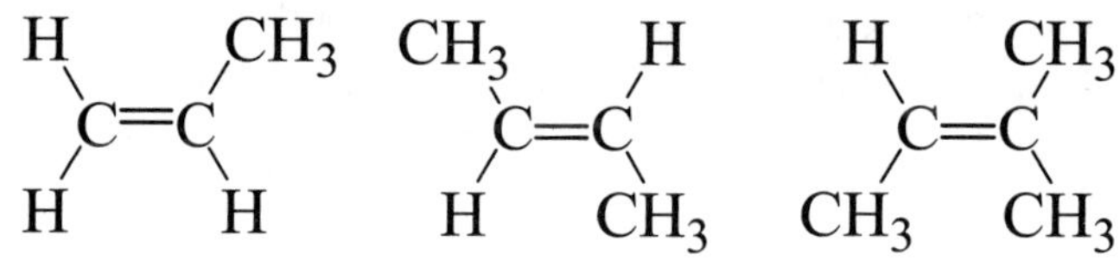

(1) (2) (3)

a) 1 only b) 2 only c) 3 only d) 1 and 2 e) 1, 2, and 3

Answer: b

30. Which of the underlined atoms (C_1, C_2, N, and O) are sp^2 hybridized?

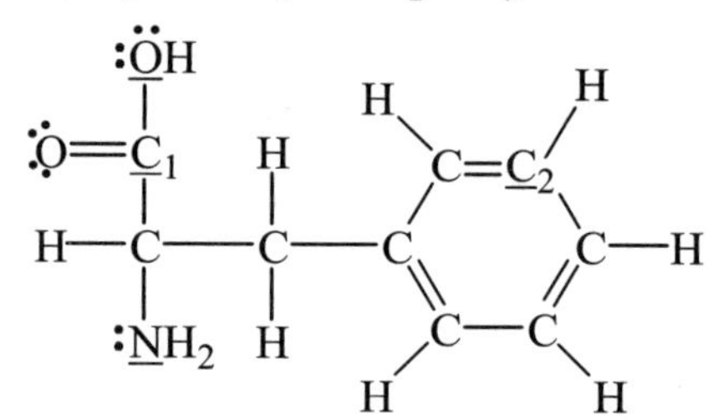

a) C_1 and C_2
b) C_1, N, and O
c) N and O
d) O and C_2
e) O only

Answer: a

10.3 Molecular Orbital Theory

31. All of the following statements concerning molecular orbital (MO) theory are correct EXCEPT
    a) the Pauli exclusion principle is obeyed.
    b) Hund's rule is obeyed.
    c) the combination of two atomic orbitals creates one molecular orbital.
    d) a bonding molecular orbital is lower in energy than its parent atomic orbitals.
    e) electrons are assigned to orbitals of successively higher energy.

    Answer: c

32. Atomic orbitals combine most effectively to form molecular orbitals when
   a) electrons in the orbitals have no spins.
   b) electrons in the orbitals have the same spin.
   c) the atomic orbitals are hybridized.
   d) the atomic orbitals have similar energies.
   e) metals combine with nonmetals.

   Answer: d

33. A molecular orbital that decreases the electron density between two nuclei is said to be ________.
   a) antibonding b) bonding c) hybridized d) dipolar e) nonpolar

   Answer: a

The following molecular orbital diagram may be used for problems 34-48. For oxygen and fluorine, the σ_{2p} orbital should be lower in energy than the π_{2p}. However, the diagram will still yield correct bond order and magnetic behavior for these molecules.

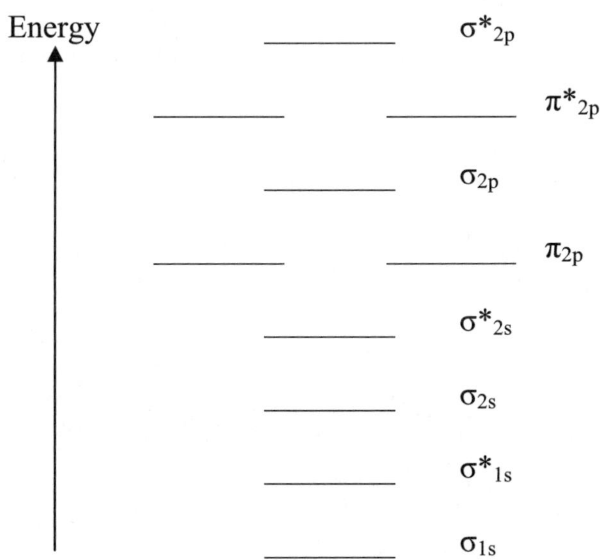

34. According to molecular orbital theory, which of the following species is the most likely to exist?
   a) H_2^{2-} b) He_2 c) Li_2 d) Li_2^{2-} e) Be_2

   Answer: c

35. According to molecular orbital theory, which of the following species is least likely to exist?
   a) H_2^- b) F_2^{2+} c) C_2^{2-} d) Li_2 e) B_2^{2+}

   Answer: e

36. According to molecular orbital theory, what is the bond order of nitrogen, N_2?
a) 1 b) 3/2 c) 2 d) 5/2 e) 3

Answer: e

37. According to molecular orbital theory, what is the bond order of superoxide, O_2^-?
a) 1 b) 3/2 c) 2 d) 5/2 e) 3

Answer: b

38. According to molecular orbital theory, which of the following lists ranks the nitrogen species in terms of increasing bond order?
a) $N_2^- < N_2^{2-} < N_2$
b) $N_2^- < N_2 < N_2^{2-}$
c) $N_2 < N_2^- < N_2^{2-}$
d) $N_2^{2-} < N_2^- < N_2$
e) $N_2 < N_2^{2-} < N_2^-$

Answer: d

39. Consider the molecules Be_2, B_2, C_2, and F_2. Which two molecules have the same bond order?
a) Be_2 and B_2 b) B_2 and C_2 c) C_2 and F_2 d) Be_2 and C_2 e) B_2 and F_2

Answer: e

40. Consider the species F_2^{2+}, N_2^{2-}, O_2^{2-}, and B_2^{2+}. Which two species have the same bond order?
a) F_2^{2+} and N_2^{2-} b) N_2^{2-} and O_2^{2-} c) O_2^{2-} and B_2^{2+} d) F_2^{2+} and O_2^{2-} e) N_2^{2-} and B_2^{2+}

Answer: a

41. Use molecular orbital theory to predict which species has the shortest bond length.
a) O_2^{2+} b) O_2^+ c) O_2 d) O_2^- e) O_2^{2-}

Answer: a

42. Use molecular orbital theory to predict which species is paramagnetic.
a) N_2 b) O_2 c) F_2 d) Li_2 e) H_2

Answer: b

43. Use molecular orbital theory to predict which ion is paramagnetic.
a) C_2^{2-} b) O_2^{2-} c) O_2^{2+} d) N_2^{2-} e) B_2^{2-}

Answer: d

44. What is the molecular orbital configuration of N_2?
a) [core electrons] $(\sigma_{2s})^2$ $(\sigma^*_{2s})^2$ $(\pi_{2p})^2$ $(\sigma_{2p})^2$ $(\sigma^*_{2p})^2$
b) [core electrons] $(\sigma_{2s})^2$ $(\sigma^*_{2s})^2$ $(\pi_{2p})^2$ $(\sigma_{2p})^2$ $(\pi^*_{2p})^2$
c) [core electrons] $(\sigma_{2s})^2$ $(\sigma^*_{2s})^2$ $(\pi_{2p})^4$ $(\sigma_{2p})^2$
d) [core electrons] $(\sigma_{2s})^2$ $(\sigma^*_{2s})^2$ $(\pi_{2p})^4$ $(\sigma_{2p})^2$ $(\pi^*_{2p})^4$
e) [core electrons] $(\sigma_{2s})^2$ $(\sigma^*_{2s})^2$ $(\pi_{2p})^4$ $(\pi^*_{2p})^2$

Answer: c

45. What is the molecular orbital configuration of B_2^{2-}?
a) [core electrons] $(\sigma_{2s})^2$ $(\sigma^*_{2s})^2$
b) [core electrons] $(\sigma_{2s})^2$ $(\sigma^*_{2s})^2$ $(\pi_{2p})^4$
c) [core electrons] $(\sigma_{2s})^2$ $(\sigma^*_{2s})^2$ $(\pi_{2p})^2$
d) [core electrons] $(\sigma_{2s})^4$ $(\sigma^*_{2s})^4$
e) [core electrons] $(\sigma_{2s})^2$ $(\sigma^*_{2s})^2$ $(\pi_{2p})^2$ $(\sigma_{2p})^2$

Answer: b

46. Assume that the molecular orbital energy diagram for a homonuclear diatomic molecule applies to a heteronuclear diatomic molecule. What is the molecular orbital configuration of CO?
a) [core electrons] $(\sigma_{2s})^2$ $(\sigma^*_{2s})^2$ $(\pi_{2p})^4$
b) [core electrons] $(\sigma_{2s})^2$ $(\sigma^*_{2s})^2$ $(\pi_{2p})^2$ $(\sigma_{2p})^2$ $(\pi^*_{2p})^2$
c) [core electrons] $(\sigma_{2s})^2$ $(\sigma^*_{2s})^2$ $(\pi_{2p})^2$ $(\sigma_{2p})^4$
d) [core electrons] $(\sigma_{2s})^2$ $(\sigma^*_{2s})^2$ $(\pi_{2p})^4$ $(\sigma_{2p})^2$
e) [core electrons] $(\sigma_{2s})^2$ $(\sigma^*_{2s})^2$ $(\pi_{2p})^2$

Answer: d

47. Assuming that the molecular orbital energy diagram for a homonuclear diatomic molecule applies to a heteronuclear diatomic molecule, determine which of the following species has the highest bond order.
a) NO^- b) OF^- c) NO^+ d) O_2 e) NO

Answer: c

48. Assuming that the molecular orbital energy diagram for a homonuclear diatomic molecule applies to a heteronuclear diatomic molecule, determine which of the following species is paramagnetic.
a) NO^+ b) CO c) CN^- d) OF^- e) NO

Answer: e

49. Which molecule will have the following valence molecular orbital level energy diagram?

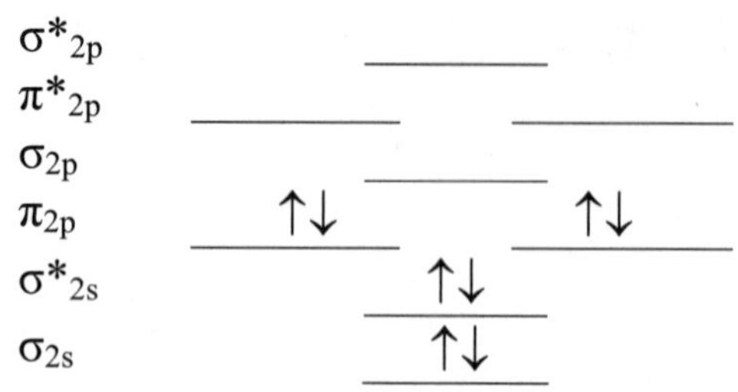

a) Li_2 b) Be_2 c) B_2 d) C_2 e) N_2

Answer: d

50. The following valence molecular orbital energy level diagram is appropriate for which one of the listed species?

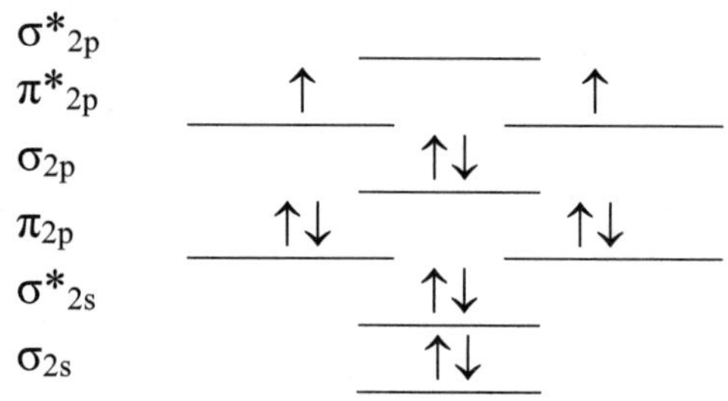

a) B_2 b) C_2 c) N_2 d) O_2 e) F_2

Answer: d

51. Which molecule will have the following valence molecular orbital energy level diagram?

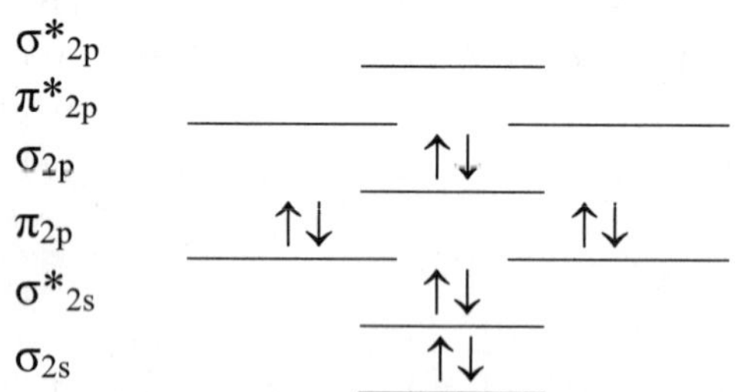

a) C_2^{2-} b) N_2^{2+} c) N_2^{2-} d) O_2^{2-} e) F_2^{2+}

Answer: a

52. Benzene, C_6H_6, consists of a six member ring of *sp*^{2} hybridized carbon atoms. How many π_{2p} molecular orbitals (both bonding and antibonding) exist for benzene?
a) 2 b) 4 c) 6 d) 8 e) 12

Answer: c

53. In the SO_2 molecule, the oxygen atoms and the sulfur atom can be viewed as sp^2 hybridized. Thus, each atom has one unhybridized *p* orbital. How many π_{2p} bonding, antibonding, and nonbonding molecular orbitals exist for SO_2?
a) Three π_{2p} molecular orbitals exist; two bonding and one antibonding.
b) Three π_{2p} molecular orbitals exist; one bonding, one antibonding, and one nonbonding.
c) Three π_{2p} molecular orbitals exist; one bonding and two antibonding.
d) Two π_{2p} molecular orbitals exist; one bonding and one antibonding.
e) Two π_{2p} molecular orbitals exist; both bonding.

Answer: b

Short Answer Questions

54. In valence bond theory, a π bond is described as the sideways overlap of two unhybridized atomic ________ orbitals.

Answer: p

55. In molecular orbital theory, the bond order is defined as 1/2(the number of electrons in ________ orbitals – the number of electrons in antibonding orbitals).

Answer: bonding

56. The ICl_2^- ion has a trigonal-bipyramidal electron-pairgeometry and a linear molecular geometry. The hybridization of the central iodine atom is ________.

Answer: sp^3d

57. The hybridization of the xenon atom in $XeOF_4$ is ________.

Answer: sp^3d^2

58. Draw a Lewis structure of xenon trioxide. What is the hybridization of the xenon atom in this molecule?

Answer: The hybridization is sp^3.

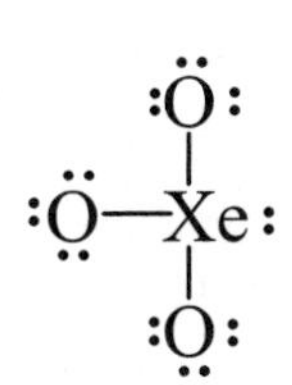

59. Draw the valence molecular orbital energy level diagram for carbon monoxide, CO.

Answer:

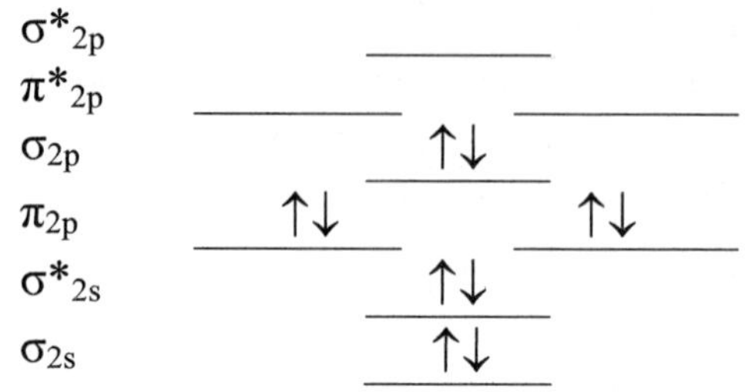

<u>Diagnostic Quiz Questions</u>

Goal #1: Understand the differences between valence bond and molecular orbital theory

1. Which of the following statements concerning valence bond (VB) and molecular orbital (MO) bonding theories is correct?
   a) In VB theory, electrons are delocalized over the molecule.
   b) VB theory predicts that oxygen is paramagnetic; MO theory does not.
   c) In VB theory, the overlap that occurs between a hydrogen 1*s* orbital and an oxygen 2*p* orbital is referred to as a pi (π) bond.
   d) VB theory describes a molecular bond as the overlap between two atomic orbitals.

   Answer: d

2. How many sigma (σ) bonds and pi (π) bonds are in the following molecule?

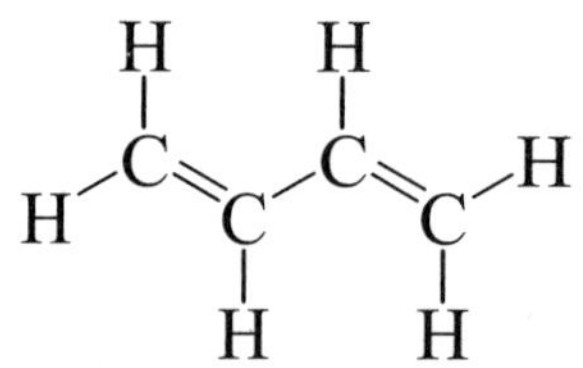

   a) two σ and nine π
   b) two σ and seven π
   c) seven σ and two π
   d) nine σ and two π

   Answer: d

3. In valence bond theory, the π bond between the two carbon atoms in C_2H_4 results from
   a) the overlap of the 2*s* orbitals from each carbon atom.
   b) the overlap of a 2*s* orbital from one carbon atom with one lobe of a 2*p* orbital from the other carbon atom.
   c) the above and below the molecular plane overlap of a 2*p* orbital from each carbon atom.
   d) the above and below the molecular plane overlap of a 2*s* orbital from one carbon atom with a 2*p* orbital from the other carbon atom.

   Answer: c

Goal #2: Identify the hybridization of an atom in a molecule or ion.

4. What is the maximum number of hybridized orbitals that can be formed by a nitrogen atom?
   a) 2
   b) 3
   c) 4
   d) 6

   Answer: c

5. In which of the following molecules does the carbon atom have sp^2 hybridization?
   a) H_2CO
   b) CH_2Cl_2
   c) CH_4
   d) CO_2

   Answer: a

6. In which of the following molecules or ions does the central atom have sp^3d^2 hybridization?
   a) XeF_4
   b) SF_4
   c) PF_5
   d) SO_4^{2-}

   Answer: a

7. What is the hybridization of the central atom in SCN^-?
   a) sp
   b) sp^2
   c) sp^3
   d) sp^3d^2

   Answer: a

8. What is the molecular geometry around an atom that is sp^3 hybridized and has two lone pairs of electrons?
   a) bent
   b) linear
   c) trigonal planar
   d) trigonal pyramidal

   Answer: a

9. One product of the combustion of ethylene, C_2H_4, is carbon dioxide. What change in hybridization of the carbon atoms occurs in this reaction?
   a) sp^3 to sp^2
   b) sp^3 to sp
   c) sp^2 to sp
   d) sp^2 to sp^3

   Answer: c

10. What are the bond angles around an sp^2 hybridized central atom?
    a) 90°
    b) 109.5°
    c) 120°
    d) 90°, 120°, and 180°

    Answer: c

Goal #3: Understand the differences between bonding and antibonding molecular orbitals

11. According to molecular orbital theory, which of the following species is the least likely to exist?
    a) Li_2
    b) Be_2
    c) B_2
    d) C_2

    Answer: b

12. Which of the following species is paramagnetic?
    a) C_2
    b) He_2^{2+}
    c) N_2
    d) B_2

    Answer: d

13. Use bond order to determine which species has the shortest bond length?
   a) N_2^{2+}
   b) N_2^{+}
   c) N_2
   d) N_2^{2-}

   Answer: c

14. Assuming that the molecular orbital energy diagram for a homonuclear diatomic molecule applies to a heteronuclear diatomic molecule, determine the bond order of NO.
   a) 1.5
   b) 2
   c) 2.5
   d) 3

   Answer: c

15. Benzene, C_6H_6, consists of a six member ring of *sp*^{2} hybridized carbon atoms. How many π_{2p} molecular orbitals (both bonding and antibonding) exist for benzene?
   a) 1
   b) 3
   c) 6
   d) 12

   Answer: c

Write the molecular orbital configuration for simple diatomic molecules

16. What is the molecular orbital configuration for N_2?
   a) [core electrons] $(\sigma_{2s})^2 (\sigma^*_{2s})^2 (\pi_{2p})^4 (\sigma_{2p})^2$
   b) [core electrons] $(\sigma_{2s})^2 (\sigma^*_{2s})^2 (\pi_{2p})^2 (\sigma_{2p})^4$
   c) [core electrons] $(\sigma_{2s})^2 (\sigma^*_{2s})^2 (\pi_{2p})^2 (\sigma_{2p})^2 (\pi^*_{2p})^2$
   d) [core electrons] $(\sigma_{2s})^2 (\sigma^*_{2s})^2 (\pi_{2p})^6$

   Answer: a

17. What is the molecular orbital configuration for CO?
   a) [core electrons] $(\sigma_{2s})^2 (\sigma^*_{2s})^2 (\pi_{2p})^6$
   b) [core electrons] $(\sigma_{2s})^2 (\sigma^*_{2s})^2 (\pi_{2p})^2 (\sigma_{2p})^4$
   c) [core electrons] $(\sigma_{2s})^2 (\sigma^*_{2s})^2 (\pi_{2p})^2 (\sigma_{2p})^2 (\pi^*_{2p})^2$
   d) [core electrons] $(\sigma_{2s})^2 (\sigma^*_{2s})^2 (\pi_{2p})^4 (\sigma_{2p})^2$

   Answer: d

CHAPTER 11: Carbon: More than just another Element

11.1 Why Carbon?

1. Structural isomers are compounds that have
   a) the same molar masses, but different elemental composition.
   b) have identical structures, but contain different isotopes of the same elements.
   c) two or more double bonds.
   d) the same elemental composition, but different molar masses.
   e) the same elemental composition, but the atoms are linked in different ways.

   Answer: e

2. Which of the following statements concerning isomers are correct?
   1. Two types of stereoisomers exist: geometric isomers and optical isomers.
   2. Enantiomers have different melting and boiling points.
   3. Optical isomers have non-superimposable mirror images.

   a) 1 only   b) 2 only   c) 3 only   d) 1 and 2   e) 1 and 3

   Answer: e

3. Which one of the following molecules has a chiral center?

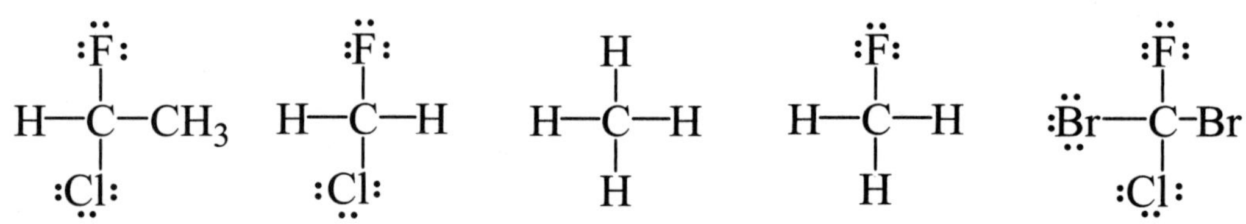

(1)   (2)   (3)   (4)   (5)

   a) (1)   b) (2)   c) (3)   d) (4)   e) (5)

   Answer: a

4. For which one of the following molecules do geometric isomers exist?
   a) H_3C-CH_3
   b) H_3C-CH_2Br
   c) BrH_2C-CH_2Br
   d) $BrHC=CHBr$
   e) $BrHC=CBr_2$

   Answer: d

11.2 Hydrocarbons

5. What is the general formula for an alkene?
   a) C_nH_{2n+2} b) C_nH_{2n} c) C_nH_{n-2} d) $C_{n+2}H_n$ e) C_nH_{2n-2}

   Answer: b

6. Which of the following molecules might be an alkyne?
   a) C_2H_6 b) C_4H_8 c) C_5H_8 d) C_6H_{12} e) C_6H_{14}

   Answer: c

7. Which of the following molecules may be a cycloalkane?
   a) C_3H_8 b) C_4H_6 c) C_5H_{12} d) C_6H_{12} e) C_7H_{16}

   Answer: d

8. Which of the following hydrocarbons has two π bonds?
   a) C_3H_6 b) C_4H_6 c) C_6H_{12} d) C_7H_{14} e) C_8H_{18}

   Answer: b

9. What is the molecular formula for heptane?
   a) C_6H_{12} b) C_6H_{14} c) C_7H_{14} d) C_7H_{16} e) C_8H_{14}

   Answer: d

10. How many structural isomers exist for C_6H_{14}?
    a) 3 b) 4 c) 5 d) 6 e) 7

    Answer: c

11. What is the name of the following compound?

    a) 2,5-dimethylheptane
    b) 2,5-diheptane
    c) 2-ethyl-5-methylhexane
    d) 3,6-dimethylheptane
    e) 1,1,4-trimethylhexane

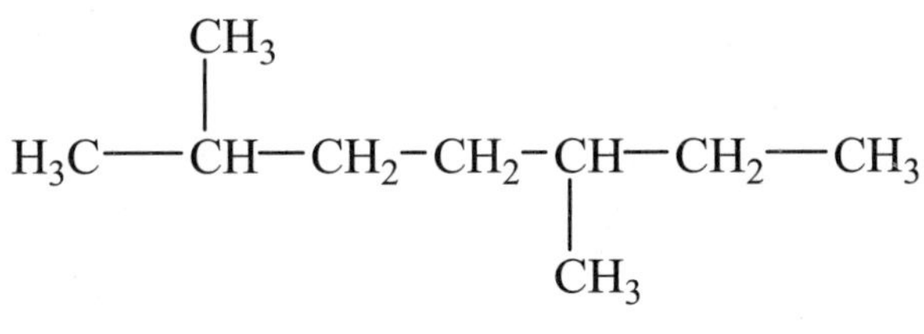

    Answer: a

12. What is the name of the following compound?
    a) pentane
    b) pentene
    c) pentanone
    d) cyclopentene
    e) cyclopentane

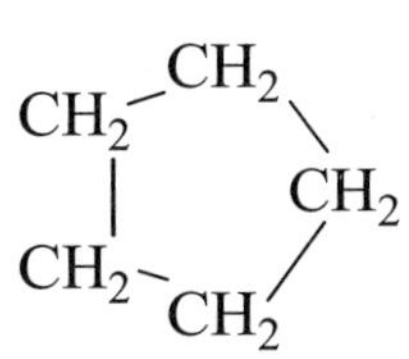

    Answer: e

13. What is the name of the following compound?
   a) 3,5-diethylmethylheptane
   b) 3-ethyl-5-methylheptane
   c) decane
   d) 2,4-diethylhexane
   e) 3-ethyl-5-methyldecane

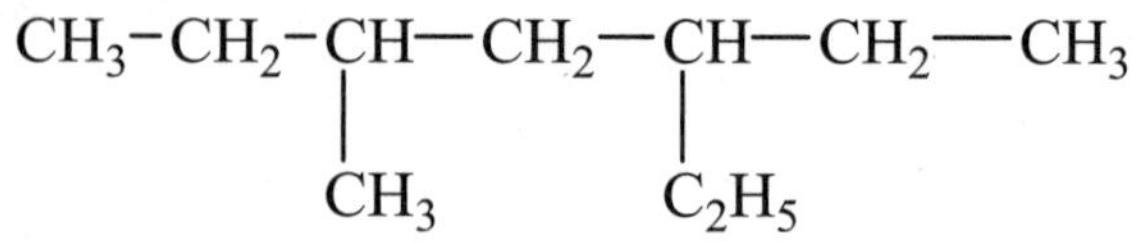

Answer: b

14. Which of the following hydrocarbons can have *cis* and *trans* isomers?
   a) 3-hexene b) 2-pentyne c) 1-butene d) 2-ethylbutane e) propene

Answer: a

15. What is the name of the following compound?
   a) 1-methyl-2-pentene
   b) *cis*-1-methyl-2-pentene
   c) *trans*-2-hexene
   d) *cis*-4-hexene
   e) *trans*-1-methyl-2-propylethylene

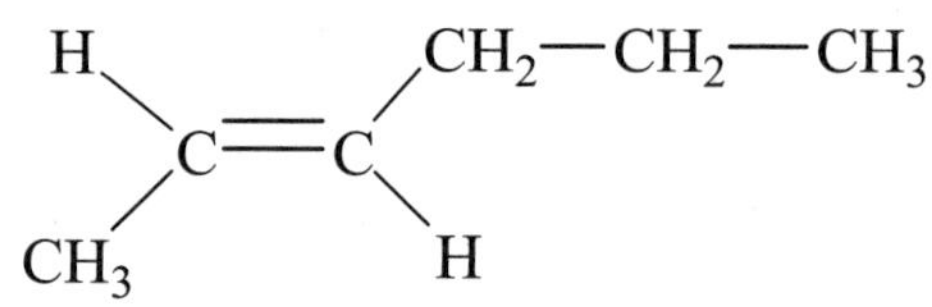

Answer: c

16. What is the name of the following compound?
   a) *trans*-acetylene
   b) *cis*-1,3-pentadiene
   c) *trans*-2,4-pentadiene
   d) *cis*-4-methyl-1,3-butadiene
   e) *cis*-4-methyl-1,3-butene

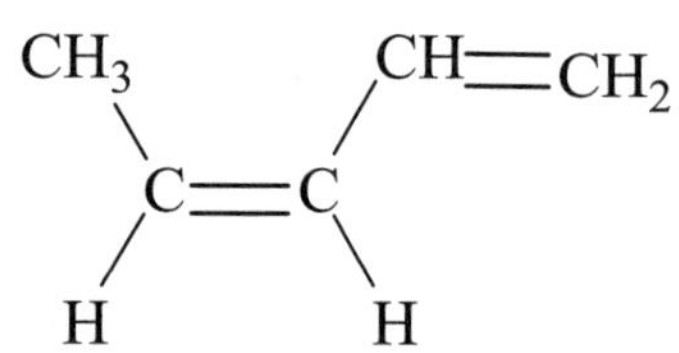

Answer: b

17. What is the name of the following compound?
   a) 2-methyl-4-pentyne
   b) 3-methyl-1-butyne
   c) 4-methyl-1-pentyne
   d) 4,4-dimethyl-1-butyne
   e) 4,4-dimethylbutyne

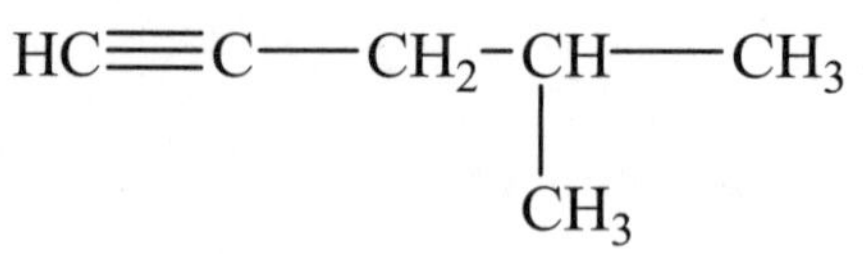

Answer: c

18. What is the molecular formula of this hydrocarbon?

a) C_6 b) C_6H_6 c) C_6H_8 d) C_6H_{12} e) C_6H_{14}

Answer: b

19. What is the molecular formula for 4,4-diethyl-1-hexyne?
a) C_6H_{10} b) C_8H_{14} c) C_8H_{18} d) $C_{10}H_{18}$ e) $C_{10}H_{22}$

Answer: d

20. What is the balanced chemical equation for the combustion of 2-methyl-1-butene?
a) $2\ C_5H_{10}(g)\ +\ 15\ O_2(g)\ \rightarrow\ 10\ CO_2(g)\ +\ 10\ H_2O(\ell)$
b) $2\ C_4H_{10}(g)\ +\ 13\ O_2(g)\ \rightarrow\ 8\ CO_2(g)\ +\ 10\ H_2O(\ell)$
c) $C_4H_8(g)\ +\ 6\ O_2(g)\ \rightarrow\ 4\ CO_2(g)\ +\ 4\ H_2O(\ell)$
d) $C_6H_{12}(g)\ +\ 9\ O_2(g)\ \rightarrow\ 6\ CO_2(g)\ +\ 6\ H_2O(\ell)$
e) $C_5H_{12}(g)\ +\ 13\ O_2(g)\ \rightarrow\ 5\ CO_2\ (g)\ +\ 6\ H_2O(\ell)$

Answer: a

21. Which of the following compounds are aromatic?

(1)

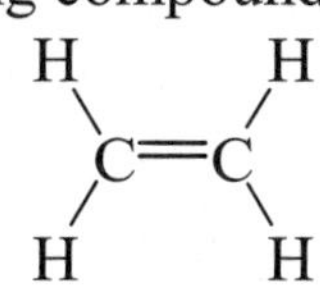

(2)

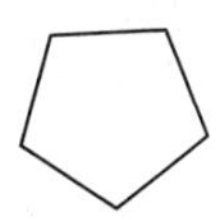

(3)

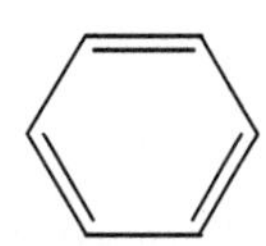

(4)

a) 1 only b) 2 only c) 3 only d) 4 only e) 1, 2 and 4

Answer: d

22. An unsaturated hydrocarbon is
a) a hydrocarbon that contains oxygen.
b) a compound in which all carbon atoms have four single bonds.
c) a compound in which one or more carbon atoms have double or triple bonds.
d) a hydrocarbon that is a gas at room temperature.
e) a cycloalkane with five or more carbons.

Answer: c

23. Which of the following molecules are polar?

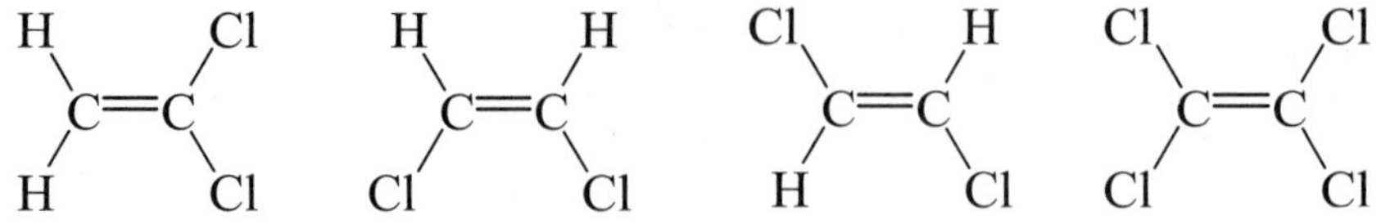

(1) (2) (3) (4)

a) 1 and 2 b) 2 and 3 c) 3 and 4 d) 1 and 3 e) 2 and 4

Answer: a

24. What is the product of the addition of Cl_2 to 2-butene?
 a) 2-chlorobutane
 b) 2-chlorobutene
 c) 2,2-dichlorobutane
 d) 2,3-dichlorobutane
 e) 1,4-dichlorobutane

Answer: d

25. What is the product(s) of the addition of HF to ethylene?
 a) fluoroethylene
 b) fluoroethane
 c) 1,2-difluoroethane
 d) 1,1-difluoroethane
 e) ethane and elemental fluorine

Answer: b

26. What is the product of the hydrogenation of *trans*-2-pentene?
 a) 2-hydroxypentane
 b) 2,3-dihydroxypentane
 c) *cis*-2-pentene
 d) 2-pentyne
 e) pentane

Answer: e

27. How many isomers exist for tribromobenzene?
 a) 3 b) 4 c) 5 d) 6 e) 7

Answer: a

28. Which of the structures below has the common name *o*-dibromobenzene (where *o*- is ortho)?

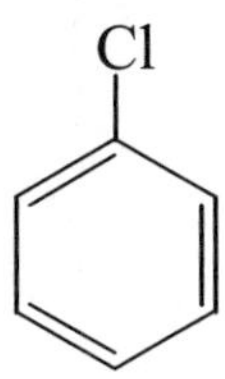

(1)

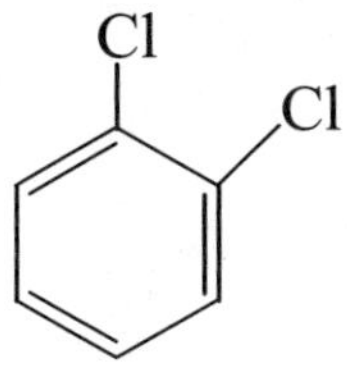

(2)

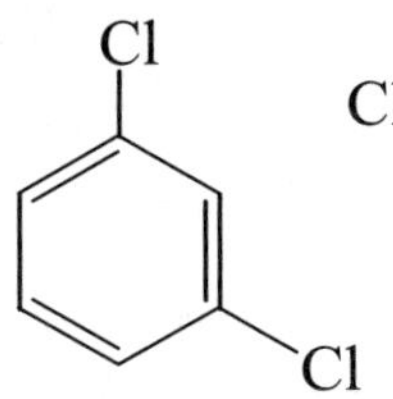

(3)

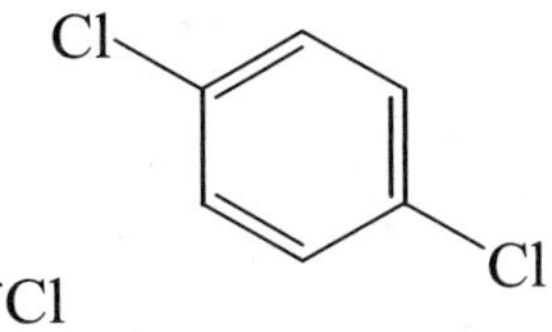

(4)

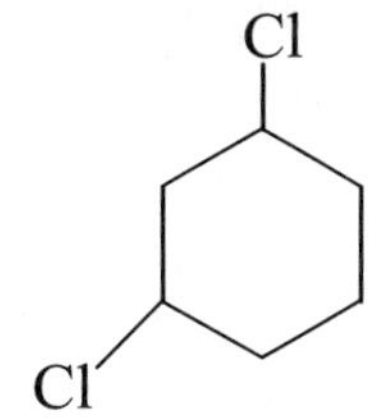

(5)

a) 1 b) 2 c) 3 d) 4 e) 5

Answer: b

29. What is the name of the following benzene derivative?

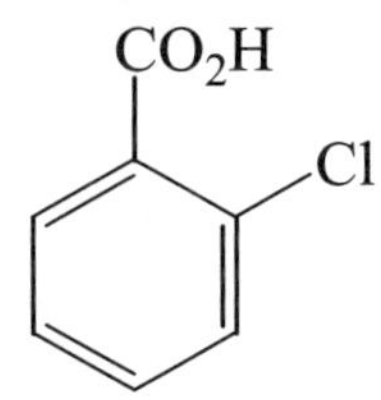

a) 1-carbonate-2-chlorobenzene
b) 3-chlorotoluene
c) 2-chlorobenzoic acid
d) 3-chlorobenzoic acid
e) 2-chlorobenzene carboxylic acid

Answer: c

30. Which of the following statements concerning hydrocarbons are CORRECT?
   1. Most hydrocarbons are gases at room temperature.
   2. The primary components of petroleum (prior to refining) are alkenes and alkynes.
   3. Reactions of hydrocarbons with oxygen generally yield carbon dioxide and water.

a) 1 only b) 2 only c) 3 only d) 1 and 3 e) 1, 2, and 3

Answer: c

31. The process of cracking, which occurs in the refining of petroleum, is
a) the method by which hydrocarbons are separated.
b) the method used to make higher boiling point hydrocarbons.
c) the method used to eliminate double or triple bonds.
d) the method used to make oxygenated fuels.
e) the fragmenting of longer chain hydrocarbons to smaller units.

Answer: e

32. Which of the following chemical equations depicts an alkylation reaction?
   a) $C_2H_2(g) + H_2(g) \rightarrow C_2H_4(g)$
   b) $CH_3CH_2OH(\ell) + 3\ O_2(g) \rightarrow 2\ CO_2(g) + 3\ H_2O(\ell)$
   c) $C_6H_{12}(\ell) \rightarrow C_6H_{10}(\ell) + H_2(g)$
   d) $C_6H_6(\ell) + CH_3Br(\ell) \rightarrow C_6H_5CH_3(\ell) + HBr(g)$
   e) $C_2H_2(g) + 2\ H_2(g) \rightarrow C_2H_6(g)$

   Answer: d

11.3 Alcohols, Ethers, and Amines

33. Formulas for derivatives of hydrocarbons may be written as R-X, where R is a hydrocarbon lacking a hydrogen atom and X is a functional group. Which of the following formulas represents an alcohol?
   a) ROH b) ROR' c) RCOR' d) RCO_2R' e) RCHO

   Answer: a

34. Which functional group does **not** contain an oxygen atom?
   a) ester b) ether c) alcohol d) amide e) amine

   Answer: e

35. Which of the following alcohols is likely to be least soluble in water?
   a) 1,2-ethanediol
   b) 2-butanol
   c) 3-pentanol
   d) 2,3-butanediol
   e) ethanol

   Answer: c

36. Compounds from which class usually have offensive odors, such as that of decaying meat or bad breath?
   a) alcohols b) amines c) ketones d) esters e) ethers

   Answer: b

37. The addition of water to ethylene has become a cheaper method than fermentation for making ________.
   a) ethanol b) ethylene glycol c) diethyl ether d) methane e) acetic acid

   Answer: a

38. Which one of the following functional groups is a base?
    a) ketone
    b) alcohol
    c) carboxylic acid
    d) amine
    e) ether

    Answer: d

<u>11.4 Compounds with a Carbonyl Group</u>

39. The functional group RCOR' is characteristic of a(n) ________.
    a) ether b) ester c) amide d) aldehyde e) ketone

    Answer: e

40. The carbonyl group occurs in molecules with all of the following functional groups EXCEPT ________.
    a) esters b) amines c) ketones d) amides e) aldehydes

    Answer: b

41. All of the following functional groups are members of the class of compounds known as carbonyls EXCEPT ________.
    a) aldehydes b) amides c) carboxylic acids d) alcohols e) ketones

    Answer: d

42. When a primary alcohol is oxidized, the product is a(n) ________.
    a) ketone b) ether c) ester d) aldehyde e) amide

    Answer: d

43. Which of the following compounds might be used to oxidize an aldehyde to a carboxylic acid?
    a) $K_2Cr_2O_7$ b) K c) NaH d) $FeCl_3$ e) $NaBH_4$

    Answer: a

44. What is the product of the reduction of butanone with sodium borohydride?
    a) butane b) butanoic acid c) acetone d) 1-butanol e) 2-butanol

    Answer: e

45. A mixture of methanol and benzoic acid is heated in the presence of acid. What is the primary product of the reaction?

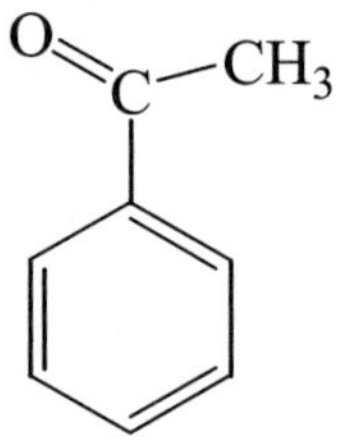
(1)

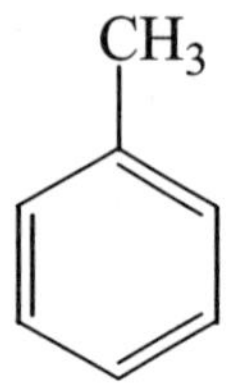
(2)

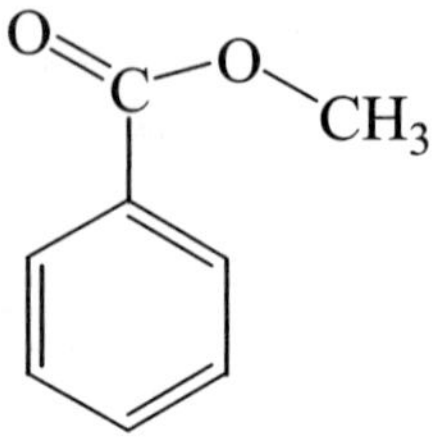
(3)

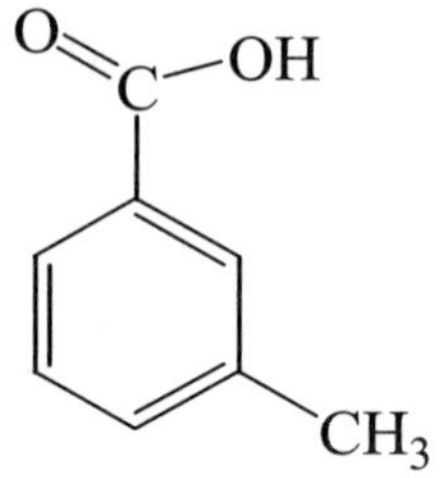
(4)

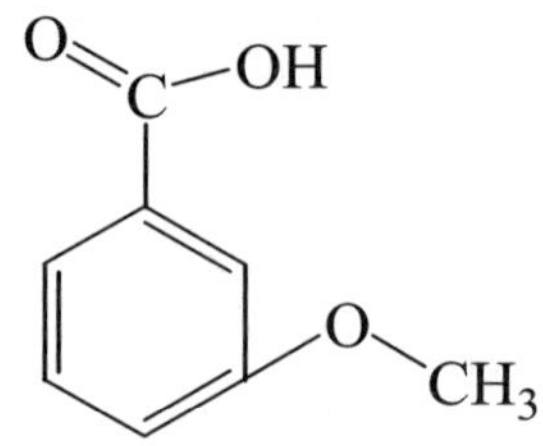
(5)

a) 1 b) 2 c) 3 d) 4 e) 5

Answer: c

46. What is the name of the product of the reaction that occurs when a mixture of ethanol and propanoic acid is heated in the presence of acid?
a) ethyl propyl ether
b) propyl ethanoate
c) 2-ethylpropanal
d) 2-ethylpropanol
e) ethyl propanoate

Answer: e

47. The hydrolysis of an ester in the presence of NaOH produces
a) a carboxylate salt of sodium and an alcohol.
b) an ether and an alcohol.
c) an alcohol and an aldehyde.
d) an aldehyde and an ether.
e) an ether and a carboxylate salt of sodium.

Answer: a

48. What class of compounds is responsible for many of the distinctive odors of artificial flavors and perfumes?
a) aldehydes b) esters c) ethers d) amides e) alcohols

Answer: b

49. What is the classification of the molecule below?

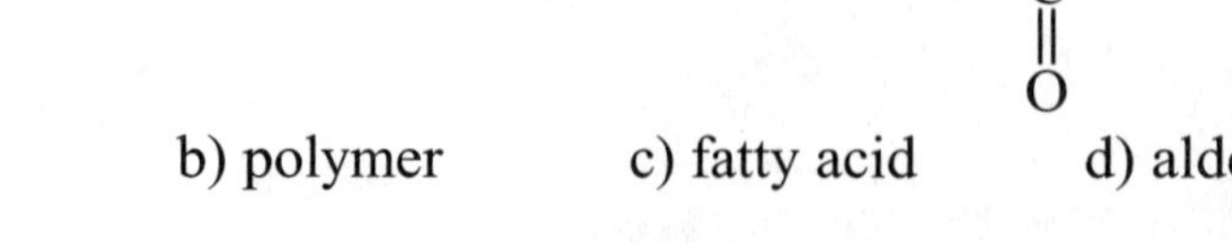

a) alcohol b) polymer c) fatty acid d) aldehyde e) triglyceride

Answer: c

11.5 Polymers

50. Which of the following statements is INCORRECT?
    a) Elastomers are materials that spring back to their original shape when stretched.
    b) Polystyrene is nonpolar and dissolves well in nonpolar solvents.
    c) Thermoplastics soften and flow when they are heated and harden when they are cooled.
    d) Polymers formed from two or more different monomers are called addition polymers.
    e) A condensation reaction involves reacting two monomers and splitting out a small molecule, often water.

    Answer: d

51. Amino acids polymerize in condensation reactions that result in the formation of an amide linkage (or peptide bond) between amino acid molecules. What dipeptide(s) may be formed in the reaction of glycine with phenylalanine?

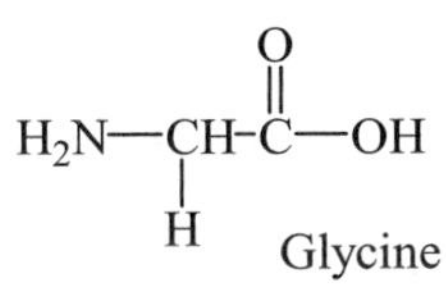

Glycine

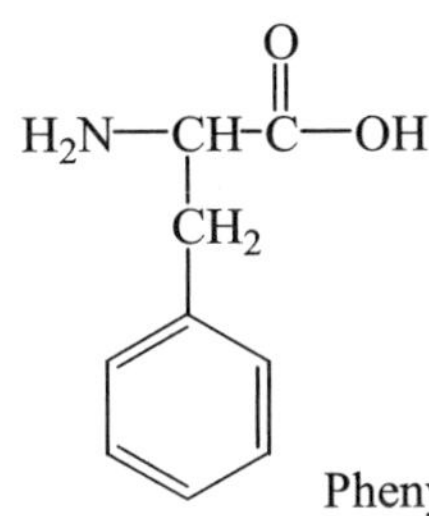

Phenylalanine

(1)

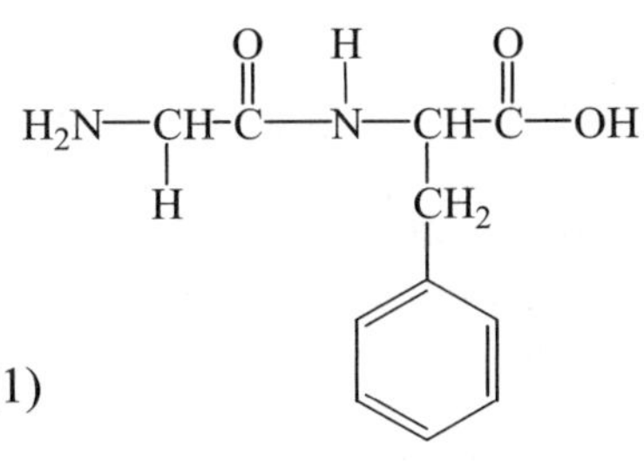

(2)

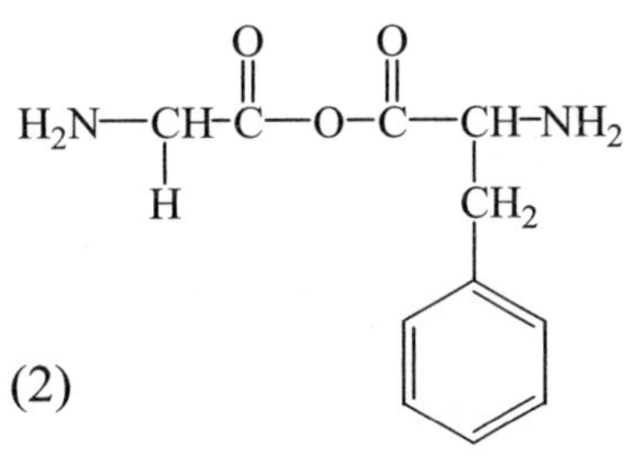

(3)

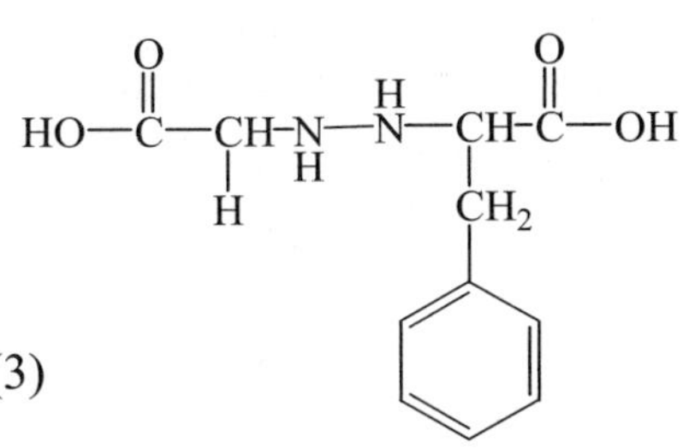

(4)

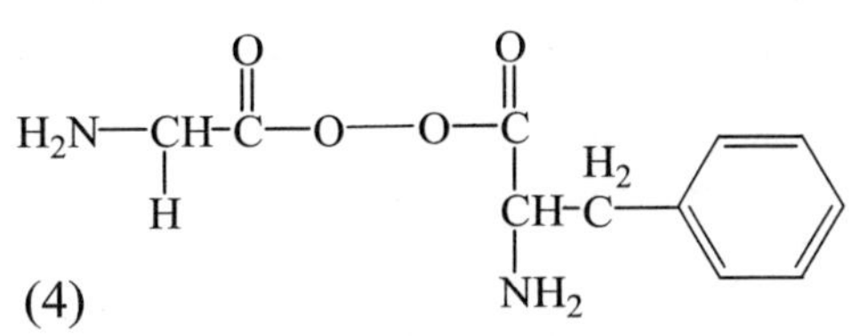

a) 1 only   b) 2 only   c) 3 only   d) 4 only   e) 1 and 3

Answer: a

52. Which of the following statements concerning polyethylene are correct?
    1. Polyethylene is made by condensation polymerization.
    2. Polyethylene contains alternating single and double carbon-carbon bonds.
    3. The monomer for polyethylene is CH_2CH_2.

a) 1 only b) 2 only c) 3 only d) 1 and 2 e) 1, 2, and 3

Answer: c

53. What is the monomer of acrylonitrile?

a) CH_2CH_2 b) CH_2CHCN c) $CH_2C(CN)_2$ d) CF_2CF_2 e) CH_2CHCH_3

Answer: b

54. Polyvinyl chloride is used in raincoats and pipes. It is produced by the addition reaction of chloroethylene, CH_2CHCl. What is the structure of the polymer produced in this reaction?

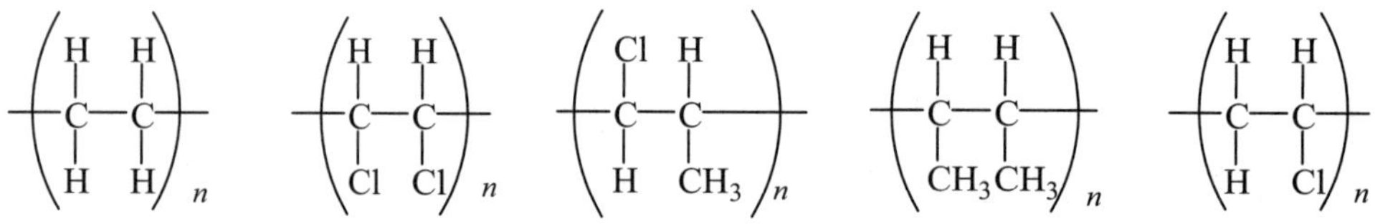

(1) (2) (3) (4) (5)

a) 1 b) 2 c) 3 d) 4 e) 5

Answer: e

Short Answer Questions

55. Hydrolysis of fats and oils in the presence of a strong base produces glycerol and the salts of fatty acids. The name for this reaction is ________, which means "soap making."

Answer: saponification

56. In vulcanized rubber, the polymer chains of natural rubber are cross-linked with sulfur atoms. The polymer chains stretch when stressed, but they return to their original shape when the stress is removed. Substances that behave this way are called __________.

Answer: elastomers

57. Examples of geometric ________ are *cis*- and *trans*-2-butene.

Answer: isomers

58. A peptide bond is the amide linkage that is formed in a condensation reaction involving the __________ group of one amino acid with the amine group of a second amino acid.

Answer: carboxylic acid

59. A pair of nonsuperimposable molecules with the same molecular formula are called ________.

Answer: enantiomers

60. Draw the Lewis structures for all the possible isomers of C_3H_6O.

Answer:

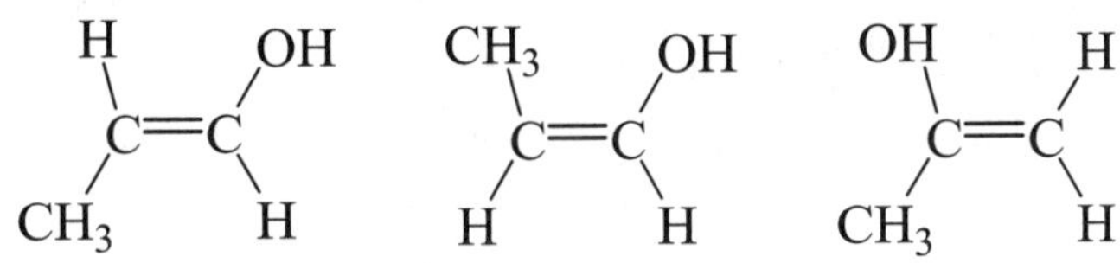

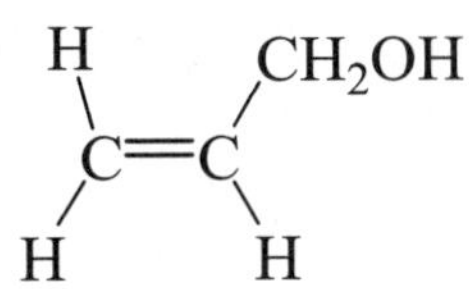

$H_2C{=}CH{-}O{-}CH_3$

61. The molecule 2,3-dibromo-4-methylhexane is made by addition of Br_2 to an alkene. Write a balanced chemical equation for this reaction.

Answer:

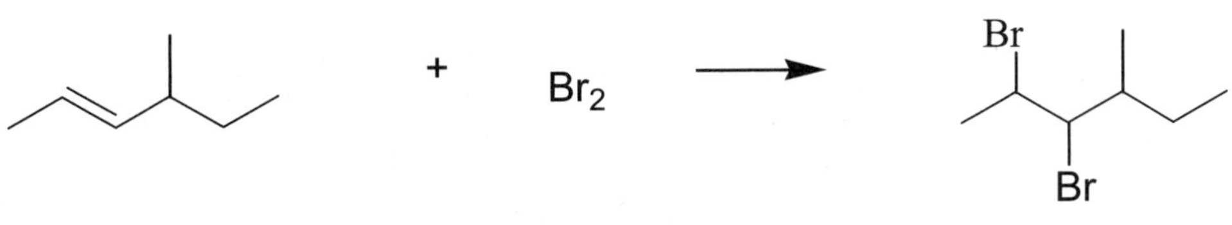

Diagnostic Quiz Questions

Goal #2: Recognize and draw structures for structural isomers and stereoisomers for carbon compounds.

1. Structural isomers are compounds that have
   a) the same elemental composition, but the atoms are linked in different ways.
   b) the same elemental composition, but different molar masses.
   c) different elemental composition, but identical physical properties.
   d) two or more resonance structures.

   Answer: a

2. How many geometric isomers exist for C_2HBrCl_2 (where a C=C bond is central to the structures)?
   a) 1
   b) 2
   c) 3
   d) 4

   Answer: c

3. Which of the following molecules has a chiral center?
   a) CH_4
   b) C_2H_6
   c) CH_3CH_2Cl
   d) $CH_3CHBrCl$

   Answer: d

Goal #3: Name and draw structures of common organic compounds

4. What is the general formula for an alkane?
   a) C_nH_{2n+2}
   b) C_nH_{2n}
   c) C_nH_{2n-2}
   d) $C_{n+2}H_{2n}$

   Answer: a

5. How many possible isomers, structural and geometric, have the formula C_4H_8?
   a) 3
   b) 4
   c) 5
   d) 6

   Answer: c

6. How many structural isomers exist for an alcohol with the formula C_4H_9OH?
   a) 2
   b) 3
   c) 4
   d) 5

   Answer: c

7. What is the name of a straight chain hydrocarbon with the formula C_8H_{18}?
   a) hexane
   b) heptane
   c) octane
   d) nonane

   Answer: c

8. What is the name of the following compound?

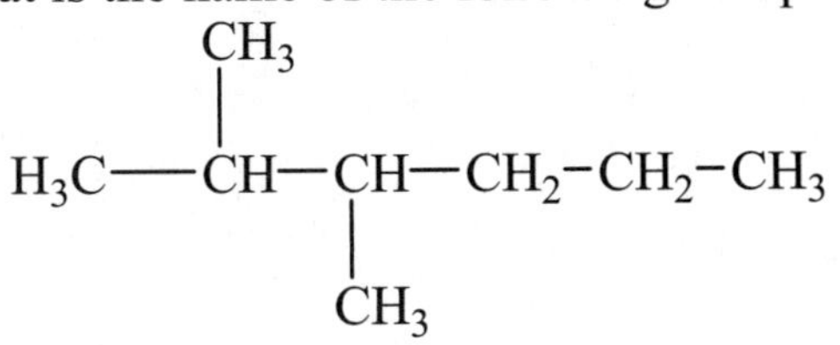

   a) 4,5-dimethylhexane
   b) 2,3-diethylhexane
   c) 2-methyl-3-propylbutane
   d) 2,3-dimethylhexane

   Answer: d

9. What is the systematic name of an alcohol with the condensed formula CH_3CH_2OH?
   a) methanol
   b) ethanol
   c) ethanone
   d) acetone

   Answer: b

10. What is the name of the following alkene?

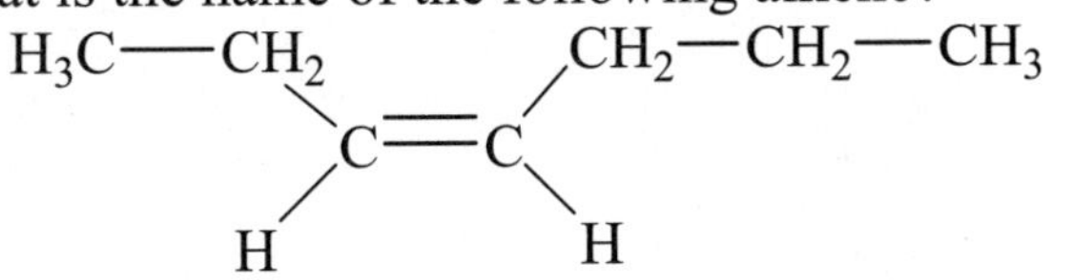

   a) *cis*-4-heptene
   b) *cis*-1-ethyle-4-methylbutene
   c) *trans*-3-heptene
   d) *cis*-3-heptene

   Answer: d

11. The functional group RCOR' is characteristic of a(n) ________.
   a) amine
   b) alcohol
   c) ketone
   d) ether

   Answer: d

12. Which of the following lists contains only compounds with carbonyl groups?
   a) amines, esters, and aldehydes
   b) ketones, ethers, and amides
   c) amides, aldehydes, and esters
   d) carboxylic acids, alcohols, and ketones

   Answer: c

Goal #3: Know the common reactions of organic functional groups

13. The oxidation of a primary alcohol produces a(n) ________.
   a) amine
   b) aldehyde
   c) ketone
   d) ether

   Answer: b

14. The oxidation of acetaldehyde, CH_3CHO, produces
   a) acetic acid, CH_3CO_2H.
   b) diethyl ether, $CH_3CH_2OCH_2CH_3$.
   c) ethanol, CH_3CH_2OH.
   d) acetone, CH_3COCH_3.

   Answer: a

15. What is the product of the addition of HCl to ethylene?
   a) chloroethane
   b) 1,2-dichloroethane
   c) *cis*-1,2-dichloroethylene
   d) 1,1-dichloroethane

   Answer: a

16. What is the product of the addition of Cl_2 to 1-butene?
   a) 1-chlorobutane
   b) 1,2-dichlorobutane
   c) 2-chlorobutane
   d) 2,2-dichlorobutane

   Answer: b

17. What two reactants combine in the presence of a strong acid to form the following ester?

$$CH_3\overset{\overset{\displaystyle O}{\|}}{C}OCH_2CH_2CH_3$$

   a) acetic acid and 1-propanol
   b) butanoic acid and methanol
   c) formic acid and butanol
   d) propanoic acid and ethanol

   Answer: a

18. The reaction of an ester with a strong base produces
   a) an alcohol and a ketone.
   b) an ether.
   c) an aldehyde and an amine.
   d) an alcohol and a carboxylate ion.

   Answer: d

Goal #4: Relate properties to molecular structure.

19. Which of the following alkanes will have the highest boiling point?
   a) hexane
   b) octane
   c) butane
   d) pentane

   Answer: b

20. Which of the following chemical equations depicts an alkylation reaction of an aromatic compound?
   a) $C_6H_6(\ell) + CH_3Br(g) \rightarrow C_6H_5CH_3(\ell) + HBr(g)$
   b) $C_2H_2(g) + Cl_2(g) \rightarrow CHClCHCl(\ell)$
   c) $C_6H_6(\ell) + Br_2(g) \rightarrow C_6H_5Br(\ell) + HBr(g)$
   d) $C_6H_6(\ell) + HNO_3(\ell) \rightarrow C_6H_5NO_2(\ell) + H_2O(\ell)$

   Answer: a

21. Which of the following statements concerning methanol are correct?
    1. Methanol, also known as wood alcohol, is toxic.
    2. Methanol and ethanol are entirely miscible with water.
    3. Methanol is used as a fuel in high-powered racing cars.

a) 1 only
b) 2 only
c) 1 and 2
d) 1, 2, and 3

Answer: d

22. What is the classification of the molecule below?

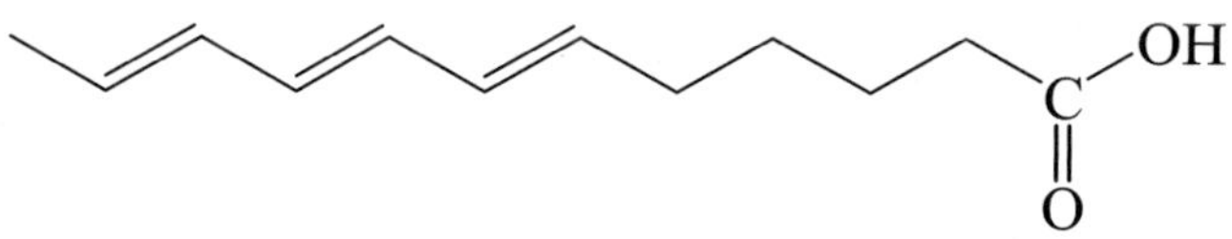

a) monosaccharide
b) oil
c) fatty acid
d) polymer

Answer: c

Goal #5: Identify common polymers

23. What is the monomer of Teflon?

a) C_2H_4
b) CH_2CF_2
c) CHFCHF
d) C_2F_4

Answer: d

24. Which of the following statements concerning polymers are correct?
    1. Amino acids polymerize by condensation reactions with water as a byproduct.
    2. Thermoplastics can withstand very high temperatures without softening or melting.
    3. An addition reaction always involves two different monomers, each with two different functional groups.

a) 1 only
b) 2 only
c) 3 only
d) 1 and 2

Answer: a

25. What two molecules copolymerize to form a polymer with the following repeating unit?

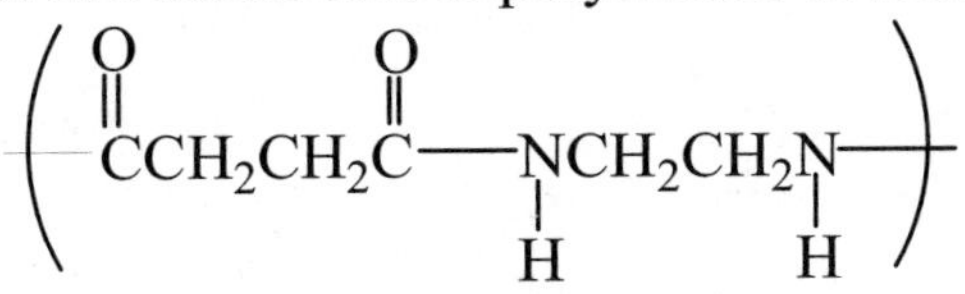

a) $CH_3CH_2CO_2H$ and $H_2NCH_2CH_3$
b) $HO_2CCH_2CH_2CO_2H$ and $H_2NCH_2CH_2NH_2$
c) CH_3CO_2H and $H_2NCH_2CH_3$
d) $HOH_2CCH_2CH_2CH_2OH$ and $H_2NCH_2CH_2NH_2$

Answer: b

CHAPTER 12: Gases & Their Properties

12.1 The Properties of Gases

1. Rank the following pressures in increasing order of magnitude: 623 mm Hg, 133 kPa, 0.155 atm, and 0.500 bar.
   a) 0.500 bar < 623 mm Hg < 133 kPa < 0.155 atm
   b) 0.155 atm < 0.500 bar < 623 mm Hg < 133 kPa
   c) 0.155 atm < 623 mm Hg < 0.500 bar < 133 kPa
   d) 623 mm Hg < 0.500 bar < 133 kPa < 0.155 atm
   e) 133 kPa < 0.155 atm < 0.500 bar < 623 mm Hg

   Answer: b

2. Convert 82.1 kPa to mm Hg. (1 atm = 760 mm Hg = 101.325 kPa)
   a) 0.810 mm Hg
   b) 10.9 mm Hg
   c) 616 mm Hg
   d) 2.40×10^3 mm Hg
   e) 6.32×10^6 mm Hg

   Answer: c

3. At 0.984 atm, the height of mercury in a barometer is 74.8 cm. If the mercury were replaced with water, what height of water (in centimeters) would be supported at this pressure? The densities of Hg and H_2O are 13.5 g/cm^3 and 1.00 g/cm^3, respectively.
   a) 5.54 cm b) 13.3 cm c) 994 cm d) 1.01×10^3 cm e) 1.84×10^5 cm

   Answer: d

12.2 Gas Laws: The Experimental Basis

4. At constant temperature, 14.0 L of O_2 at 0.882 atm is compressed to 1.75 L. What is the final pressure of O_2?
   a) 0.111 atm b) 0.142 atm c) 7.06 atm d) 21.6 atm e) 27.8 atm

   Answer: c

5. If the volume of a confined gas is quadrupled while its temperature remains constant, what change will be observed?
   a) The pressure of the gas will decrease to 1/4 its original value.
   b) The pressure of the gas will quadruple.
   c) The density of the gas will decrease to 1/2 its original value.
   d) The average velocity of the molecules will quadruple.
   e) The average velocity of the molecules will double.

   Answer: a

6. The lid is tightly sealed on a rigid flask containing 3.50 L H_2 at 17 °C and 0.913 atm. If the flask is heated to 71 °C, what is the pressure in the flask?
a) 0.770 atm b) 1.08 atm c) 2.95 atm d) 3.81 atm e) 13.3 atm

Answer: b

7. A tightly sealed 5.0-L flask contains 781 mm Hg of Ar at 19 °C. The flask is heated until the pressure is doubled. What is the temperature of the gas?
a) -127 °C b) 38 °C c) 95 °C d) 149 °C e) 311 °C

Answer: e

8. A balloon is filled with He gas to a volume of 3.22 L at 32 °C. The balloon is placed in liquid nitrogen until its temperature reaches –132 °C. Assuming the pressure remains constant, what is the volume of the cooled balloon?
a) 0.628 L b) 0.781 L c) 1.24 L d) 1.49 L e) 6.96 L

Answer: d

9. A 0.225-L flask contains CH_4 at 27 °C and 318 mm Hg. What is the pressure of the CH_4 if the volume is increased to 0.500 L and the temperature increased to 95 °C?
a) 117 mm Hg b) 176 mm Hg c) 503 mm Hg d) 508 mm Hg e) 867 mm Hg

Answer: b

10. You have a 30.0 L cylinder of helium at a pressure of 132 atm and a temperature of 24 °C. The He is used to fill balloons to a volume of 3.00 L at 1.07 atm and 29 °C. How many balloons can be filled with He? Assume that the cylinder can provide He until its internal pressure reaches 1.00 atm (i.e., there are 131 atmospheres of usable He in the cylinder).
a) 1.24×10^3 b) 3.73×10^3 c) 3.64×10^3 d) 4.18×10^3 e) 1.12×10^4

Answer: a

11. Avogadro's hypothesis states that equal volumes of gases under the same conditions of temperature and pressure have equal ________.
a) densities
b) particle velocities
c) molar masses
d) numbers of particles
e) masses

Answer: d

12. Which of the following relationships are true for gases?
    1. The number of moles of a gas is directly proportional to its volume (at constant pressure).
    2. The pressure of a gas is inversely proportional to its temperature in kelvins (at constant volume).
    3. The volume of a gas is directly proportional to its pressure (at constant temperature).

    a) 1 only b) 2 only c) 3 only d) 1 and 2 e) 2 and 3

    Answer: a

13. What volume of CO_2 at 15 °C and 1.50 atm contains the same number of molecules as 0.410 L of O_2 at 35 °C and 3.00 atm?
    a) 0.192 L b) 0.219 L c) 0.767 L d) 0.820 L e) 1.14 L

    Answer: c

14. If a 3.44-L flask of SO_2 at 1.65 atm contains the same number of molecules as a 5.00-L flask of H_2 at -7 °C and 1.00 atm, what is the temperature (in °C) of the SO_2?
    a) -39 °C b) 29 °C c) 88 °C d) 249 °C e) 575 °C

    Answer: b

12.3 The Ideal Gas Law

15. If 3.67 g CO_2 gas is introduced into an evacuated 2.50 L flask at 65 °C, what is the pressure inside the flask? (R = 0.08206 L·atm/mol·K)
    a) 0.178 atm b) 0.445 atm c) 0.925 atm d) 40.7 atm e) 496 atm

    Answer: c

16. The pressure in a 20.0 L flask is 0.512 atm at 72 °C. How many moles of gas are in the flask? (R = 0.08206 L·atm/mol·K)
    a) 0.0297 mol b) 0.142 mol c) 0.276 mol d) 0.362 mol e) 1.73 mol

    Answer: d

17. A 50.0 L cylinder of Ar has a pressure of 137 atm at 25 °C. What mass of Ar is inside the cylinder? (R = 0.08206 L·atm/mol·K)
    a) 0.143 g b) 4.48 g c) 918 g d) 2.80×10^2 g e) 1.12×10^4 g

    Answer: e

18. What volume is occupied by 8.50 g C_2H_2 at STP (standard temperature and pressure)? (R = 0.08206 L·atm/mol·K)
a) 0.670 L b) 7.31 L c) 7.98 L d) 68.7 L e) 191 L

Answer: b

19. Which of the following gases has the greatest density at 25 °C and 5.0 atm?
a) CH_4 b) O_2 c) N_2 d) F_2 e) CO_2

Answer: e

20. Calculate the density (in g/L) of Kr at 308 K and 527 mm Hg. (R = 0.08206 L·atm/mol·K)
a) 2.30 g/L b) 4.78 g/L c) 31.1 g/L d) 1.75×10^3 g/L e) 2.36×10^4 g/L

Answer: a

21. If the density of oxygen in air is 0.263 g/L at 25 °C, what is the pressure of oxygen at 25 °C? (R = 0.08206 L·atm/mol·K)
a) 0.201 atm b) 0.344 atm c) 0.402 atm d) 2.91 atm e) 6.43 atm

Answer: a

22. At what temperature does 1.00 atm of Ar gas have the same density as 1.00 atm of He gas at 273.2 K?
a) 27.4 K b) 117 K c) 367 K d) 2.73×10^3 K e) 4.71×10^3 K

Answer: d

23. A mass of 1.663 g of an unknown gas is introduced into an evacuated 2.00 L flask. If the pressure in the flask is 0.544 atm at 78 °C, which of the following gases might be in the flask? (R = 0.08206 L·atm/mol·K)
a) N_2 b) C_2H_2 c) NH_3 d) HCl e) N_2O

Answer: e

12.4 Gas Laws and Chemical Reactions

24. At 453 K and 755 mm Hg, what volume of N_2 will react completely with 22.2 L H_2 to produce NH_3?

$$N_2(g) + 3\ H_2(g) \rightarrow 2\ NH_3(g)$$

a) 7.40 L b) 14.8 L c) 22.2 L d) 44.4 L e) 66.6 L

Answer: a

25. What volume of O_2, measured at 91.2 °C and 743 mm Hg, will be produced by the decomposition of 4.88 g $KClO_3$? (R = 0.08206 L·atm/mol·K)

$$2\ KClO_3(s) \rightarrow 2\ KCl(s) + 3\ O_2(g)$$

a) 0.305 L b) 1.22 L c) 1.83 L d) 24.0 L e) 37.4 L

Answer: c

26. What volume of $O_2(g)$, measured at 17.7 °C and 0.978 atm reacts with 15.1 g C_4H_{10} to produce $CO_2(g)$ and $H_2O(\ell)$? (R = 0.08206 L·atm/mol·K)

a) 2.51 L b) 6.34 L c) 23.5 L d) 41.2 L e) 239 L

Answer: d

27. If 6.46 L of gaseous ethanol is burned in excess oxygen, what is the maximum volume of carbon dioxide produced? Assume that the temperature of the reactants and products is 425 °C and the pressure remains constant at 1.00 atm.

$$CH_3CH_2OH(g) + 3\ O_2(g) \rightarrow 2\ CO_2(g) + 3\ H_2O(g)$$

a) 6.46 L b) 3.23 L c) 12.9 L d) 19.4 L e) 32.3 L

Answer: c

28. Ammonia gas is synthesized according to the balanced equation below.

$$N_2(g) + 3\ H_2(g) \rightarrow 2\ NH_3(g)$$

If 2.75 L N_2 react with 7.75 L H_2, what is the theoretical yield (in liters) of NH_3? Assume that the volumes of reactants and products are measured at the same temperature and pressure.

a) 2.75 L b) 5.17 L c) 5.50 L d) 10.5 L e) 15.5 L

Answer: b

29. If 4.00 g S_8 react with 4.25 L O_2 at 39 °C and 1.00 atm, what is the maximum mass of SO_3 produced? (R = 0.08206 L·atm/mol·K)

$$S_8(s) + 12\ O_2(g) \rightarrow 8\ SO_3(g)$$

a) 2.83 g b) 8.86 g c) 9.99 g d) 13.3 g e) 19.9 g

Answer: b

30. Hydrochloric acid reacts with zinc to produce hydrogen gas according to the reaction below.

$$2\ HCl(aq) + Zn(s) \rightarrow ZnCl_2(aq) + H_2(g)$$

If 750.0 mL of 0.250 M HCl is combined with 5.98 g Zn, what volume of dry hydrogen gas can be produced? Assume the temperature and pressure of the gas are 25 °C and 742 mm Hg, respectively. (R = 0.08206 L·atm/mol·K)

a) 1.97 L b) 2.29 L c) 2.35 L d) 4.58 L e) 4.70 L

Answer: b

31. The empirical formula of a certain hydrocarbon is CH_2. When 0.120 mole of the hydrocarbon is completely combusted with excess oxygen, 17.7 L CO_2 gas is produced at 27 °C and 1.00 atm. What is the molecular formula of the hydrocarbon? (R = 0.08206 L·atm/mol·K)
a) C_2H_2 b) C_2H_4 c) C_3H_6 d) C_5H_{10} e) C_6H_{12}

Answer: e

32. An unknown gaseous hydrocarbon contains 85.63 % C. Its density is 0.426 g/L at 0.465 atm and 373 K. What is the molecular formula of the gas? (R = 0.08206 L·atm/mol·K)
a) C_2H_4 b) C_3H_6 c) C_4H_8 d) C_5H_{10} e) C_6H_{12}

Answer: a

12.5 Gas Mixtures and Partial Pressures

33. Nitroglycerin (227.1 g/mol) decomposes according to the reaction below.

$$4\ C_3H_5N_3O_9(\ell) \rightarrow 6\ N_2(g) + 12\ CO_2(g) + 10\ H_2O(g) + O_2(g)$$

What total volume of gases is produced at 65 °C and 744 mm Hg by the decomposition of 5.00 g nitroglycerin? (R = 0.08206 L·atm/mol·K)
a) 5.95×10^{-3} L b) 0.120 L c) 0.624 d) 4.52 L e) 18.1 L

Answer: d

34. Water can be decomposed by electrolysis into hydrogen gas and oxygen gas. What mass of water must decompose to fill a 4.00 L flask to a total pressure of 2.63 atm at 298 K with a mixture hydrogen and oxygen? (R = 0.08206 L·atm/mol·K)

$$2\ H_2O(\ell) \rightarrow 2\ H_2(g) + O_2(g)$$

a) 5.17 g b) 6.22 g c) 7.75 g d) 11.6 g e) 27.9 g

Answer: a

35. Carbon monoxide reacts with oxygen to form carbon dioxide.

$$2\ CO(g) + O_2(g) \rightarrow 2\ CO_2(g)$$

In a 1.00 L flask, 2.40 atm of CO reacts with 4.50 atm of O_2. Assuming that the temperature remains constant, what is the final pressure in the flask?
a) 1.20 atm b) 2.40 atm c) 3.30 atm d) 3.60 atm e) 5.70 atm

Answer: e

36. A mixture of He and O_2 is placed in a 4.00 L flask at 32 °C. The partial pressure of the He is 2.7 atm and the partial pressure of the O_2 is 1.4 atm. What is the mole fraction of O_2?
a) 0.224 b) 0.341 c) 0.481 d) 0.518 e) 0.659

Answer: b

37. A 1.00 L flask at 298 K contains a mixture of Kr and N_2 with a total pressure of 0.940 atm. If the mole fraction of Kr is 0.455, what is the mass fraction of Kr?
a) 0.218 b) 0.286 c) 0.455 d) 0.714 e) 0.782

Answer: d

12.6 The Kinetic-Molecular Theory of Gases

38. Which of the following are postulates of kinetic-molecular theory of gases?
1. The distance between gas molecules is large in comparison to their size.
2. Gas molecules are in constant, random motion.
3. The kinetic energy of gas particles is proportional to the kelvin temperature.

a) 1 only b) 2 only c) 3 only d) 1 and 2 e) 1, 2, and 3

Answer: e

39. What is the root-mean-square velocity of ethane molecules (C_2H_6) at 273 K?
a) 15.0 m/s b) 47.2 m/s c) 275 m/s d) 476 m/s e) 652 m/s

Answer: d

40. Place the following gases in order of increasing average velocity at 325 K: Ar, Ne, N_2, and SO_2.
a) Ar < Ne < N_2 < SO_2
b) Ne < Ar < N_2 < SO_2
c) N_2 < SO_2 < Ne < Ar
d) SO_2 < Ar < N_2 < Ne
e) Ar = Ne = N_2 = SO_2

Answer: d

41. The average velocity of a gas molecule is
a) inversely proportional to its kinetic energy.
b) directly proportional to the gas constant, *R*.
c) directly proportional to the square root of its temperature in K.
d) inversely proportional to the square of its mass.
e) directly proportional to the square root of its mass.

Answer: c

12.7 Effusion and Diffusion

42. Carbon monoxide gas diffuses through a barrier at a rate of 0.480 mL/minute. If an unknown gas diffuses through the same barrier at a rate of 0.277 mL/minute, what is the molar mass of the gas?
a) 9.33 g/mol b) 21.3 g/mol c) 36.9 g/mol d) 44.01 g/mol e) 84.1 g/mol

Answer: e

43. How long will it take 10.0 mL of Cl_2 gas to effuse through a porous barrier if it has been observed that 76.3 minutes are required for 10.0 mL of HCl gas to effuse through the same barrier?
a) 54.7 min. b) 106 min. c) 148 min. d) 202 min. e) 289 min.

Answer: b

44. If a gas effuses 1.618 times faster than Kr, what is its molar mass?
a. 18.02 g/mol b. 28.01 g/mol c. 32.01 g/mol d. 51.79 g/mol e. 65.88 g/mol

Answer: c

12.9 Nonideal Behavior: Real Gases

45. Non-ideal behavior for a gas is most likely to be observed under conditions of
a) high temperature and low pressure.
b) high temperature and high pressure.
c) low temperature and low pressure.
d) standard temperature and pressure.
e) low temperature and high pressure.

Answer: e

46. One way in which real gases differ from ideal gases is that the molecules of a real gas
a) are attracted to each other.
b) are liquids or solids above 273.15 K.
c) are always polar.
d) move in circles rather than straight paths.
e) have no kinetic energy.

Answer: a

Short Answer Questions

47. The rate of effusion of a gas is inversely proportional to the ________ of its molar mass.

Answer: square root

48. Boyle learned that at constant temperature and number of molecules, the volume of a gas is inversely proportional to its ________.

    Answer: pressure

49. A statement of Avogadro's hypothesis is that equal volumes of gases under the same conditions of pressure and temperature will contain equal numbers of ________.

    Answer: particles (or molecules, or moles)

50. The gas constant, *R*, has a value of 0.08206. The units of the constant are ________.

    Answer: L·atm/mol·K

51. The ideal gas law can be modified to correct for the errors arising from nonideality. The modified equation is known as the ________ equation of state for a real gas.

    Answer: van der Waals

52. The pressure exerted by gas molecules on the inside of a container is due to gas molecules colliding with the walls of the container. Why does one mole of oxygen molecules apply the same pressure to the inside of a flask as one mole of hydrogen molecules, even though the oxygen has sixteen times the mass of hydrogen?

    Answer: At the same temperature, hydrogen molecules have a greater average velocity than oxygen molecules. The force applied to the container walls due to a collision is both a function of the mass and the velocity of the colliding molecule. In addition, hydrogen molecules collide more frequently with the walls of the container, due to their higher velocity. So, the combination of hydrogen's higher collision frequency and greater velocity compensates for its lower mass.

53. What are two ways in which real gases differ from ideal gases?

    Answer: Gas molecules occupy a finite volume of space. Gas molecules are attracted to each other through a variety of intermolecular forces.

Diagnostic Quiz Questions

Goal #1: Understand the basis of the gas laws and how to use those laws.

1. Convert 357 mm Hg to atmospheres.
   a) 0.357 atm
   b) 0.470 atm
   c) 2.13 atm
   d) 2.71×10^5 atm

   Answer: b

2. At constant temperature, 10.0 L of N_2 at 0.983 atm is compressed to 2.88 L. What is the final pressure of N_2?
   a) 0.283 atm
   b) 1.51 atm
   c) 3.41 atm
   d) 29.3 atm

   Answer: c

3. The lid is tightly sealed on a rigid flask containing 2.00 L O_2 at 15 °C and 0.723 atm. If the flask is heated to 55 °C, what is the pressure in the flask?
   a) 0.635 atm
   b) 0.823 atm
   c) 1.13 atm
   d) 2.65 atm

   Answer: b

4. A balloon is inflated with hydrogen to a volume of 21 L at 27 °C. If the balloon is cooled to 14 °C, what is the new volume of the balloon?
   a) 11 L
   b) 2.0×10^1 L
   c) 22 L
   d) 41 L

   Answer: b

5. Acetylene burns in oxygen to give CO_2 and H_2O, according to the equation below.

$$2\ C_2H_2(g)\ +\ 5\ O_2(g)\ \rightarrow\ 4\ CO_2(g)\ +\ 2\ H_2O(g)$$

What volume of oxygen will react completely with 26 L C_2H_2?
   a) 11 L
   b) 26 L
   c) 52 L
   d) 65 L

   Answer: d

Goal #2 Use the ideal gas law

6. If 3.25 g N_2 gas is introduced into an evacuated 1.50 L flask at 325 K, what is the pressure inside the flask? ($R = 0.08206$ L·atm/mol·K)
   a) 2.06 atm
   b) 3.09 atm
   c) 14.1 atm
   d) 57.8 atm

   Answer: a

7. A mass of 22.8 g of an unknown gas exerts a pressure of 0.912 atm at 78 °C in a 10.0 L flask. What is the molar mass of the gas?
   a) 1.56 g/mol
   b) 16.0 g/mol
   c) 44.0 g/mol
   d) 72.0 g/mol

   Answer: d

8. Calculate the density (in g/L) of Ar at 32 °C and 342 mm Hg. ($R = 0.08206$ L·atm/mol·K)
   a) 0.0615 g/L
   b) 0.718 g/L
   c) 6.85 g/L
   d) 546 g/L

   Answer: b

9. What volume of O_2, measured at 27.2 °C and 735 mm Hg, will be produced by the decomposition of 5.22 g $KClO_3$? (R = 0.08206 L·atm/mol·K)

$$2\ KClO_3(s) \rightarrow 2\ KCl(s) + 3\ O_2(g)$$

a) 1.09 L
b) 1.63 L
c) 14.7 L
d) 2.00×10^2 L

Answer: b

10. A mixture of H_2 and Ne is placed in a 5.00 L flask at 20 °C. The partial pressure of the H_2 is 1.4 atm and the partial pressure of the Ne is 2.1 atm. What is the mole fraction of H_2?
a) 0.014
b) 0.40
c) 0.60
d) 0.67

Answer: b

11. A 1.00 L flask at 298 K contains a mixture of He and O_2 with a total pressure of 1.00 atm. If the mole fraction of O_2 is 0.50, what is the mass fraction of O_2?
a) 0.11
b) 0.50
c) 0.73
d) 0.89

Answer: d

12. Sulfur dioxide reacts with oxygen to form sulfur trioxide.

$$2\ SO_2(g) + O_2(g) \rightarrow 2\ SO_3(g)$$

In a 1.00 L flask, 2.00 atm of SO_2 react with 5.00 atm of O_2. Assuming that the temperature remains constant, what is the final pressure in the flask?
a) 2.00 atm
b) 5.00 atm
c) 6.00 atm
d) 7.00 atm

Answer: c

Goal #3: Understand the kinetic molecular theory applied to gases, especially the distribution of molecular speeds.

13. Which of the following are postulates of the kinetic-molecular theory of gases?
    1. The distance between gas molecules is large in comparison to their size.
    2. Gas molecules are in constant, random motion.
    3. The kinetic energy of a gas molecule is inversely proportional to its temperature.

    a) 1 only
    b) 2 only
    c) 3 only
    d) 1 and 2

    Answer: d

14. Calculate the rms speed of H_2 molecules at 15 °C.
    a) 59.7 m/s
    b) 188 m/s
    c) 431 m/s
    d) 1.89×10^3 m/s

    Answer: d

Goal #4: Apply kinetic-molecular theory of gas behavior at the molecular level.

15. Which of the following statements concerning gas pressure are correct?
    1. Gas pressure arises from collisions of gas molecules with the walls of the container holding the gas.
    2. Increasing the number of gas molecules within a container increases the number of collisions with the walls of the container, thereby increasing the gas pressure.
    3. As the temperature of a gas increases, gas molecules exert more force on the walls of their container.

    a) 1 only
    b) 2 only
    c) 3 only
    d) 1, 2, and 3

    Answer: d

16. Carbon dioxide gas diffuses through a barrier at a rate of 0.40 mL/minute. If an unknown gas diffuses through the same barrier at a rate of 0.31 mL/minute, what is the molar mass of the gas?
   a) 26 g/mol
   b) 39 g/mol
   c) 51 g/mol
   d) 73 g/mol

   Answer: d

17. How long will it take 10.0 mL of nitrogen gas to effuse through a porous barrier if it has been observed that 92.1 minutes are required for 10.0 mL of Cl_2 gas to effuse through the same barrier?
   a) 14.4 min.
   b) 36.4 min.
   c) 57.9 min.
   d) 147 min.

   Answer: c

Goal #5: Recognize why gases do not behave as ideal gases

18. Which of the following statements concerning real gases are correct?
    1. The measured pressure of a container of chlorine gas is generally higher than the pressure calculated by the ideal gas law.
    2. Nonideality in gas behavior becomes more pronounced as the temperature of a gas is decreased.
    3. Nonideality in gas behavior becomes more pronounced as the pressure of a gas is decreased.

   a) 1 only
   b) 2 only
   c) 3 only
   d) 1 and 2

   Answer: b

CHAPTER 13: Intermolecular Forces, Liquids, and Solids

13.2 Intermolecular Forces

1. Which of the following statements concerning the attraction of ions to polar molecules are CORRECT?
    1. The energy of attraction between an ion and a polar molecule is inversely proportional to the distance between the center of the ion and the oppositely charged pole of the dipole.
    2. The higher the ion charge, the greater the attraction between the ion and a polar molecule.
    3. The greater the dipole moment, the greater the attraction between the ion and a polar molecule.

    a) 1 only b) 2 only c) 3 only d) 2 and 3 e) 1, 2, and 3

    Answer: e

2. Place the following cations in order from the lowest to the highest hydration energy.
    a) $K^+ < Ca^{2+} < Cs^+$
    b) $K^+ < Cs^+ < Ca^{2+}$
    c) $Ca^{2+} < K^+ < Cs^+$
    d) $Ca^{2+} < Cs^+ < K^+$
    e) $Cs^+ < K^+ < Ca^{2+}$

    Answer: e

3. Which of the following statements concerning intermolecular forces are CORRECT?
    1. London dispersion forces exist in all molecular solids.
    3. Dipole-dipole attractions occur in all molecules that contain polar bonds.
    4. Hydrogen bonding only occurs for molecules containing OH bonds.

    a) 1 only b) 2 only c) 3 only d) 1 and 2 e) 1, 2, and 3

    Answer: a

4. Which one of the following substances will exhibit dipole-dipole intermolecular forces?
    a) C_2H_4 b) SO_2 c) CO_2 d) F_2 e) Ar

    Answer: b

5. What intermolecular force or bond is primarily responsible for the solubility of H_2S in water?
   a) ion-dipole force
   b) dipole-dipole force
   c) ionic bonding
   d) covalent bonding
   e) hydrogen bonding

   Answer: b

6. What intermolecular force or bond is primarily responsible for the solubility of chlorine (Cl_2) in water?
   a) dipole/dipole force
   b) hydrogen bonding
   c) dipole/induced dipole force
   d) ion-dipole force
   e) ion-induced dipole force

   Answer: c

7. As molecular solids, which of the following will exhibit dipole-dipole intermolecular forces: PH_3, SO_3, Br_2, and $SiCl_4$?
   a) PH_3 only
   b) PH_3 and SO_3
   c) SO_3 and $SiCl_4$
   d) Br_2 and $SiCl_4$
   e) PH_3, SO_3, and $SiCl_4$

   Answer: a

8. Which of the following molecules will have the highest boiling point?
   a) H_2 b) CH_4 c) Ne d) Cl_2 e) O_2

   Answer: d

9. Which one of the following molecules will have the lowest boiling point?
   a) CCl_4 b) $CHCl_3$ c) CH_2Cl_2 d) CH_3Cl e) CH_4

   Answer: e

10. Which intermolecular forces are present in $PCl_3(s)$?
    1. London dispersion
    2. dipole-dipole
    3. hydrogen bonding

    a) 1 only b) 2 only c) 3 only d) 1 and 2 e) 1 and 3

    Answer: d

11. For which of the following pure solids is it necessary to break covalent bonds to make a liquid or gas?
a) NaCl b) NH_3 c) CH_3CO_2H d) Kr e) SiO_2

Answer: e

13.3 Hydrogen Bonding

12. Which is the dominant intermolecular force present in formic acid, $HCO_2H(\ell)$?
a) London dispersion
b) ionic bonding
c) hydrogen bonding
d) dipole-dipole
e) dipole/induced dipole

Answer: c

13. Which of the following molecules would be expected to form hydrogen bonds in the liquid or solid state: ethylene glycol ($HOCH_2CH_2OH$), methanol (CH_3OH), and formaldehyde (H_2CO)?
a) ethylene glycol only
b) methanol only
c) formaldehyde only
d) ethylene glycol and methanol
e) ethylene glycol, methanol, and formaldehyde

Answer: d

14. Arrange H_2O, H_2S, and H_2Se in order from lowest to highest boiling point.
a) $H_2O < H_2S < H_2Se$
b) $H_2O < H_2Se < H_2S$
c) $H_2S < H_2Se < H_2O$
d) $H_2S < H_2O < H_2Se$
e) $H_2Se < H_2O < H_2S$

Answer: c

15. Which intermolecular force or bond is responsible for the density of $H_2O(s)$ being less than that of $H_2O(\ell)$?
a) London dispersion forces
b) hydrogen bonding
c) ionic bonding
d) covalent bonding
e) dipole/induced dipole forces

Answer: b

16. Hydrogen bonding is present in all of the following molecular solids EXCEPT _______ .
a) CH_3OCH_3 b) H_2O c) NH_3 d) HF e) H_2O_2

Answer: a

17. Arrange NaCl, HCl, CCl_4, and $GeCl_4$ in order of increasing boiling point.
a) HCl < CCl_4 < $GeCl_4$ < NaCl
b) HCl < $GeCl_4$ < CCl_4 < NaCl
c) CCl_4 < NaCl < HCl < $GeCl_4$
d) $GeCl_4$ < NaCl < HCl < CCl_4
e) NaCl < $GeCl_4$ < CCl_4 < HCl

Answer: a

13.5 Properties of Liquids

18. Which of the following liquids will have the largest enthalpy of vaporization?
a) C_3H_8 b) C_4H_{10} c) C_4H_8 d) C_5H_{12} e) C_6H_{14}

Answer: e

19. Ammonia, NH_3, is used as a refrigerant. At its boiling point of –33 °C, the enthalpy of vaporization of ammonia is 23.3 kJ/mol. How much heat is required to vaporize 355 g of ammonia at –33 °C?
a) 1.12 kJ b) 152 kJ c) 251 kJ d) 486 kJ e) 8.27×10^3 kJ

Answer: d

20. At its boiling point of –42.1 °C, 14.6 kJ of heat is required to vaporize 33.9 g of propane (C_3H_8). What is the molar enthalpy of vaporization of propane?
a) 2.32 kJ/mol b) 11.2 kJ/mol c) 19.0 kJ/mol d) 97.8 kJ/mol e) 473 kJ/mol

Answer: c

21. At 50.0 °C, water has an equilibrium vapor pressure of 92.5 mm Hg. If 4.700 g H_2O is sealed in an evacuated 10.0-L flask and heated to 50.0 °C, what mass of H_2O will evaporate? Assume any liquid remaining in the flask has a negligible volume. (R = 0.08206 L·atm/mol·K, 1 atm = 760 mm Hg)
a) 0.0459 g b) 0.827 g c) 1.21 g d) 3.873 g e) 4.700 g

Answer: b

22. If 3.01 g of water is sealed in an evacuated 2.00-L flask and heated to the normal boiling point of 373 K, what is the pressure in the flask? (R = 0.08206 L·atm/mol·K)
a) 0.391 atm b) 0.678 atm c) 1.00 atm d) 2.56 atm e) 46.1 atm

Answer: c

23. A linear relationship exists between the natural logarithm of the vapor pressure of a gas and the reciprocal of its temperature (in kelvin). What is the slope of the line?
a) $-\frac{R}{\Delta H_{vap}}$ b) $-R \times \Delta H_{vap}$ c) R d) $-\frac{\Delta H_{vap}}{R}$ e) $-\Delta H_{vap}$

Answer: d

24. Diethyl ether has an enthalpy of vaporization of 29.0 kJ/mol. The compound has a vapor pressure of 1.00 atm at 35.0 °C. At what temperature is the vapor pressure equal to 225 mm Hg? (R = 8.314 J/K·mol)
a) 5.1 °C b) 11.2 °C c) 15.2 °C d) 72.1 °C e) 83.0 °C

Answer: a

25. Diethyl ether has a vapor pressure of 760 mm Hg at 34.6 °C and a vapor pressure of 534 mm Hg at 25.0 °C. What is the enthalpy of vaporization of diethyl ether?
a) 0.406 kJ/mol b) 12.1 kJ/mol c) 16.0 kJ/mol d) 28.1 kJ/mol e) 113 kJ/mol

Answer: d

26. Which of the following properties of water can be attributed to hydrogen bonding?
   1. high melting point
   2. high heat of vaporization
   3. low vapor pressure
   4. high surface tension

a) 1 and 3 b) 2 and 3 c) 2, 3, and 4 d) 1, 3, and 4 e) 1, 2, 3, and 4

Answer: e

27. Which of the following phase transitions is endothermic?
a) gas to solid
b) liquid to gas
c) liquid to solid
d) gas to liquid
e) none of the above

Answer: b

28. The normal boiling point is defined as
a) the temperature at which the vapor pressure of a liquid equals 1 atm.
b) the pressure at which any liquid boils at 273.15 K.
c) the temperature at which water always boils.
d) the pressure of a gas when its temperature reaches 373.15 K.
e) the temperature at which the enthalpy of vaporization equals 40.7 kJ/mol.

Answer: a

29. Which of the following statements are correct?
1. Above the critical pressure, only the solid phase of a pure substance can exist.
2. Above the critical temperature, a gas may not be condensed to a liquid.
3. At the critical point, three phases (gas, liquid, and solid) may coexist at equilibrium.

a) 1 only b) 2 only c) 3 only d) 1 and 2 e) 1 and 3

Answer: b

30. Which of the following statements concerning a supercritical fluid (SCF) are correct?
1. A SCF has a viscosity similar to that of a liquid.
2. A SCF has a density similar to that of a gas.
3. A SCF fills its entire container, much like a gas.

a) 1 only b) 2 only c) 3 only d) 1 and 3 e) 1, 2, and 3

Answer: c

31. Which of the following gases can be liquefied at 25 °C?

Gas	boiling pt.	critical temp.
CH_4	-161.5 °C	-82.6 °C
C_3H_8	-42.1 °C	+96.7 °C
O_2	-183 °C	-119 °C

a) CH_4 only b) C_3H_8 only c) O_2 only d) CH_4 and O_2 e) C_3H_8 and O_2

Answer: b

32. The toughness of the skin of a liquid is a measure of its ________.
a) meniscus quality
b) adhesive forces
c) dispersion forces
d) viscosity
e) surface tension

Answer: e

33. Which of the following are validify why vegetable oil has a greater viscosity than diethyl ether, CH_3OCH_3?
   1. Oil molecules have long chains that become entangled.
   2. Oil molecules are not held together by hydrogen bonds.
   3. Intermolecular forces are greater for the larger oil molecules.

   a) 1 only b) 2 only c) 3 only d) 1 and 3 e) 2 and 3

   Answer: d

13.6 The Solid State: Metals

34. An atom located on a face of a unit cell is shared equally between ________ unit cells.
   a) 2 b) 4 c) 6 d) 8 e) 10

   Answer: a

35. Which equation represents the number of atoms in a body-centered cubic unit cell?
   a) # atoms $= \frac{1}{8}(8)$
   b) # atoms $= \frac{1}{4}(8)$
   c) # atoms $= 1 + \frac{1}{8}(8)$
   d) # atoms $= \frac{1}{2}(6) + \frac{1}{8}(8)$
   e) # atoms $= 1 + \frac{1}{2}(6) + \frac{1}{8}(8)$

   Answer: c

36. If a metal crystallizes in a face-centered cubic lattice, each metal atom has ________ "nearest neighbors."
   a) 6 b) 8 c) 10 d) 12 e) 16

   Answer: d

37. The radius of a copper atom is 90.2 pm. If copper crystallizes in a face-centered cubic lattice, what is the length (in pm) of an edge of a unit cell.
   a) 128 pm b) 1.80×10^2 pm c) 208 pm d) 255 pm e) 312 pm

   Answer: d

38. Sodium crystallizes in a body-centered cubic unit cell. If the length of an edge of the unit cell is 0.430 nm, what is the atomic radius (in nm) of a sodium atom?
a) 0.186 nm b) 0.304 nm c) 0.372 nm d) 0.430 nm e) 0.745 nm

Answer: a

39. Aluminum (atomic mass 26.98 g/mol) crystallizes in a face-centered cubic unit cell. In addition, aluminum has an atomic radius of 143 pm. What is the density of aluminum?
a) 1.35 g/cm^3 b) 2.71 g/cm^3 c) 4.98 g/cm^3 d) 7.66 g/cm^3 e) 10.5 g/cm^3

Answer: b

13.7 The Solid State: Structures and Formulas of Ionic Compounds

40. Cesium chloride forms a simple cubic crystal with chloride ions at the lattice points. The cesium ions occupy cubic holes. How many chloride ions surround each cesium ion in cesium chloride?
a) 1 b) 2 c) 4 d) 8 e) 12

Answer: d

41. Which of the following statements are correct? If an ionic compound with the formula MX forms a face-centered cubic lattice with the anions (X^{n-}), the cations (M^{n+}) will occupy
    1. all of the tetrahedral holes in each unit cell.
    2. all of the octahedral holes in each unit cell.
    3. the center of each face in each unit cell.

a) 1 only b) 2 only c) 3 only d) 1 and 2 e) 1, 2, and 3

Answer: b

42. If an ionic compound with the formula MX_2 forms a face-centered cubic lattice with the anions (X^{n-}), the cations (M^{2n+}) will occupy
a) all of the tetrahedral holes in each unit cell.
b) half of the tetrahedral holes in each unit cell.
c) all of the octahedral holes in each unit cell.
d) the center of each face in each unit cell.
e) the cubic holes in the center of the each unit cell.

Answer: a

13.8 Other Kinds of Solid Materials

43. Which of the following is NOT a network solid?
    a) graphite
    b) diamond
    c) buckminster fullerene (C_{60})
    d) silicon dioxide
    e) elemental silicon

    Answer: c

44. Which of the following solids are always amorphous: water, polyethylene, glass, potassium nitrate?
    a) water and glass
    b) polyethylene and potassium nitrate
    c) water, polyethylene, and glass
    d) water and potassium nitrate
    e) polyethylene and glass

    Answer: e

13.9 The Physical Properties of Solids

45. Which process requires the greatest endothermic change in enthalpy for water?
    a) freezing b) condensation c) vaporization d) melting e) sublimation

    Answer: e

13.10 Phase diagrams

46. A phase diagram of a pure compound has a triple point at 27 °C and 65 mm Hg, a normal melting point at 24 °C, and a normal boiling point at 118 °C. Which of the following statements regarding this compound are CORRECT?
    1. The density of the solid is greater than that of the liquid.
    2. Sublimation occurs if the temperature of the solid phase is increased from 11 °C to 34 °C at a constant pressure of 61 mm Hg.
    3. Evaporation occurs if the pressure is raised from 65 mm Hg to 820 mm Hg at a constant temperature of 24 °C.

    a) 1 only b) 2 only c) 3 only d) 2 and 3 e) 1, 2, and 3

    Answer: b

47. The phase diagram for CO_2 has a triple point at -56.6 °C and 5.19 atm, and a critical point at 31.0 °C and 73 atm. The solid and liquid phases are in equilibrium at -78.7 °C and 1.00 atm. Which of the following statements regarding CO_2 are CORRECT?
    1. Sublimation occurs if the temperature of the solid phase is increased from -79.0 °C to 0.0 °C at a constant pressure of 2.5 atm.
    2. CO_2 is a supercritical fluid at 55 °C and 75 atm.
    3. Condensation occurs if the pressure is decreased from 73 atm to 5.0 atm at 34 °C.

    a) 1 only b) 2 only c) 3 only d) 1 and 2 e) 1, 2, and 3

    Answer: d

48. Which of the following statements concerning the phase diagram below are CORRECT?

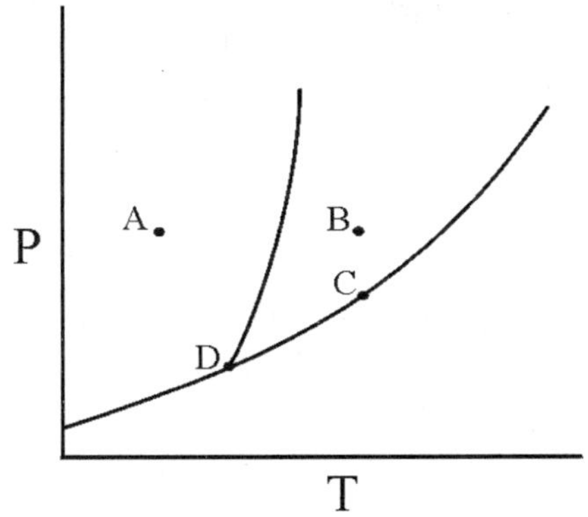

1. Moving from point B to A results in a phase transition from liquid to gas.
2. Point D lies at the critical point.
3. At point C, liquid and gas phases coexist at equilibrium.

a) 1 only b) 2 only c) 3 only d) 1 and 3 e) 2 and 3

Answer: c

Short Answer Questions

49. Sodium chloride crystallizes in a ________ cubic unit cell with chloride ions occupying the lattice points. The sodium ions occupy interstitial regions, with each cation in contact with six chloride ions.

    Answer: face-centered

50. A cup of oil takes longer to pour from a glass than a cup of water. A liquid's resistance to flow is referred to as its ________.

    Answer: viscosity

51. ________ is a measure of the degree to which the electron cloud surrounding an atom or molecule can be distorted in an electric field.

Answer: Polarizability

52. On a 3-phase diagram, the ________ point is located at the temperature and pressure where gas, liquid, and solid phases of a substance all coexist at equilibrium.

Answer: triple

53. Which ion, K^+ or Ca^{2+}, is expected to have the more negative enthalpy of hydration? Why?

Answer: Ca^{2+} will have the more negative enthalpy of hydration. A calcium ion is smaller than a potassium ion. The smaller ion will bond more closely to water's dipole, allowing for a stronger ion-dipole interaction. In addition, the higher the charge on the ion, the greater the ion-dipole interaction.

54. Two important allotropes of phosphorus are white phosphorus and red phosphorus. White phosphorus, P_4, has a melting point of 44.1 °C and spontaneously reacts with oxygen. Red phosphorus, P, melts at 280 °C and is stable in air. Use your knowledge of intermolecular and intramolecular bonding to explain why these two forms of the same element have such different properties.

Answer: White phosphorus is a molecular solid composed of tetrahedral P_4 molecules. The P_4 molecules are held together by London dispersion forces. Red phosphorus is a network solid; all phosphorus atoms are connected through covalent bonds. It is easier to disrupt the weaker London dispersion forces of P_4 than it is to break the covalent bonds of red phosphorus.

Diagnostic Quiz Questions

Goal #1: Describe intermolecular forces and their effects.

1. Which of the following ions will have the greatest enthalpy of hydration?
   a) Na^+
   b) Mg^{2+}
   c) K^+
   d) Ca^{2+}

   Answer: b

2. What intermolecular force or bond is responsible for the solubility of CO_2 in water?
   a) London dispersion force
   b) dipole-dipole force
   c) dipole/induced dipole force
   d) hydrogen bonding

   Answer: c

3. What intermolecular forces or bonds must be overcome in converting SO_2 from a liquid to a gas?
   1. London dispersion forces
   2. dipole-dipole forces
   3. hydrogen bonds

   a) 1 only
   b) 2 only
   c) 3 only
   d) 1 and 2

   Answer: d

4. Which of the following molecules will have the highest boiling point?
   a) H_2
   b) N_2
   c) O_2
   d) Cl_2

   Answer: d

5. What is the dominant intermolecular force or bond that must be overcome in converting CH_3OH from a liquid to a gas?
   a) London dispersion force
   b) hydrogen bonding
   c) dipole-dipole force
   d) covalent bonding

   Answer: b

6. Which of the following molecules is expected to form hydrogen bonds in the liquid state or solid state: H_2SO_4, HF, CH_3OH, and CH_2O (formaldehyde)?
   a) H_2SO_4
   b) CH_3OH and CH_2O
   c) H_2SO_4, HF, and CH_3OH
   d) H_2SO_4, HF, CH_3OH, and CH_2O

   Answer: c

Goal #2: Understand the importance of hydrogen bonding

7. Which of the following properties of water can be attributed to hydrogen bonding?
   1. high heat of vaporization
   2. high surface tension
   3. higher density of liquid versus solid

   a) 1 only
   b) 2 only
   c) 3 only
   d) 1, 2, and 3

   Answer: d

Goal #3: Understand the properties of liquids

8. The molar enthalpy of vaporization of NH_3 is 23.3 kJ/mol at its normal boiling point of -33.3 °C. How much energy is required to evaporate 1.00 kg of ammonia at -33.3 °C?
   a) 0.397 kJ
   b) 2.52 kJ
   c) 1.37×10^3 kJ
   d) 2.33×10^4 kJ

   Answer: c

9. A mass of 1.50 g H_2O is sealed in an evacuated 3.0-L flask and heated to 55 °C. What mass of liquid water remains in the flask when equilibrium is established? The equilibrium vapor pressure of water at 55 °C is 118 mm Hg.
   a) 0.00 g
   b) 0.31 g
   c) 1.19 g
   d) 1.48 g

   Answer: c

10. Calculate the enthalpy of vaporization of Br_2. This element has vapor pressures of 271 mm Hg and 565 mm Hg at 30.0 °C and 50.0 °C, respectively.
    a) 0.458 kJ/mol
    b) 29.9 kJ/mol
    c) 131 kJ/mol
    d) 1.95×10^4 kJ/mol

    Answer: b

11. Diethyl ether has an enthalpy of vaporization of 29.0 kJ/mol. The compound has a vapor pressure of 524 mm Hg at 25 °C. At what temperature is the vapor pressure equal to 115 mm Hg?
    a) -17 °C
    b) -9 °C
    c) 9 °C
    d) 25 °C

    Answer: b

12. Which of the following statements are correct?
    1. A liquid boils when its vapor pressure is equal to the pressure above its surface.
    2. Above the critical pressure and critical temperature, only the solid phase of a pure substance can exist.
    3. The gas, liquid, and solid phases can all coexist at the critical point.

    a) 1 only
    b) 2 only
    c) 3 only
    d) 1 and 3

    Answer: a

13. The toughness of the "skin" of a liquid is a measure of its ________.
    a) capillary action
    b) viscosity
    c) adhesive force
    d) surface tension

    Answer: d

Goal #4: Understand cubic unit cells

14. An atom located on a lattice point is shared equally by _____ unit cells.
    a) 2
    b) 4
    c) 8
    d) 12

    Answer: c

15. Copper crystallizes in a face-centered cubic lattice with a unit cell length of 361 pm. What is the radius (in pm) of a copper atom?
    a) 108 pm
    b) 128 pm
    c) 157 pm
    d) 181 pm

    Answer: b

16. An element crystallizes in a face-centered cubic lattice. If the length of an edge of the unit cell is 0.408 nm, and the density of the element is 19.3 g/cm^3, what is the identity of the element?
    a) Au
    b) Ag
    c) W
    d) Ir

    Answer: a

17. If an ionic compound with the formula MX forms a simple cubic lattice with the anions (X^{n-}) at the lattice points, the cations (M^{n+}) must
    a) occupy all the cubic holes in the lattice.
    b) occupy half of the cubic holes in the lattice.
    c) occupy all of the octahedral holes in the lattice.
    d) occupy half of the tetrahedral holes in the lattice.

    Answer: a

Goal #5: Describe the properties of solids

18. Which one of the following compounds forms a network solid composed of covalently bonded atoms?
    a) SiO_2
    b) CO_2
    c) NH_4Cl
    d) NH_3

    Answer: a

19. Which of the following compounds is likely to have the lowest melting point?
    a) SiO_2
    b) Si
    c) NaCl
    d) NH_3

    Answer: d

Goal #6: Understand the nature of phase diagrams

20. Which of the following statements concerning the phase diagram below are correct?

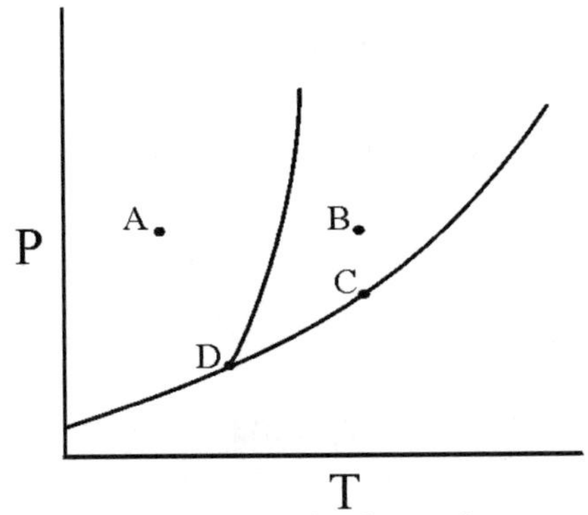

1. Moving from point A to B results in a phase transition from solid to liquid.
2. Point D lies at the critical point.
3. At point C, liquid and gas phases coexist at equilibrium.

a) 1 only
b) 2 only
c) 3 only
d) 1 and 3

Answer: d

21. A phase diagram of a pure compound has a triple point at 13 °C and 205 mm Hg, a normal melting point at 17 °C, and a normal boiling point at 87 °C. Which of the following statements regarding this compound are correct?
    1. The density of the solid is greater than that of the liquid.
    2. If the pressure is reduced from 535 mm Hg to 25 mm Hg at a constant temperature of 12 °C, sublimation occurs.
    3. At a constant pressure of 835 mm Hg, complete evaporation will occur if the temperature is raised from 13 °C to 81 °C.

a) 1 only
b) 2 only
c) 3 only
d) 1 and 2

Answer: d

CHAPTER 14: Solutions and Their Behavior

14.1 Units of Concentration

1. What is the definition of molality?
   a) moles of solute per liter of solution
   b) grams of solute per kg of solution
   c) milligrams of solute per liter of solution
   d) moles of solute per liter of solvent
   e) moles of solute per kg of solvent

   Answer: e

2. To prepare 1.000 kg of 5.5% by mass KCl(aq), one may dissolve
   a) 994.5 g KCl in 5.5 g of water.
   b) 5.5 g KCl in 994.5 g of water.
   c) 5.5 g KCl in 1000.0 g of water.
   d) 55 g KCl in 945 g of water.
   e) 55.0 g KCl in 1000.0 g of water.

   Answer: d

3. What is the molality of 13.82% by mass sulfuric acid? The molar mass of H_2SO_4 is 98.08 g/mol.
   a) 0.1409 *m* b) 1.409 *m* c) 1.635 *m* d) 13.55 *m* e) 15.73 *m*

   Answer: c

4. What is the molality of Cl^- ions in 5.11% by mass $CaCl_2$(aq)? The molar mass of calcium chloride is 110.98 g/mol.
   a) 0.0921 *m* b) 0.485 *m* c) 0.921 *m* d) 0.970 *m* e) 1.46 *m*

   Answer: d

5. What is the mole fraction of 1.98 *m* $Fe(NO_3)_3$(aq)? The molar mass of $Fe(NO_3)_3$ is 241.9 g/mol and the molar mass of water is 18.02 g/mol.
   a) 0.0345 b) 0.0641 c) 0.324 d) 0.479 e) 0.863

   Answer: a

6. What is the mole fraction of $MgCl_2$ in a solution that is 11.04% by mass $MgCl_2$(aq)? The molar mass of $MgCl_2$ is 95.21 g/mol.
   a) 0.02295 b) 0.02349 c) 0.1104 d) 0.1160 e) 1.303

   Answer: a

7. What is the weight percent of ammonia in 3.33 *m* $NH_3(aq)$?
a) 0.332% b) 5.37% c) 5.67% d) 6.01% e) 56.7%

Answer: b

8. The weight percent of concentrated H_2SO_4 is 96.0% and its density is 1.84 g/mL. What is the molarity of concentrated H_2SO_4?
a) 9.79 M b) 12.0 M c) 18.0 M d) 53.2 M e) 245 M

Answer: c

9. Concentrated hydrofluoric acid is 28.9 M and has a density of 1.18 g/mL. What is the weight percent of concentrated HF?
a) 24.5% b) 49.0% c) 51.0% d) 68.2% e) 75.5%

Answer: b

10. Concentrated sodium hydroxide is 19.4 M and 50.5% NaOH by mass. What is the density of concentrated NaOH?
a) 0.392 g/mL b) 1.26 g/mL c) 1.54 g/mL d) 2.55 g/mL e) 9.80 g/mL

Answer: c

11. A solution is prepared by mixing 66.7 g H_2O with 33.3 g $HOCH_2CH_2OH$ (ethylene glycol). What is the concentration of ethylene glycol in units of molality? The molar masses of water and ethylene glycol are 18.02 g/mol and 62.07 g/mol, respectively.
a) 0.127 *m* b) 0.333 *m* c) 4.99 *m* d) 5.36 *m* e) 8.04 *m*

Answer: e

12. If the concentration of potassium chloride in water is 113 ppm, what is the molarity of potassium chloride in water? The molar mass of KCl is 74.55 g/mol. Assume the density of the solution is 1.00 g/mL.
a) 1.52×10^{-9} M b) 1.19×10^{-4} M c) 1.52×10^{-3} M d) 8.42×10^{-3} M e) 0.660 M

Answer: c

13. A cylindrical reservoir of water has a radius of 75 m and a depth of 16 m. What mass of fluoride ion is present in the reservoir if the concentration of fluoride ion is 2.3 ppm? (Volume = $\pi r^2 h$; assume the density of the solution is 1.00 g/mL)
a) 8.1 g b) 3.4×10^4 g c) 1.2×10^5 g d) 6.5×10^5 g e) 6.5×10^8 g

Answer: d

14. What concentration of chloride ion (in ppb) is present in 3.1×10^{-8} M $CaCl_2$(aq)? For very dilute aqueous solutions, you can assume the solution's density is 1.0 g/mL. The molar mass of $CaCl_2$ is 110.98 g/mol.
a) 0.28 ppb b) 2.2 ppb c) 3.4 ppb d) 6.8 ppb e) 62 ppb

Answer: b

15. What mass of $Cu(NO_3)_2$ must be added to 1.00 kg H_2O to prepare 97 ppm Cu^{2+}(aq)?
a) 3.3×10^{-4} g b) 5.2×10^{-4} g c) 9.7×10^{-3} g d) 0.033 g e) 0.29 g

Answer: e

16. Which of the following concentration conversions requires a knowledge of the density of the solution?
a) ppm to weight percent
b) molarity to mole fraction
c) molality to ppm
d) weight percent to mole fraction
e) mole fraction to ppb

Answer: b

14.2 The Solution Process

17. Which of the following statements are CORRECT?
   1. Solubility is the concentration of solute in equilibrium with undissolved solute in a saturated solution.
   2. If two liquids mix to an appreciable extent to form a solution, they are immiscible.
   3. If two liquids mix completely in any proportion to form a solution, the resulting solution is supersaturated.

a) 1 only b) 2 only c) 3 only d) 1 and 2 e) 2 and 3

Answer: a

18. Which of the following liquids are likely to be miscible with water: 1-propanol ($CH_3CH_2CH_2OH$), carbon tetrachloride (CCl_4), hexane (C_6H_{14}), and acetic acid (CH_3CO_2H)?
a) 1 propanol and carbon tetrachloride
b) 1-propanol and hexane
c) carbon tetrachloride and hexane
d) carbon tetrachloride and acetic acid
e) 1-propanol and acetic acid

Answer: e

19. Two nonpolar solvents, such as hexane and carbon tetrachloride, may be miscible even though the enthalpy of mixing of these liquids might be small. A reason that mixing occurs is that mixtures have greater disorder than pure solvents. The tendency toward disorder is a thermodynamic function called ________.
a) entropy b) enthalpy c) saturation d) adhesion e) cohesion

Answer: a

20. In which solvent should Na^+ have the most negative enthalpy of solvation?
a) CCl_4 b) H_2O c) C_6H_6 d) CS_2 e) CH_3OCH_3

Answer: b

21. Determine the heat of solution of NaOH given the thermodynamic equations below.

$$Na(s) + 1/2\ O_2(g) + 1/2\ H_2(g) \rightarrow NaOH(s) \qquad \Delta H^\circ_f = -425.9 \text{ kJ/mol}$$

$$Na(s) + 1/2\ O_2(g) + 1/2\ H_2(g) \rightarrow NaOH(aq, 1\ m) \qquad \Delta H^\circ_f = -469.2 \text{ kJ/mol}$$

a) -895.1 kJ/mol b) -43.3 kJ/mol c) +1.102 kJ/mol d) +43.3 kJ/mol e) +895.1 kJ/mol

Answer: b

22. The heat of formation of KCl(s) is -436.7 kJ/mol and the heat of formation of KCl(aq, 1 *m*) is -419.5 kJ/mol. Determine the heat of solution of KCl.
a) -856.2 kJ/mol b) -17.2 kJ/mol c) +1.041 kJ/mol d) +17.2 kJ/mol e) +856.2 kJ/mol

Answer: d

14.3 Factors Affecting Solubility: Pressure and Temperature

23. Which substance will most likely NOT quantitatively follow Henry's law when dissolved in water?
a) He b) O_2 c) CH_4 d) N_2 e) NH_3

Answer: e

24. The Henry's law constant for N_2 in water at 37 °C is 8.2×10^{-7} M/mm Hg. What is the equilibrium concentration of N_2 in water when the partial pressure of N_2 is 634 mm Hg?
a) 1.3×10^{-9} M b) 5.2×10^{-4} M c) 1.9×10^{-2} M d) 1.9×10^{3} M e) 7.7×10^{8} M

Answer: b

25. Which action(s) will decrease the equilibrium concentration of a non-reacting gas in water?
    1. decreasing the temperature of the water
    2. decreasing the volume of water
    3. decreasing the pressure of the gas above the liquid

    a) 1 only b) 2 only c) 3 only d) 1 and 3 3) 1, 2, and 3

    Answer: c

26. The Henry's law constant for O_2 in water at 25 °C is 1.66×10^{-6} M/mm Hg. What partial pressure of O_2 is necessary to achieve an equilibrium concentration of 0.150 M O_2?
    a) 2.49×10^{-7} mm Hg
    b) 6.23×10^{-6} mm Hg
    c) 1.11×10^{-5} mm Hg
    d) 9.04×10^{4} mm Hg
    e) 4.02×10^{6} mm Hg

    Answer: d

14.4 Colligative Properties

27. Ideally, colligative properties depend only on the
    a) concentration of solute particles in a solution.
    b) identity of the solute particles in a solution.
    c) color of a solution.
    d) hydrated radii of the molecules or ions dissolved in a solution.
    e) partial pressure of the gases above the surface of a solution.

    Answer: a

28. What is the equilibrium partial pressure of water vapor above a mixture of 24.0 g H_2O and 48.0 g CH_3CH_2OH at 25 °C. The partial pressure of pure water at 25.0 °C is 23.8 mm Hg. Assume ideal behavior for the solution.
    a) 0.561 mm Hg b) 7.93 mm Hg c) 10.4 mm Hg d) 11.9 mm Hg e) 13.4 mm Hg

    Answer: e

29. The vapor pressure of pure water at 45 °C is 71.9 mm Hg. What is the vapor pressure of a mixture of 21.0 g sucrose ($C_{12}H_{22}O_{11}$, molar mass 342.3 g/mol) and 79.0 g water?
    a) 0.992 mm Hg b) 15.1 mm Hg c) 19.1 mm Hg d) 55.8 mm Hg e) 70.9 mm Hg

    Answer: e

30. What mass of ethylene glycol, when mixed with 90.5 g H_2O, will reduce the equilibrium vapor pressure of H_2O from 1.00 atm to 0.500 atm at 100 °C. The molar masses of water and ethylene glycol are 18.02 g/mol and 62.07 g/mol, respectively. Assume ideal behavior for the solution.
a) 90.5 g b) 156 g c) 181 g d) 263 g e) 312 g

Answer: e

31. Which of the following aqueous solutions should have the lowest freezing point?
a) pure H_2O b) 1 *m* NaBr c) 1 *m* $Ca(NO_3)_2$ d) 1 *m* NH_3 e) 1 *m* $C_6H_{12}O_6$

Answer: c

32. Which of the following aqueous solutions should have the lowest boiling point?
a) 0.50 *m* KCl b) 0.50 *m* Na_2SO_4 c) 0.50 *m* $CaBr_2$ d) 1.5 *m* KI e) 2.0 *m* $C_6H_{12}O_6$

Answer: a

33. The freezing point depression constant for water is –1.86 °C/*m*. At what temperature will a solution containing 7.99 g $BaCl_2$ and 52.0 g H_2O begin to freeze? Assume that no ion-pairing occurs between Ba^{2+} and Cl^-.
a) -4.12 °C b) -2.75 °C c) -1.37 °C d) -0.857 °C e) -0.248 °C

Answer: a

34. What is the freezing point of a solution containing 4.78 grams naphthalene (molar mass = 128.2 g/mol) dissolved in 32.0 grams paradichlorobenzene? The freezing point of pure paradichlorobenzene is 53.0 °C and the freezing point depression constant, K_{fp}, is -7.10 °C/*m*.
a) -1.63 °C b) 44.7 °C c) 46.9 °C d) 51.9 °C e) 52.1 °C

Answer: b

35. What is the molar mass of a nonpolar molecular compound if 5.52 grams dissolved in 36.0 grams of benzene begins to freeze at –1.87 °C? The freezing point of pure benzene is 5.50 °C and the freezing point depression constant, K_{fp}, is –5.12 °C/*m*.
a) 46.1 g/mol b) 107 g/mol c) 216 g/mol d) 221 g/mol e) 419 g/mol

Answer: b

36. What is the boiling point of a solution containing 2.33 g of caffeine, $C_8H_{10}N_4O_2$, dissolved in 15.0 g of benzene? The boiling point of pure benzene is 80.1 °C and the boiling point elevation constant, K_{bp}, is 2.53 °C/*m*.
a) 2.02 °C b) 78.1 °C c) 80.2 °C d) 82.1 °C e) 83.3 °C

Answer: d

37. What is the molar mass of a compound if 6.21 grams is dissolved in 24.0 grams of chloroform to form a solution that has a boiling point of 68.04 °C? The boiling point of pure chloroform is 61.70 °C and the boiling point elevation constant, K_{bp}, is 3.63 °C/*m*.
a) 1.75 g/mol b) 13.8 g/mol c) 148 g/mol d) 452 g/mol e) 573 g/mol

Answer: c

38. What mass of NaCl must be dissolved in 65.0 grams of water to lower the freezing point by 7.50 °C? The freezing point depression constant, K_{fp}, for water is –1.86 °C/*m*. Assume the van't Hoff factor for NaCl is 1.87.
a) 0.140 g b) 1.76 g c) 4.55 g d) 8.19 g e) 15.3 g

Answer: d

39. A 0.561 *m* solution of an unknown electrolyte depresses the freezing point of water by 2.93 °C. What is the van't Hoff factor for this electrolyte? The freezing point depression constant, K_{fp}, for water is –1.86 °C/*m*.
a) 0.103 b) 1.92 c) 2.81 d) 3.05 e) 3.99

Answer: c

40. The freezing point depression constant for water is –1.86 °C/*m*. If 5.00 g Na_2SO_4 is dissolved in 45.0 g H_2O, the freezing point is changed by –3.82 °C. Calculate the van't Hoff factor for Na_2SO_4.
a) 0.381 b) 2.05 c) 2.41 d) 2.63 e) 2..97

Answer: d

41. What concentration unit is used in the calculation of osmotic pressure for a dilute solution?
a) molality
b) weight percent
c) mass fraction
d) mole fraction
e) molarity

Answer: e

42. Which of the following is an application of osmotic pressure?
a) salting meats for preservation
b) mixing water and ethylene glycol in antifreeze
c) pressurizing beverages with carbon dioxide
d) salting icy roads
e) using a pressure cooker to increase the boiling point of water

Answer: a

43. At 25 °C, what is the osmotic pressure of 8.65 g urea (CON_2H_4) diluted with water to 1.50 L? (R = 0.08206 L·atm/mol·K)
a) 0.197 atm b) 2.35 atm c) 3.52 atm d) 141 atm e) 212 atm

Answer: b

44. The osmotic pressure of blood is 7.65 atm at 37 °C. What mass of glucose ($C_6H_{12}O_6$, molar mass = 180.2 g/mol) is needed to prepare 5.00 L of solution for intravenous injection? The osmotic pressure of the glucose solution must equal the osmotic pressure of blood. (R = 0.08206 L·atm/mol·K)
a) 1.50 g b) 54.2 g c) 126 g d) 271 g e) 2270 g

Answer: d

45. A solution is prepared by dissolving 4.78 g of an unknown nonelectrolyte in enough water to make 0.500 L of solution. The osmotic pressure of the solution is 1.98 atm at 27 °C. What is the molar mass of the solute? (R = 0.08206 L·atm/mol·K)
a) 24.9 g/mol b) 59.4 g/mol c) 119 g/mol d) 233 g/mol e) 466 g/mol

Answer: c

14.5 Colloids

46. All of the following are colloidal dispersions EXCEPT
a) ice cubes. b) cheese. c) butter. d) clouds. e) shaving cream.

Answer: a

47. Cheese and butter are colloidal dispersions that contain a liquid dispersed in a solid medium. This type of colloid is called a(n) ________.
a) emulsion b) sol c) foam d) aerosol e) gel

Answer: e

48. A(n) ________ is a colloidal dispersion of a liquid in another liquid.
a) sol b) emulsion c) aerosol d) foam e) gel

Answer: b

Short Answer Questions

49. Intravenous fluids must have the same overall solute partial concentration as blood. If two fluids have the same solute particle concentration, they are said to be ________.

Answer: isotonic

50. A surfactant used for cleaning is called a(n) ________.

Answer: detergent

51. For an aqueous ionic solution, the ratio of the experimentally determined change in freezing point divided by the calculated change in freezing point (assuming no ionization) is called the ________ factor.

Answer: van't Hoff

52. If one of the factors determining the equilibrium of a system is changed, the system adjusts to counteract that change. This is known as ________ principle.

Answer: Le Chatelier's

53. A solution in which there is more dissolved solute than in a saturated solution is known as a(n) ________ solution.

Answer: supersaturated

54. Pickles are made by soaking cucumbers in a concentrated brine solution. Why does the cucumber shrivel up when it is soaked in brine?

Answer: The cucumber's skin acts as a semi-permeable membrane. Since the solute concentration is greater outside the cucumber, water flows out of the cucumber. This is due to the higher osmotic pressure of the brine solution.

55. Henry's law states that the solubility of a gas in a liquid is directly proportional to its pressure. This law holds true for gases such as nitrogen and oxygen. However, Henry's law does not hold true for hydrogen chloride gas. Why?

Answer: Henry's law is only true for gases that do not interact with the solvent. Hydrogen chloride dissociates in water to form H^+ and Cl^- ions.

Diagnostic Quiz Questions

Goal #1: Learn additional methods of expressing solution concentration

1. What is the molality of a solution prepared by dissolving 21.7 g KCl in 225 g H_2O?
   a) 0.0880 *m*
   b) 0.0964 *m*
   c) 0.773 *m*
   d) 1.29 *m*

   Answer: d

2. What is the molality of a 13.82% by mass glucose solution? The molar mass of glucose, $C_6H_{12}O_6$, is 180.16 g/mol.
   a) 0.7671 *m*
   b) 0.8901 *m*
   c) 1.549 *m*
   d) 3.461 *m*

   Answer: b

3. The molality of an aqueous sodium nitrate solution is 1.44 *m*. What is the mole fraction of $NaNO_3$? The molar mass of $NaNO_3$ is 85.00 g/mol; the molar mass of water is 18.02 g/mol.
   a) 0.0253
   b) 0.0287
   c) 0.0296
   d) 0.284

   Answer: a

4. Concentrated NaOH is 19.3 M and has a density of 1.53 g/mL. What is the percent NaOH by mass of concentrated NaOH?
   a) 31.5%
   b) 49.5%
   c) 50.5%
   d) 73.8%

   Answer: c

5. A mass of 0.250 g calcium chloride is diluted with 975 g H_2O. What is the chloride ion concentration (in ppm)?
   a) 81.8 ppm
   b) 164 ppm
   c) 251 ppm
   d) 502 ppm

   Answer: b

6. Which of the following pairs of liquids will be miscible?
   a) CCl_4 and CH_3OH
   b) H_2O and C_6H_6
   c) C_6H_{14} and C_8H_{18}
   d) C_8H_{18} and H_2O

   Answer: c

Goal #2: Understand the solution process

7. The standard heat of formation of NaF(s) is -573.6 kJ/mol and the standard heat of formation of NaF(aq, 1 *m*) is -572.8 kJ/mol. Determine the standard heat of solution of NaF.
   a) +1146.6 kJ/mol
   b) +0.8 kJ/mol
   c) -0.8 kJ/mol
   d) -1146.6 kJ/mol

   Answer: b

8. What is the equilibrium concentration of carbon dioxide in water that is in contact with air at 25 °C and 0.982 atm. The mole fraction of CO_2 in air is 3.01×10^{-4}. The Henry's law constant for carbon dioxide is 4.48×10^{-5} M/mm Hg.
   a) 1.32×10^{-8} M
   b) 1.01×10^{-5} M
   c) 8.68×10^{-3} M
   d) 3.34×10^{-2} M

   Answer: b

9. Which of the following statements concerning the solubility of a solute are correct?
   1. The solubility of a gas in a liquid increases with increasing temperature.
   2.) The solubility of a gas in a liquid is directly proportional to the partial pressure of the gas above the liquid.
   3. The dissolution of a gas in a liquid is an endothermic process.

   a) 1 only
   b) 2 only
   c) 3 only
   d) 2 and 3

   Answer: b

Goal #3: Understand and use the colligative properties of solution

10. What concentration unit is necessary for the calculation of boiling point elevation?
   a) mole fraction of solute
   b) mass fraction of solvent
   c) molarity of solute
   d) molality solute

   Answer: d

11. The vapor pressure of pure water at 50.0 °C is 92.5 mm Hg. Calculate the ideal partial pressure of water at this temperature for a mixture of 45.0 g H_2O and 55.0 g CH_3OH?
   a) 41.6 mm Hg
   b) 45.2 mm Hg
   c) 50.9 mm Hg
   d) 54.8 mm Hg

   Answer: d

12. Predict which of the following compounds will have the lowest freezing point.
   a) pure H_2O
   b) 1 *m* NH_3
   c) 1 *m* NaCl
   d) 1 *m* $Ca(NO_3)_2$

   Answer: d

13. What is the boiling point of a mixture composed of 85 g $HOCH_2CH_2OH$ (ethylene glycol) and 125 g H_2O? The boiling point elevation constant for H_2O is 0.512 °C/*m*.
   a) 100.2 °C
   b) 105.6 °C
   c) 107.1 °C
   d) 135.2 °C

   Answer: b

14. An aqueous potassium nitrate solution has a freezing point of -4.5 °C. What is the molality of the solution? The freezing point depression constant for water is -1.86 °C/*m*.
   a) 0.21 *m*
   b) 0.41 *m*
   c) 1.2 *m*
   d) 2.4 *m*

   Answer: c

15. What is the molar mass of a nonelectrolyte if 6.02 grams dissolved in 30.0 grams of benzene freezes at –1.55 °C? The freezing point of pure benzene is 5.50 °C and the freezing point depression constant, K_{fp}, is –5.12 °C/*m*.
   a) 146 g/mol
   b) 187 g/mol
   c) 276 g/mol
   d) 663 g/mol

   Answer: a

16. What is the osmotic pressure of 0.250 M aqueous urea (CON_2H_4) at 25 °C? (R = 0.08206 L·atm/mol·K)
   a) 0.513 atm
   b) 4.99 atm
   c) 6.11 atm
   d) 8.21 atm

   Answer: c

Goal #4: Describe colloids and their applications.

17. An example of a colloidal solution where the dispersing medium is a gas and the dispersed phase is a liquid is ________.
   a) clouds
   b) milk
   c) smoke
   d) butter

   Answer: a

18. A(n) ________ is a colloidal dispersion of a liquid in another liquid.
   a) sol
   b) gel
   c) emulsion
   d) aerosol

   Answer: c

CHAPTER 15: Principles of Reactivity: Chemical Kinetics

15.1 Rates of Chemical Reactions

1. Which statement concerning relative rates of reaction is correct for the decomposition of phosphine?

$$4\ PH_3(g) \rightarrow P_4(g) + 6\ H_2(g)$$

a) The rate of disappearance of PH_3 is 6/4 the rate of appearance of H_2.
b) The rate of appearance of P_4 is equal to the rate of appearance of H_2.
c) The rate of disappearance of PH_3 is 1/4 the rate of appearance of P_4.
d) The rate of disappearance of PH_3 is 4 times the rate of appearance of H_2.
e) The rate of appearance of H_2 is 6 times the rate of appearance of P_4.

Answer: e

2. Which relationship is correct for the rate of decomposition of dinitrogen pentaoxide?

$$2\ N_2O_5(g) \rightarrow 4\ NO_2(g) + O_2(g)$$

a) $\frac{\Delta[N_2O_5]}{\Delta t} = \frac{\Delta[NO_2]}{\Delta t} = \frac{\Delta[O_2]}{\Delta t}$

b) $-\frac{\Delta[N_2O_5]}{\Delta t} = \frac{\Delta[NO_2]}{\Delta t} = \frac{\Delta[O_2]}{\Delta t}$

c) $-\frac{1}{2}\left(\frac{\Delta[N_2O_5]}{\Delta t}\right) = \frac{1}{4}\left(\frac{\Delta[NO_2]}{\Delta t}\right)$

d) $2\left(\frac{\Delta[N_2O_5]}{\Delta t}\right) = \frac{\Delta[O_2]}{\Delta t}$

e) $-2\left(\frac{\Delta[N_2O_5]}{\Delta t}\right) = \frac{\Delta[O_2]}{\Delta t}$

Answer: c

3. Ozone decomposes to oxygen according to the balanced chemical equation below.

$$2\ O_3(g) \rightarrow 3\ O_2(g)$$

If the rate of disappearance of ozone is -7.2×10^{-4} M/s, what is the rate of formation of oxygen?
a) 4.8×10^{-4} M/s b) 7.2×10^{-4} M/s c) 1.1×10^{-3} M/s d) 1.4×10^{-3} M/s e) 2.2×10^{-3} M/s

Answer: c

4. Which of the following expressions relates the rates of reaction of NOCl(g) and Cl_2(g)?

$$2\,NOCl(g) \rightarrow 2\,NO(g) + Cl_2(g)$$

a) $-\frac{1}{2}\left(\frac{\Delta[NOCl]}{\Delta t}\right) = \frac{\Delta[Cl_2]}{\Delta t}$

b) $-2\left(\frac{\Delta[NOCl]}{\Delta t}\right) = \frac{\Delta[Cl_2]}{\Delta t}$

c) $-\frac{\Delta[NOCl]}{\Delta t} = \frac{\Delta[Cl_2]}{\Delta t}$

d) $-\frac{\Delta t}{\Delta[NOCl]} = \frac{\Delta t}{\Delta[Cl_2]}$

e) $\frac{\Delta[NOCl]}{\Delta t} = \frac{\Delta[Cl_2]}{\Delta t}$

Answer: a

15.2 Reaction Conditions and Rate

5. Which of the following changes are likely to increase the rate of a chemical reaction?
   1. Adding a catalyst
   2. Increasing the temperature of the reactants
   3. Increasing the concentrations of the reactants

   a) 1 only b) 2 only c) 3 only d) 1 and 3 e) 1, 2, and 3

Answer: e

6. What is the name given to a substance that increases the rate of a chemical reaction but is not itself consumed?

   a) reactant b) product c) intermediate d) catalyst e) element

Answer: d

15.3 Effect of Concentration on Reaction Rate

7. What is the overall order of the reaction below

$$2\,NO(g) + 2\,H_2(g) \rightarrow N_2(g) + 2\,H_2O(g)$$

if it proceeds via the following rate expression?

$$-\frac{\Delta[NO]}{\Delta t} = k[NO]^2[H_2]$$

   a) zero-order b) first-order c) second-order d) third-order e) fourth-order

Answer: d

8. Carbon monoxide reacts with oxygen according to the equation below.

$$2\ CO(g) + O_2(g) \rightarrow 2\ CO_2(g)$$

What is the overall order of the reaction?

a) first-order
b) second-order
c) third-order
d) fourth-order
e) not enough information is given to determine the reaction order

Answer: e

9. Given the initial rate data for the decomposition reaction,

$$A \rightarrow B + C$$

determine the rate expression for the reaction.

[A], M	$-\Delta[A]/\Delta t$ M/s
0.110	7.81×10^{-3}
0.220	3.12×10^{-2}
0.330	7.03×10^{-2}

a) $\dfrac{-\Delta[A]}{\Delta t} = 7.10 \times 10^{-2}\ s^{-1}[A]$

b) $\dfrac{-\Delta[A]}{\Delta t} = 0.645\ M^{-1}s^{-1}[A]^2$

c) $\dfrac{-\Delta[A]}{\Delta t} = 3.63 \times 10^{-2}\ M^{-1}s^{-1}[A]^2$

d) $\dfrac{-\Delta[A]}{\Delta t} = 3.63 \times 10^{-2}\ M\ s^{-1}$

e) $\dfrac{-\Delta[A]}{\Delta t} = 5.87\ M^{-2}s^{-1}[A]^3$

Answer: b

10. Given the initial rate data for the reaction A + B → C, determine the rate expression for the reaction.

[A], M	[B], M	Δ[C]/Δ*t* (initial) M/s
0.250	0.150	8.90×10^{-6}
0.250	0.300	1.78×10^{-5}
0.500	0.300	7.12×10^{-5}

a) $\frac{\Delta[C]}{\Delta t} = 2.37 \times 10^{-4}\ M^{-1}s^{-1}[A][B]$

b) $\frac{\Delta[C]}{\Delta t} = 9.49 \times 10^{-4}\ M^{-2}s^{-1}[A]^2[B]$

c) $\frac{\Delta[C]}{\Delta t} = 1.58 \times 10^{-3}\ M^{-2}s^{-1}[A][B]^2$

d) $\frac{\Delta[C]}{\Delta t} = 6.33 \times 10^{-3}\ M^{-3}s^{-1}[A]^2[B]^2$

e) $\frac{\Delta[C]}{\Delta t} = 3.80 \times 10^{-3}\ M^{-3}s^{-1}[A]^3[B]$

Answer: b

11. Given the initial rate data for the reaction A + B → C, determine the rate expression for the reaction.

[A], M	[B], M	Δ[C]/Δ*t* (initial) M/s
0.100	0.100	1.26×10^{-3}
0.100	0.150	1.89×10^{-3}
0.170	0.150	5.46×10^{-3}

a) $\frac{\Delta[C]}{\Delta t} = 1.26\ M^{-2}s^{-1}[A]^2[B]$

b) $\frac{\Delta[C]}{\Delta t} = 0.126\ M^{-1}s^{-1}[A][B]$

c) $\frac{\Delta[C]}{\Delta t} = 1.26\ M^{-2}s^{-1}[A][B]^2$

d) $\frac{\Delta[C]}{\Delta t} = 12.6\ M^{-3}s^{-1}[A]^2[B]^2$

e) $\frac{\Delta[C]}{\Delta t} = 0.0126\ s^{-1}[B]$

Answer: a

12. Given the initial rate data for the reaction A + B → C, determine the rate expression for the reaction.

[A], M	[B], M	$\Delta[C]/\Delta t$ (initial) M/s
0.334	0.134	4.11×10^{-9}
0.334	0.187	8.00×10^{-9}
0.668	0.134	4.11×10^{-9}

a) $\frac{\Delta[C]}{\Delta t} = 2.75 \times 10^{-7}\ M^{-2}s^{-1}[A]^2[B]$

b) $\frac{\Delta[C]}{\Delta t} = 3.07 \times 10^{-8}\ s^{-1}[B]$

c) $\frac{\Delta[C]}{\Delta t} = 2.29 \times 10^{-7}\ M^{-1}s^{-1}[B]^2$

d) $\frac{\Delta[C]}{\Delta t} = 6.85 \times 10^{-7}\ M^{-2}s^{-1}[A][B]^2$

e) $\frac{\Delta[C]}{\Delta t} = 1.23 \times 10^{-8}\ s^{-1}[A]$

Answer: c

13. For the reaction A + 2B → C, the rate law is

$$\frac{\Delta[C]}{\Delta t} = k[A][B]^2 .$$

What are the units of the rate constant where time is measured in seconds?

a) $\frac{1}{M \cdot s}$ b) $\frac{1}{M^2 \cdot s}$ c) $M \cdot s$ d) $\frac{M^2}{s}$ e) $\frac{M}{s}$

Answer: b

14. For the reaction A → B, the rate law is

$$\frac{\Delta[B]}{\Delta t} = k[A] .$$

What are the units of the rate constant where time is measured in seconds?

a) $\frac{1}{s}$ b) $M^2 \cdot s$ c) $\frac{s}{M}$ d) $\frac{M}{s}$ e) $\frac{1}{M \cdot s}$

Answer: a

15. For the reaction 2A + B → C, the rate law is

$$\frac{\Delta[C]}{\Delta t} = k[A]^2[B] .$$

Which of the factor(s) will affect the value of the rate constant for this reaction?

1. Increasing the concentration of reactant A
2. Adding a catalyst
3. Increasing the temperature

a) 1 only b) 2 only c) 3 only d) 2 and 3 e) 1, 2, and 3

Answer: d

16. The reaction of NO and O_2 produces NO_2.

$$2\ NO(g) + O_2(g) \rightarrow 2\ NO_2(g)$$

The reaction is second-order with respect to NO(g) and first-order with respect to O_2(g). At a given temperature, the rate constant, k, equals $6.1 \times 10^3\ M^{-2}s^{-1}$. What is the rate of reaction when the initial concentrations of NO and O_2 are 0.020 M and 0.015 M, respectively?

a) 9.8×10^{-10} M/s b) 0.027 M/s c) 0.037 M/s d) 1.0 M/s e) 1.8 M/s

Answer: c

17. How are the exponents in a rate law determined?

1. They are equal to the concentrations of the reactants.
2. They are equal to the coefficients in the balanced chemical equation.
3. They are determined by experimentation.

a) 1 only b) 2 only c) 3 only d) 1 and 2 e) 2 and 3

Answer: c

15.4 Concentration-Time Relationships: Integrated Rate Laws

18. For a second-order decomposition reaction,

$$2A \rightarrow B \quad \text{rate} = k[A]^2$$

which of the following functions can be plotted versus time to give a straight line?

a) [A] b) $[A]^2$ c) $\ln\frac{1}{[A]}$ d) ln[A] e) $\frac{1}{[A]}$

Answer: e

19. A student analyzed a first-order reaction and obtained the graph below. Unfortunately, the student forgot to label the axes. What are the correct labels for the x and y axes?
a) x axis = time, y axis = ln[A]
b) x axis = time, y axis = [A]
c) x axis = time, y axis = 1/[A]
d) x axis = 1/time, y axis = [A]
e) x axis = 1/time, y axis = 1/[A]

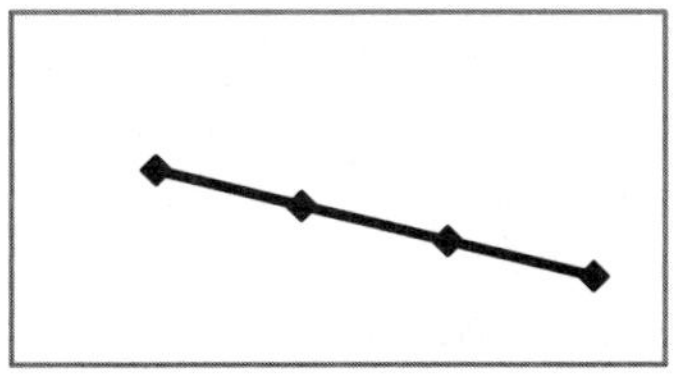

Answer: a

20. Which of the following equations corresponds to the integrated expression for a second-order decomposition reaction?
a) $[A] = -kt + [A]_0$
b) $\frac{1}{[A]} = kt + \frac{1}{[A]_0}$
c) $\ln[A] = -kt + \ln[A]_0$
d) $\frac{[A]}{[A]_0} = -kt$
e) $\frac{[A]_0}{[A]} = -kt$

Answer: b

21. The half-life of a first-order decomposition reaction is 2.36 hours. If the initial concentration of reactant is 0.52 M, what is the concentration of reactant after 752 seconds?
a) 0.046 M b) 0.16 M c) 0.26 M d) 0.47 M e) 0.49 M

Answer: e

22. What is the rate constant for a first-order reaction if the half-life of the reaction is 6.43×10^{-5} seconds?
a) 9.28×10^{-5} s^{-1} b) 0.0967 s^{-1} c) 10.3 s^{-1} d) 1.08×10^{4} s^{-1} e) 1.56×10^{4} s^{-1}

Answer: d

23. For the first-order decomposition of N_2O_5 at 340 K, where $k = 5.8 \times 10^{-3}$ s^{-1}, calculate the original concentration if the concentration of N_2O_5 is 0.41 M after 209 seconds.
a) 0.12 M b) 0.23 M c) 0.72 M d) 1.4 M e) 1.7 M

Answer: d

24. Which of the following equations corresponds to the integrated expression for a zero-order decomposition reaction?

a) $[A] = -kt + [A]_0$

b) $\ln[A] = -kt + \ln[A]_0$

c) $\dfrac{1}{[A]} = kt + \dfrac{1}{[A]_0}$

d) $\dfrac{[A]}{[A]_0} = -kt$

e) $\dfrac{[A]_0}{[A]} = -kt$

Answer: a

25. The reaction A → B follows first-order kinetics with a half-life of 21.7 seconds. If the concentration of A is 0.430 M after 14.5 seconds, what is the initial concentration of A?

a) 0271 M b) 0.287 M c) 0.644 M d) 0.683 M e) 1.59 M

Answer: d

26. Hydrogen peroxide decomposes into water and oxygen in a first-order process.

$$H_2O_2(aq) \rightarrow H_2O(\ell) + 1/2\ O_2(g)$$

At 20.0 °C, the half-life for the reaction is 3.92×10^4 seconds. If the initial concentration of hydrogen peroxide is 0.88 M, what is the concentration after 2.20 days?

a) 0.00 M b) 0.031 M c) 018 M d) 0.44 M e) 0.88 M

Answer: b

27. The decomposition of formic acid follows first-order kinetics.

$$HCO_2H(g) \rightarrow CO_2(g) + H_2(g)$$

The half-life for the reaction at 550 °C is 24 seconds. How many seconds does it take for the formic acid concentration to decrease by 75%?

a) 6.0 s b) 18 s c) 24 s d) 36 s e) 48 s

Answer: e

28. The decomposition of phosphine, PH_3, follows first-order kinetics.

$$4\ PH_3(g) \rightarrow P_4(g) + 6\ H_2(g)$$

The half-life for the reaction at 550 °C is 81.3 seconds. How long does it take for 89.0% of a phosphine sample to decompose?

a) 0.0296 s b) 123 s c) 259 s d) 903 s e) 732 s

Answer: c

29. What is the half-life of a first-order reaction if it takes 4.8×10^{-3} seconds for the concentration to decrease from 1.33 M to 0.51 M?
a) 1.8×10^{-3} s b) 3.5×10^{-3} s c) 1.3×10^{-2} s d) 2.0×10^{2} s e) 3.8×10^{2} s

Answer: b

30. What is the half-life of a first-order reaction with a rate constant of 0.457 s^{-1}?
a) 0.659 s b) 0.783 s c) 1.48 s d) 1.52 s e) 2.19 s

Answer: d

31. For the second-order reaction below, the concentration of the product, B, after 444 seconds is 0.0730 M. If no B is initially present, and the initial concentration of A is 0.744 M, what is the rate constant?

$$2A \rightarrow B \quad \text{rate} = k[A]^2$$

a) 3.3×10^{-4} $M^{-1}s^{-1}$
b) 7.4×10^{-4} $M^{-1}s^{-1}$
c) 1.2×10^{-2} $M^{-1}s^{-1}$
d) 1.3×10^{-3} $M^{-1}s^{-1}$
e) 2.8×10^{-2} $M^{-1}s^{-1}$

Answer: b

32. For the second-order reaction below, the initial concentration of reactant A is 0.24 M. If the rate constant for the reaction is 5.5×10^{-3} $M^{-1}s^{-1}$, what is the concentration of A after 265 seconds?

$$2A \rightarrow B + C \quad \text{rate} = k[A]^2$$

a) 0.056 M b) 0.035 M c) 0.18 M d) 0.21 M e)0.37 M

Answer: c

33. For the zero-order reaction below, a graph of ________ versus time will generate a straight line.

$$A \rightarrow B + C \quad \text{rate} = k[A]^0$$

a) [A] b) $\frac{1}{[A]^2}$ c) ln[A] d) $\frac{1}{[A]}$ e) $\frac{\ln(2)}{[A]}$

Answer: a

15.5 A Microscopic View of Reaction Rates

34. According to collision theory, which condition(s) must be met in order for molecules to react?
    1. The reacting molecules must collide with each other to react.
    2. A catalyst must be in contact with the reacting molecules for a reaction to occur.
    3. The reacting molecules must collide with an orientation that can lead to rearrangement of the atoms.

    a) 1 only b) 2 only c) 3 only d) 1 and 2 e) 1 and 3

    Answer: e

35. In general, as temperature increases, the rate of a chemical reaction
    a) decreases due to fewer collisions with proper molecular orientation.
    b) decreases for endothermic reactions.
    c) decreases due to an increase in the activation energy.
    d) increases due to a decrease in the activation energy.
    e) increases due to a greater number of effective collisions.

    Answer: e

36. The Arrhenius equation, $k = Ae^{-E_a/RT}$, relates the rate of reaction and temperature. A plot of $\ln(k)$ versus $1/T$ will yield a straight line with a slope of ________.
    a) $-E_a/R$
    b) $-E_a$
    c) A
    d) $e^{-E_a/R}$
    e) $1/RT$

    Answer: a

37. Molecules must overcome a barrier called the activation energy if they are to react. The highest energy point reached during the progress of a reaction is called the ________.
    a) transition state
    b) Arrhenius state
    c) half-life
    d) elementary step
    e) intermediate state

    Answer: a

38. For a chemical reaction, the activation energy for the forward reaction is +112 kJ and the activation energy for the backward reaction is +187 kJ. What is the overall energy change for the forward reaction?
a) -299 kJ b) -75 kJ c) -0.599 kJ d) +75 kJ e) +299 kJ

Answer: b

39. Calculate the activation energy, E_a for
$$N_2O_5(g) \rightarrow 2\ NO_2(g) + \tfrac{1}{2}\ O_2(g)$$
given k (at 25.0 °C) = $3.46 \times 10^{-5}\ s^{-1}$ and k (at 45.0 °C) = $5.79 \times 10^{-4}\ s^{-1}$. ($R$ = 8.314 J/K·mol)
a) 1.32 kJ/mol b) 1.61 kJ/mol c) 14.0 kJ/mol d) 48.2 kJ/mol e) 111 kJ/mol

Answer: e

40. For a given reaction, the rate constant doubles when the temperature is increased from 37 °C to 51 °C. What is the activation energy for this reaction? (R = 8.314 J/K·mol)
a) 18 kJ/mol b) 41 kJ/mol c) 59 kJ/mol d) 86 kJ/mol e) 780 kJ/mol

Answer: b

41. For a given reaction, the activation energy is 21.7 kJ/mol. If the reaction rate constant is $6.4 \times 10^{-4}\ M^{-1}s^{-1}$ at 25.0 °C, what is the reaction rate constant at 72.0 °C? (R = 8.314 J/K·mol)
a) $1.9 \times 10^{-4}\ M^{-1}s^{-1}$
b) $6.4 \times 10^{-4}\ M^{-1}s^{-1}$
c) $2.1 \times 10^{-3}\ M^{-1}s^{-1}$
d) $3.6 \times 10^{-3}\ M^{-1}s^{-1}$
e) $2.5 \times 10^{26}\ M^{-1}s^{-1}$

Answer: c

42. The rate constant at 305 K for a first-order reaction is $4.1 \times 10^{-4}\ s^{-1}$ and the activation energy is 15.3 kJ/mol. What is the value of the frequency factor, A, in the Arrhenius equation? (R = 8.314 J/K·mol)
a) $4.1 \times 10^{-4}\ s^{-1}$ b) $0.17\ s^{-1}$ c) $1.3\ s^{-1}$ d) $5.8\ s^{-1}$ e) $6.3\ s^{-1}$

Answer: b

43. The effect of adding a catalyst to a reaction is to
a) increase the number of collisions between reactants.
b) increase the energy of the products.
c) increase the equilibrium constant of a reaction.
d) lower the activation energy of a reaction.
e) decrease the enthalpy change of a reaction.

Answer: d

15.6 Reaction Mechanisms

44. The elementary steps for the catalyzed decomposition of dinitrogen monoxide are shown below.

$$N_2O(g) + NO(g) \rightarrow N_2(g) + NO_2(g)$$
$$2\ NO_2(g) \rightarrow 2\ NO(g) + O_2(g)$$

Which of the following statements are CORRECT?

1. The overall balanced reaction is $2\ N_2O(g) \rightarrow 2\ N_2(g) + O_2(g)$.
2. $NO(g)$ is a catalyst for the reaction.
3. $N_2(g)$ is a reaction intermediate.

a) 1 only b) 2 only c) 3 only d) 1 and 2 e) 1, 2, and 3

Answer: d

45. In basic solution, $(CH_3)_3CCl$ reacts according to the equation below.

$$(CH_3)_3CCl + OH^- \rightarrow (CH_3)_3COH + Cl^-$$

The accepted mechanism for the reaction is

$$(CH_3)_3CCl \rightarrow (CH_3)_3C^+ + Cl^- \quad \text{(slow)}$$
$$(CH_3)_3C^+ + OH^- \rightarrow (CH_3)_3COH \quad \text{(fast)}$$

What is a rate law that is consistent with the mechanism for this reaction?

a) rate = $k[(CH_3)_3CCl]$
b) rate = $k[(CH_3)_3CCl][OH^-]$
c) rate = $k[(CH_3)_3C^+][OH^-]$
d) rate = $k[(CH_3)_3CCl][OH^-]/[Cl^-]$
e) rate = $k[(CH_3)_3CCl]\ [OH^-]/[Cl^-]$

Answer: a

46. Nitrogen dioxide reacts with carbon monoxide to produce nitrogen monoxide and carbon dioxide.

$$NO_2(g) + CO(g) \rightarrow NO(g) + CO_2(g)$$

A proposed mechanism for this reaction is

$$2\ NO_2(g) \rightleftharpoons NO_3(g) + NO(g) \quad \text{(fast, equilibrium)}$$
$$NO_3(g) + CO(g) \rightarrow NO_2(g) + CO_2(g) \quad \text{(slow)}$$

What is a rate law that is consistent with the proposed mechanism?

a) rate = $k[NO_2][CO]$
b) rate = $k[[NO_2]^2[CO]$
c) rate = $k[[NO_2]^2[CO]/[NO]$
d) rate = $k[NO_3][CO]$
e) rate = $k[NO_2]^2$

Answer: c

47. For the overall reaction

$$A + 2B \rightarrow C$$

which of the following mechanisms is consistent with the rate equation below?

$$\text{rate} = k[A]\,[B]^2$$

a) $2A + B \rightleftharpoons I$ (fast)
   $I + B \rightarrow C + A$ (slow)

b) $A + B \rightarrow I$ (slow)
   $I + B \rightarrow C$ (fast)

c) $2B \rightarrow I$ (slow)
   $A + I \rightarrow C$ (fast)

d) $2B \rightleftharpoons I$ (fast)
   $I + A \rightarrow C$ (slow)

e) $A + 2B \rightarrow I$ (fast)
   $I + B \rightarrow C + B$ (slow)

Answer: d

Short Answer Questions

48. If a catalyst is present in a different phase than the reactants and products, it is referred to as a ________ catalyst.

Answer: heterogeneous

49. The ________ of an elementary step is defined as the number of reactant molecules that come together in the reaction.

Answer: molecularity

50. The pre-exponential, *A*, in the Arrhenius equation is called the ________ factor.

Answer: frequency

51. Radioactive isotopes decay by ________ -order kinetics.

Answer: first

52. Elementary steps in a reaction mechanism often include reaction ________. These (usually) short-lived species, which are at one point produced and then later consumed, do not appear in the overall chemical reaction.

Answer: intermediates

53. Termolecular elementary steps are rare. Why?

Answer: A termolecular elementary step requires three reactant particles to simultaneously interact. The probability is small that three particles will effectively collide with the necessary energy and orientation.

54. The decay of radioactive elements is a first-order process. The half-life of carbon-14 is 5730 years. How many years will it take for 5.0 g of carbon-14 to decay to 1.0 mg?

Answer: 7.04×10^4 years

Diagnostic Quiz Questions

Goal #1: Understand rates of reaction and the conditions affecting rates.

1. Which of the statements concerning relative rates of reaction is correct for the decomposition of dinitrogen pentaoxide?

$$2\ N_2O_5(g) \rightarrow 4\ NO_2(g) + O_2(g)$$

a) The rate of disappearance of N_2O_5 is equal to the rate of appearance of NO_2.
b) The rate of appearance of NO_2 is equal to the rate of appearance of O_2.
c) The rate of disappearance of N_2O_5 is twice the rate of appearance of NO_2.
d) The rate of disappearance of N_2O_5 is twice the rate of appearance of O_2.

Answer: d

2. In the reaction below, the rate of formation of O_2 is 3.4×10^{-3} mol/L·s. What is the rate of formation of HF?

$$2\ HOF(g) \rightarrow 2\ HF(g) + O_2(g)$$

a) 1.7×10^{-3} mol/L·s
b) 3.4×10^{-3} mol/L·s
c) 6.8×10^{-3} mol/L·s
d) 14×10^{-3} mol/L·s

Answer: c

3. Which of the following statements are correct?
   1. Catalysts are substances that increase the rate of reaction but are not consumed in the reaction.
   2. Changes in temperature affect reaction rates.
   3. Changes in reactant concentrations affect reaction rates.

a) 1 only
b) 2 only
c) 3 only
d) 1, 2, and 3

Answer: d

Goal #2: Derive the rate equation, rate constant, and reaction order from experimental data

4. What is the overall order of the reaction

$$CO(g) + NO_2(g) \rightarrow CO_2(g) + NO(g)$$

if it proceeds via the following rate expression?

$$\frac{\Delta[CO_2]}{\Delta t} = k[CO][NO_2]$$

a) 1
b) 2
c) 3
d) 4

Answer: b

5. For the reaction 2 NO(g) + Cl_2(g) → 2 NOCl(g), the rate law is

$$\frac{\Delta[NOCl]}{\Delta t} = k[NO]^2[Cl_2].$$

What are the units of the rate constant where time is measured in seconds?

a) 1/s
b) M/s
c) 1/M·s
d) $1/M^2 \cdot s$

Answer: d

6. Given the initial rate data for the reaction A + B → C, determine the order of the reaction with respect to each reactant.

[A], M	[B], M	$\Delta[C]/\Delta t$ (initial) M/s
0.10	0.20	6.80×10^{-6}
0.10	0.40	2.72×10^{-5}
0.20	0.40	5.44×10^{-5}

a) The reaction is first-order with respect to A and first-order with respect to B.
b) The reaction is first-order with respect to A and second-order with respect to B.
c) The reaction is second-order with respect to A and first-order with respect to B.
d) The reaction is fourth-order with respect to A and first-order with respect to B.

Answer: b

7. Given the initial rate data for the reaction A + B → C, determine the rate expression for the reaction.

[A], M	[B], M	Δ[C]/Δ*t* (initial) M/s
0.25	0.10	2.16×10^{-3}
0.25	0.25	5.40×10^{-3}
0.40	0.10	5.53×10^{-3}

a) Rate = 0.086 $M^{-1}s^{-1}$[A] [B]
b) Rate = 0.86 $M^{-2}s^{-1}$[A][B]^2
c) Rate = 0.35 $M^{-2}s^{-1}$[A]^2[B]
d) Rate = 3.5 $M^{-3}s^{-1}$[A]^2[B]^2

Answer: c

Goal #3: Use integrated rate laws

8. For the reaction R → products, which of the following equations corresponds to the integrated expression for a first-order decomposition reaction?

a) $\frac{[R]_t}{[R]_0} = -kt$

b) $\ln\frac{[R]_t}{[R]_0} = -kt$

c) $\frac{1}{[R]_t} - \frac{1}{[R]_0} = kt$

d) $[R]_t - [R]_0 = -kt$

Answer: b

9. Hydrogen peroxide decays into water and oxygen in a first-order process.

$$H_2O_2(aq) \rightarrow H_2O(\ell) + \tfrac{1}{2} O_2(g)$$

At 20.0 °C, the rate constant is $1.77 \times 10^{5}\ s^{-1}$. If the initial concentration of hydrogen peroxide is 1.2 M, what is the concentration after 7.0 days?

a) 2.4×10^{-11} M
b) 2.7×10^{-5} M
c) 8.7×10^{-2} M
d) 1.2 M

Answer: b

10. For the second-order reaction below, the concentration of the product, B, after 539 seconds is 0.0811 M. If no B is initially present, and the initial concentration of A is 0.744 M, what is the rate constant?

$$2A \rightarrow B \quad \text{rate} = k[A]^2$$

a) 3.1×10^{-4} $M^{-1}s^{-1}$
b) 4.6×10^{-4} $M^{-1}s^{-1}$
c) 7.0×10^{-4} $M^{-1}s^{-1}$
d) 2.0×10^{-2} $M^{-1}s^{-1}$

Answer: c

11. What is the half-life for a first-order reaction with a rate constant of 0.291 s^{-1}?
a) 0.420 s
b) 1.93 s
c) 2.38 s
d) 6.87 s

Answer: c

12. The decomposition of HCO_2H follows first-order kinetics.

$$HCO_2H(g) \rightarrow CO_2(g) + H_2(g)$$

The half-life for the reaction at 550 °C is 24 seconds. How many seconds are needed for the formic acid concentration to decrease by 33%?
a) 7.9 s
b) 14 s
c) 16 s
d) 38 s

Answer: b

Goal #4: Understand the collision theory of reaction rates and the role of activation energy

13. Which the following statements concerning collision theory and the Arrhenius equation are correct?
    1. The frequency factor, A, in the Arrhenius equation relates to the number of collisions and the fraction of those collisions that have the proper geometry for a reaction to occur.
    2. As the temperature increases, the rate of a reaction increases due a greater number of effective collisions.
    3. The activation energy is the energy barrier that must be overcome by reactants for a reaction to occur.

    a) 1 only
    b) 2 only
    c) 3 only
    d) 1, 2, and 3

    Answer: d

14. For a chemical reaction, the activation energy for the forward reaction is +235 kJ and the activation energy for the reverse reaction is +51 kJ. What is the overall energy change for the forward reaction?
    a) -286 kJ
    b) -184 kJ
    c) +184 kJ
    d) +286 kJ

    Answer: c

15. For a given reaction, the activation energy is 29.1 kJ/mol. If the reaction rate constant is 8.12 $\times 10^{-4}$ $M^{-1}s^{-1}$ at 298 K, what is the reaction rate constant at 318 K?
    a) 2.46×10^{-4} $M^{-1}s^{-1}$
    b) 8.13×10^{-4} $M^{-1}s^{-1}$
    c) 1.70×10^{-3} $M^{-1}s^{-1}$
    d) 5.88×10^{2} $M^{-1}s^{-1}$

    Answer: c

16. For a given reaction, the rate constant doubles when the temperature is increased from 27 °C to 49 °C. What is the activation energy for this reaction?
    a) 0.347 kJ/mol
    b) 25.3 kJ/mol
    c) 31.5 kJ/mol
    d) 52.7 kJ/mol

    Answer: b

Goal #5: Relate reaction mechanisms and rate laws.

17. Nitrogen monoxide reacts with chlorine to produce NOCl.

$$2\,NO(g) + Cl_2(g) \rightarrow 2\,NOCl(g)$$

A proposed mechanism for this reaction is

$$NO(g) + NO(g) \rightleftharpoons N_2O_2(g) \qquad \text{(fast, equilibrium)}$$
$$N_2O_2(g) + Cl_2(g) \rightarrow 2\,NOCl(g) \qquad \text{(slow)}$$

What is a rate law that is consistent with this mechanism?

a) rate = $k[N_2O_2][Cl_2]$
b) rate = $k[NO][Cl_2]$
c) rate = $k[NO]^2[Cl_2]$
d) rate = $k[NO]^2$

Answer: c

18. The elementary steps for the catalyzed decomposition for dinitrogen monoxide are shown below. Identify the catalyst.

$$NO(g) + N_2O(g) \rightarrow N_2(g) + NO_2(g)$$
$$2NO_2(g) \rightarrow 2NO(g) + O_2(g)$$

a) NO
b) N_2O
c) N_2
d) NO_2

Answer: a

19. The elementary steps for the formation of nitrogen and water from nitric oxide and hydrogen are given below. Identify any intermediates.

$$2\,NO(g) \rightarrow N_2O_2(g)$$
$$N_2O_2(g) + H_2(g) \rightarrow N_2O(g) + H_2O(g)$$
$$N_2O(g) + H_2(g) \rightarrow N_2(g) + H_2O(g)$$

a) N_2O_2 only
b) N_2O only
c) N_2O and H_2O
d) N_2O_2 and N_2O

Answer: d

CHAPTER 16: Principles of Reactivity: Chemical Equilibria

16.2 The Equilibrium Constant and the Reaction Quotient

1. Which of the following statements are CORRECT?
   1. For gas phase equilibria, the partial pressures of reactants and products are equal.
   2. For a chemical system at equilibrium, the forward and reverse rates of reaction are equal.
   3. For an aqueous chemical system at equilibrium, the concentrations of products divided by the concentrations of reactants equals one.

   a) 1 only b) 2 only c) 3 only d) 1 and 2 e) 1, 2, and 3

   Answer: b

2. Which of the following statements are always CORRECT?
   1. Product concentrations appear in the numerator of an equilibrium constant expression.
   2. A reaction favors the formation of products if K >> 1.
   3. Equilibrium constants have units of atmospheres for gas phase reactions.

   a) 1 only b) 2 only c) 3 only d) 1 and 2 e) 1, 2, and 3

   Answer: d

3. Write the expression for K_c for the reaction below.

   $$Cu^{2+}(aq) + 4\,NH_3(aq) \rightleftharpoons Cu(NH_3)_4^{2+}(aq)$$

   a) $K_c = \dfrac{[Cu^{2+}][NH_3]}{[Cu(NH_3)_4^{2+}]}$

   b) $K_c = \dfrac{[Cu(NH_3)_4^{2+}]}{[Cu^{2+}]}$

   c) $K_c = \dfrac{[Cu(NH_3)_4^{2+}]}{[Cu^{2+}][NH_3]}$

   d) $K_c = \dfrac{[Cu^{2+}][NH_3]^4}{[Cu(NH_3)_4^{2+}]}$

   e) $K_c = \dfrac{[Cu(NH_3)_4^{2+}]}{[Cu^{2+}][NH_3]^4}$

   Answer: e

4. Write the expression for K for the reaction below.

$$Al_2S_3(s) \rightleftharpoons 2\ Al^{3+}(aq) + 3\ S^{2-}(aq)$$

a) $K = [Al^{3+}]^2[S^{2-}]^3$

b) $K = [Al^{3+}][S^{2-}]$

c) $K = [2\ Al^{3+}][3\ S^{2-}]$

d) $K = \dfrac{[Al_2S_3]}{[Al^{3+}]^2[S^{2-}]^3}$

e) $K = \dfrac{[Al^{3+}]^2[S^{2-}]^3}{[Al_2S_3]}$

Answer: a

5. Write the expression for K for the reaction of acetate ion with water.

$$CH_3CO_2^-(aq) + H_2O(\ell) \rightleftharpoons CH_3CO_2H(aq) + OH^-(aq)$$

a) $K = \dfrac{[CH_3CO_2H]}{[CH_3CO_2^-]}$

b) $K = \dfrac{[CH_3CO_2^-]}{[CH_3CO_2H][OH^-]}$

c) $K = \dfrac{[CH_3CO_2H][OH^-]}{[CH_3CO_2^-]}$

d) $K = \dfrac{[CH_3CO_2H][OH^-]}{[CH_3CO_2^-][H_2O]}$

e) $K = \dfrac{[CH_3CO_2^-][H_2O]}{[CH_3CO_2H][OH^-]}$

Answer: c

6. Write the expression for K_p for the reaction below.

$$2\ HBr(g) \rightleftharpoons H_2(g) + Br_2(\ell)$$

a) $K_p = \dfrac{P_{HBr}^2}{P_{Br_2}P_{H_2}}$ b) $K_p = \dfrac{P_{H_2}}{P_{HBr}^2}$ c) $K_p = P_{HBr}^2$ d) $K_p = \dfrac{P_{HBr}^2}{P_{H_2}}$ e) $K_p = \dfrac{P_{H_2}P_{Br_2}}{P_{HBr}^2}$

Answer: b

7. Write the expression for K_p for the reaction below.

$$2\ Hg_2O(s) \rightleftharpoons 4\ Hg(\ell) + O_2(g)$$

a) $K_p = \dfrac{P_{Hg}P_{O_2}}{P_{Hg_2O}}$

b) $K_p = \dfrac{P_{Hg}^4 P_{O_2}}{P_{Hg_2O}^2}$

c) $K_p = P_{Hg}^4 P_{O_2}$

d) $K_p = \dfrac{1}{P_{Hg}^4 P_{O_2}}$

e) $K_p = P_{O_2}$

Answer: e

8. Write a balanced chemical equation which corresponds to the following equilibrium constant expression.

$$K_p = \frac{P_{NH_3}}{P_{N_2}^{1/2} P_{H_2}^{3/2}}$$

a) $1/2\ N_2(g) + 3/2\ H_2(g) \rightleftharpoons NH_3(g)$
b) $N_2(g) + 3\ H_2(g) \rightleftharpoons 2\ NH_3(g)$
c) $2\ NH_3(g) \rightleftharpoons N_2(g) + 3\ H_2(g)$
d) $NH_3(g) \rightleftharpoons 1/2\ N_2(g) + 3/2\ H_2(g)$
e) $2\ N_2(g) + 6\ H_2(g) \rightleftharpoons 4\ NH_3(g)$

Answer: a

9. Write a balanced chemical equation which corresponds to the following equilibrium constant expression.

$$K = [Fe^{3+}][OH^-]^3$$

a) $FeOH^{2+}(s) \rightleftharpoons Fe^{3+}(aq) + OH^-(aq)$
b) $3\ Fe^{3+}(aq) + 3\ OH^-(aq) \rightleftharpoons 3\ Fe(OH)_3(aq)$
c) $Fe(OH)_3(aq) \rightleftharpoons Fe^{3+}(aq) + 3\ OH^-(aq)$
d) $Fe(OH)_3(s) \rightleftharpoons Fe^{3+}(aq) + 3\ OH^-(aq)$
e) $Fe^{3+}(aq) + 3\ OH^-(aq) \rightleftharpoons 3\ Fe(OH)_3(s)$

Answer: d

10. Write a balanced chemical equation which corresponds to the following equilibrium constant expression.

$$K = \frac{[F^-][H_3O^+]}{[HF]}$$

a) $HF(aq) \rightleftharpoons F^-(aq) + H_3O^+(aq)$

b) $F^-(aq) + H_3O^+(aq) \rightleftharpoons HF(aq) + H_2O(\ell)$

c) $HF(aq) + H_2O(\ell) \rightleftharpoons F^-(aq) + H_3O^+(aq)$

d) $H^+(aq) + OH^-(aq) \rightleftharpoons H_2O(\ell)$

e) $F^-(aq) + H_3O^+(aq) \rightleftharpoons HF(aq)$

Answer: c

11. For which one of the following reactions does K_p equal K_c?

a) $2\ SO_2(g) + O_2(g) \rightleftharpoons 2\ SO_3(g)$

b) $2\ O_3(g) \rightleftharpoons 3\ O_2(g)$

c) $2\ Ag_2O(s) \rightleftharpoons 4\ Ag(s) + O_2(g)$

d) $2\ NH_3(g) \rightleftharpoons 3\ H_2(g) + N_2(g)$

e) $CO(g) + NO_2(g) \rightleftharpoons NO(g) + CO_2(g)$

Answer: e

12. What is the relationship between K_p and K_c for the reaction below?

$$CS_2(g) + 3\ Cl_2(g) \rightleftharpoons S_2Cl_2(g) + CCl_4(g)$$

a) $K_c = \dfrac{K_p}{(RT)^2}$

b) $K_c = \dfrac{(RT)^2}{K_p}$

c) $K_c = (RT)^2 K_p$

d) $K_c = \left(\dfrac{RT}{K_p}\right)^2$

e) $K_c = \left(\dfrac{K_p}{RT}\right)^2$

Answer: c

13. Dinitrogen tetraoxide decomposes to nitrogen dioxide.

$$N_2O_4(g) \rightleftharpoons 2\ NO_2(g)$$

Calculate the value of K_p, given that $K_c = 5.88 \times 10^{-3}$ at 273 K. ($R = 0.08206$ L·atm/mol·K)
a) 2.62×10^{-4} b) 0.132 c) 7.59 d) 1.70×10^{2} e) 3.81×10^{-3}

Answer: b

14. The oxidation of sulfur dioxide produces sulfur trioxide.

$$2\ SO_2(g) + O_2(g) \rightleftharpoons 2\ SO_3(g)$$

Calculate the value of K_c, given that $K_p = 2.8 \times 10^{2}$ at 999 K. ($R = 0.08206$ L·atm/mol·K)
a) 4.4×10^{-5} b) 0.29 c) 3.4 d) 2.3×10^{4} e) 6.8×10^{3}

Answer: d

15. If the reaction quotient, Q, is greater than K_p, then
a) the chemical system has reached equilibrium.
b) the temperature must be increased for the reaction to proceed in the forward direction.
c) the reaction will proceed in the direction that results in fewer gas phase particles.
d) the reaction will proceed to the right until equilibrium is established.
e) the reaction will proceed to the left until equilibrium is established.

Answer: e

16. A 4.00 L flask is filled with 0.75 mol SO_3, 2.50 mol SO_2, and 1.30 mol O_2, and allowed to reach equilibrium. Predict the effect on the concentrations of SO_3 as equilibrium is achieved by using Q, the reaction quotient. Assume the temperature of the mixture is chosen so that $K_c = 12$.

$$2\ SO_3(g) \rightleftharpoons 2\ SO_2(g) + O_2(g)$$

a) $[SO_3]$ will decrease because $Q > K$.
b) $[SO_3]$ will decrease because $Q < K$.
c) $[SO_3]$ will increase because $Q < K$.
d) $[SO_3]$ will increase because $Q > K$.
e) $[SO_3]$ will remain the same because $Q = K$.

Answer: b

17. Consider the reaction $A(g) \rightleftharpoons 2\ B(g)$ where $K_p = 5.0$ at 25 °C. If 0.50 mol A and 2.0 mol B are introduced into a 1.0 L flask at 25 °C, what change in concentrations (if any) will occur in time?
a) [A] will decrease and [B] will decrease.
b) [A] will decrease and [B] will increase.
c) [A] will increase and [B] will increase.
d) [A] will increase and [B] will decrease.
e) [A] and [B] remain unchanged.

Answer: d

16.3 Determining an Equilibrium Constant

18. The reaction below is studied at a high temperature.

$$PCl_5(g) \rightleftharpoons PCl_3(g) + Cl_2(g)$$

At equilibrium, the partial pressures of the gases are as follows: $PCl_5 = 1.8 \times 10^{-2}$ atm, $PCl_3 = 5.6 \times 10^{-2}$ atm, and $Cl_2 = 3.8 \times 10^{-4}$ atm. What is the value of K_p for the reaction?

a) 3.8×10^{-7} b) 1.2×10^{-3} c) 3.1 d) 8.5×10^{2} e) 2.6×10^{6}

Answer: b

19. At a given temperature, an equilibrium mixture contains the following concentrations of gases: $[SO_3] = 0.054$ M, $[SO_2] = 0.0047$ M, and $[O_2] = 0.58$ M. What is the equilibrium constant, K_c, for the reaction below?

$$2\ SO_3(g) \rightleftharpoons 2\ SO_2(g) + O_2(g)$$

a) 4.4×10^{-3} b) 5.0×10^{-2} c) 11 d) 2.0×10^{2} e) 2.3×10^{2}

Answer: a

20. Excess $Ca(IO_3)_2(s)$ is placed in 1.5 L of water. At equilibrium, the solution contains 0.011 M $IO_3^-(aq)$. What is the equilibrium constant for the reaction below?

$$Ca(IO_3)_2(s) \rightleftharpoons Ca^{2+}(aq) + 2\ IO_3^-(aq)$$

a) 3.3×10^{-7} b) 6.7×10^{-7} c) 1.3×10^{-6} d) 5.3×10^{-6} e) 6.1×10^{-5}

Answer: b

21. At 25 °C, only 1.9 g $CaSO_4$ will dissolve in 2.00 L of water. What is the equilibrium constant for the reaction below?

$$CaSO_4(s) \rightleftharpoons Ca^{2+}(aq) + SO_4^{2-}(aq)$$

a) 4.9×10^{-5} b) 1.9×10^{-4} c) 1.4×10^{-2} d) 7.0×10^{-3} e) 0.90

Answer: a

22. We place 0.0774 mol $N_2O_4(g)$ in a 1.00 L flask at a given temperature. After reaching equilibrium, the concentration of $NO_2(g)$ is 0.0068 M. What is K_c for the reaction below?

$$N_2O_4(g) \rightleftharpoons 2\ NO_2(g)$$

a) 8.8×10^{-5} b) 5.9×10^{-4} c) 6.2×10^{-4} d) 7.2×10^{-4} e) 9.6×10^{-2}

Answer: c

23. When 0.20 mole NH_4Cl is dissolved in water to a volume of 1.00 L, 0.0053% of the NH_4^+ dissociates to form NH_3. What is the equilibrium constant for the reaction?

$$NH_4^+(aq) + H_2O(\ell) \rightleftharpoons NH_3(aq) + H_3O^+(aq)$$

a) 2.2×10^{-11} b) 1.1×10^{-10} c) 5.6×10^{-10} d) 1.4×10^{-8} e) 5.6×10^{-6}

Answer: c

24. A sealed tube is prepared with 1.07 atm PCl_5 at 500 K. The PCl_5 decomposes until equilibrium is established.

$$PCl_5(g) \rightleftharpoons PCl_3(aq) + Cl_2(g)$$

The equilibrium pressure in the tube is 1.54 atm. Calculate K_p.

a) 0.052 b) 0.20 c) 0.27 d) 0.37 e) 2.2

Answer: d

25. A mixture of 0.200 mol NO_2 and 0.200 mol CO is placed in a 1.00 L flask and given time to equilibrate. Analysis of the equilibrium mixture indicates that 0.134 mol of CO_2 is present. Calculate K_c for the reaction.

$$NO_2(g) + CO(g) \rightleftharpoons NO(g) + CO_2(g)$$

a) 0.27 b) 0.45 c) 0.67 d) 2.0 e) 4.1

Answer: e

26. Nitrosyl bromide decomposes according to the chemical equation below.

$$2\ NOBr(g) \rightleftharpoons 2\ NO(g) + Br_2(g)$$

When 0.260 atm of NOBr is sealed in a flask and allowed to reach equilibrium, 22% of the NOBr decomposes. What is the equilibrium constant, K_p, for the reaction?

a) 2.3×10^{-3} b) 4.5×10^{-3} c) 3.5×10^{-2} d) 4.8×10^{-2} e) 8.0×10^{-2}

Answer: a

16.4 Using Equilibrium Constants in Calculations

27. At 2010 K, the equilibrium constant, K_c, for the following reaction is 4.0×10^{-4}.

$$N_2(g) + O_2(g) \rightleftharpoons 2\ NO(g)$$

If the equilibrium concentrations of N_2 and O_2 are 0.28 mol/L and 0.38 mol/L at 2010 K, what is the equilibrium concentration of NO?

a) 1.8×10^{-9} M b) 2.1×10^{-5} M c) 4.3×10^{-5} M d) 6.5×10^{-3} M e) 6.1×10^{-2} M

Answer: d

28. A gaseous mixture of NO_2 and N_2O_4 is in equilibrium. If the concentration of N_2O_4 is 7.1×10^{-4} M, what is the concentration of NO_2?

$$2\ NO_2(g) \rightleftharpoons N_2O_4(g) \quad K_c = 170$$

a) 1.7×10^{-11} M b) 4.2×10^{-6} M c) 2.0×10^{-3} M d) 4.9×10^{2} M e) 2.4×10^{5} M

Answer: c

29. At 25 °C, the decomposition of dinitrogen tetraoxide

$$N_2O_4(g) \rightleftharpoons 2\ NO_2(g)$$

has an equilibrium constant (K_p) of 0.144. At equilibrium, the total pressure of the system is 0.48 atm. What is the partial pressure of each gas?
a) 0.030 atm $NO_2(g)$ and 0.42 $N_2O_4(g)$
b) 0.060 atm $NO_2(g)$ and 0.42 $N_2O_4(g)$
c) 0.16 atm $NO_2(g)$ and 0.18 $N_2O_4(g)$
d) 0.20 atm $NO_2(g)$ and 0.28 $N_2O_4(g)$
e) 0.28 atm $NO_2(g)$ and 0.20 $N_2O_4(g)$

Answer: d

30. The equilibrium constant at 25 °C for the dissolution of silver bromide is 5.4×10^{-13}.

$$AgBr(s) \rightleftharpoons Ag^+(aq) + Br^-(aq)$$

If an excess quantity of AgBr(s) is added to water and allowed to equilibrate, what is the equilibrium concentration of Ag^+?
a) 2.9×10^{-25} M b) 2.7×10^{-13} M c) 5.4×10^{-13} M d) 1.1×10^{-12} M e) 7.3×10^{-7} M

Answer: e

31. Hydrogen iodide can decompose into hydrogen and iodine gases.

$$2\ HI(g) \rightleftharpoons H_2(g) + I_2(g) \qquad K_p = 0.016$$

If 0.820 atm HI(g) is sealed in a flask, what is the pressure of each gas when equilibrium is established?
a) HI = 0.576 atm, H_2 = 0.096 atm, I_2 = 0.096 atm
b) HI = 0.654 atm, H_2 = 0.083 atm, I_2 = 0.083 atm
c) HI = 0.728 atm, H_2 = 0.092 atm, I_2 = 0.092 atm
d) HI = 0.737 atm, H_2 = 0.083 atm, I_2 = 0.083 atm
e) HI = 0.768 atm, H_2 = 0.111 atm, I_2 = 0.111 atm

Answer: b

32. We place 1.32 mol PCl_5 in a 1.0 L flask and allow it to reach equilibrium at a given temperature. What is the final concentration of Cl_2 in the flask?

$$PCl_5(g) \rightleftharpoons PCl_3(aq) + Cl_2(g) \qquad K_c = 0.47$$

a) 0.36 M b) 0.59 M c) 0.62 M d) 0.79 M e) 85 M

Answer: b

33. The equilibrium constant, K_c, for the following reaction is 1.0×10^{-5} at 1500 K.

$$N_2(g) + O_2(g) \rightleftharpoons 2\,NO(g)$$

If 0.750 M N_2 and 0.750 M O_2 are allowed to equilibrate at 1500 K, what is the concentration of NO?

a) 6.7×10^{-4} M b) 1.2×10^{-3} M c) 2.4×10^{-3} M d) 2.7×10^{-3} M e) 5.5×10^{-3} M

Answer: c

34. At 800 K, the equilibrium constant, K_p, for the following reaction is 3.2×10^{-7}.

$$2\,H_2S(g) \rightleftharpoons 2\,H_2(g) + S_2(g)$$

A reaction vessel at 800 K initially contains 3.50 atm of H_2S. If the reaction is allowed to equilibrate, what is the equilibrium pressure of H_2?

a) 1.0×10^{-3} atm b) 2.1×10^{-3} atm c) 5.0×10^{-3} atm d) 9.9×10^{-3} atm e) 2.0×10^{-2} atm

Answer: e

35. Carbonyl bromide decomposes to carbon monoxide and bromine.

$$COBr_2(g) \rightleftharpoons CO(g) + Br_2(g)$$

K_c is 0.19 at 73 °C. If an initial concentration of 0.63 M $COBr_2$ is allowed to equilibrate, what are the equilibrium concentrations of $COBr_2$, CO, and Br_2?

a) $[COBr_2] = 0.11$ M, $[CO] = 0.26$ M, $[Br_2] = 0.26$ M
b) $[COBr_2] = 0.28$ M, $[CO] = 0.35$ M, $[Br_2] = 0.35$ M
c) $[COBr_2] = 0.30$ M, $[CO] = 0.33$ M, $[Br_2] = 0.33$ M
d) $[COBr_2] = 0.37$ M, $[CO] = 0.26$ M, $[Br_2] = 0.26$ M
e) $[COBr_2] = 0.63$ M, $[CO] = 0.35$ M, $[Br_2] = 0.35$ M

Answer: d

16.5 More about Balanced Equations and Equilibrium Constants

36. For the following reaction,

$$2\,SO_2(g) + O_2(g) \rightleftharpoons 2\,SO_3(g)$$

the equilibrium constant, K_p, is 0.758 at 627 °C. What is the equilibrium constant, at 627 °C, for the reaction below?

$$SO_3(g) \rightleftharpoons SO_2(g) + 1/2\,O_2(g)$$

a) 0.660 b) 0.871 c) 1.15 d) 1.32 e) 1.74

Answer: c

37. The equilibrium constant (K_p) for the following reaction is 3.94×10^{-3} at a given temperature.

$$N_2(g) + 2\ H_2O(g) \rightleftharpoons 2\ NO(g) + 2\ H_2(g)$$

What is the equilibrium constant for the reaction below at the same temperature?

$$3\ N_2(g) + 6\ H_2O(g) \rightleftharpoons 6\ NO(g) + 6\ H_2(g)$$

a) 6.12×10^{-8} b) 1.31×10^{-3} c) 3.94×10^{-3} d) 1.18×10^{-2} e) 1.58×10^{-1}

Answer: a

38. Given the following chemical equilibria,

$$N_2(g) + 3\ H_2(g) \rightleftharpoons 2\ NH_3(g) \qquad K_1$$
$$4\ NH_3(g) + 5\ O_2(g) \rightleftharpoons 4\ NO(g) + 6\ H_2O(g) \qquad K_2$$
$$H_2(g) + 1/2\ O_2(g) \rightleftharpoons H_2O(g) \qquad K_3$$

Determine the equilibrium constant for the reaction below.

$$N_2(g) + O_2(g) \rightleftharpoons 2\ NO(g) \qquad K_4$$

a) $K_4 = K_1 \times K_2 \times K_3$

b) $K_4 = K_1 + K_2 + K_3$

c) $K_4 = K_1 + \dfrac{K_2}{2} + 3K_3$

d) $K_4 = \dfrac{3(K_1 \times K_2)}{2K_3}$

e) $K_4 = \dfrac{K_1 \times \sqrt{K_2}}{K_3^3}$

Answer: e

39. Determine the equilibrium constant for the following reaction

$$Ca^{2+}(aq) + 2\ H_2O(\ell) \rightleftharpoons Ca(OH)_2(s) + 2\ H^+(aq)$$

given the chemical reactions below.

$$Ca(OH)_2(s) \rightleftharpoons Ca^{2+}(aq) + 2\ OH^-(aq) \qquad K = 6.5 \times 10^{-6}$$
$$H_2O(\ell) \rightleftharpoons H^+(aq) + OH^-(aq) \qquad K = 1.0 \times 10^{-14}$$

a) 1.5×10^{-23} b) 6.5×10^{-20} c) 1.3×10^{-19} d) 1.5×10^{-9} e) 1.5×10^{19}

Answer: a

40. Given the following equilibria,

$$MgCO_3(s) \rightleftharpoons Mg^{2+}(aq) + CO_3^{2-}(aq) \qquad K_1 = 6.8 \times 10^{-6}$$
$$MgF_2(s) \rightleftharpoons Mg^{2+}(aq) + 2\ F^-(aq) \qquad K_2 = 5.2 \times 10^{-11}$$

determine the equilibrium constant for the following reaction.

$$MgCO_3(s) + 2\ F^-(aq) \rightleftharpoons MgF_2(s) + CO_3^{2-}(aq) \qquad K_3$$

a) 3.5×10^{-16} b) 6.8×10^{-6} c) 7.6×10^{-6} d) 1.3×10^{5} e) 2.8×10^{15}

Answer: d

41. Given the following equilibria,

$Ni^{2+}(aq) + 2\ OH^-(aq) \rightleftharpoons Ni(OH)_2(s)$ $K_1 = 1.8 \times 10^{15}$

$Ni(NH_3)_6^{2+}(aq) \rightleftharpoons Ni^{2+}(aq) + 6\ NH_3(aq)$ $K_2 = 5.6 \times 10^8$

determine the equilibrium constant for the following reaction.

$Ni(OH)_2(s) + 6\ NH_3(aq) \rightleftharpoons Ni(NH_3)_6^{2+}(aq) + 2\ OH^-(aq)$ K_3

a) 9.9×10^{-25} b) 1.8×10^{-9} c) 3.1×10^{-7} d) 3.2×10^{6} e)1.0×10^{24}

Answer: a

42. Assume that the following chemical reaction is at equilibrium.

$I_2(g) + Cl_2(g) \rightleftharpoons 2\ ICl(g)$ $\Delta H° = -26.9$ kJ

At 25 °C, $K_p = 2.0 \times 10^5$. If the temperature is decreased to 5 °C, which statement applies?
a) K_p will decrease and the reaction will proceed in the forward direction.
b) K_p will decrease and the reaction will proceed in the backward direction.
c) K_p will remain unchanged and the reaction will proceed in the forward direction.
d) K_p will remain unchanged and the reaction will proceed in the backward direction.
e) K_p will increase and the reaction will proceed in the forward direction.

Answer: e

43. Assume that the following endothermic chemical reaction is at equilibrium.

$C(s) + H_2O(g) \rightleftharpoons H_2(g) + CO(g)$

Which of the following statements are CORRECT?

1. Increasing the amount of C(s) will increase the equilibrium concentration of CO(g).
2. Increasing the temperature will increase the equilibrium concentration of H_2(g).
3. Decreasing the concentration of H_2O(g) will increase the equilibrium concentration of CO(g).

a) 1 only b) 2 only c) 3 only d) 1 and 2 e) 1, 2, and 3

Answer: b

44. The thermochemical equation for the formation of ammonia from elemental nitrogen and hydrogen is as follows.

$N_2(g) + 3\ H_2(g) \rightleftharpoons 2\ NH_3(g)$ $\Delta H = -92.2$ kJ

Which of the following will drive the equilibrium system to the right?
a) adding NH_3(g)
b) removing N_2(g)
c) increasing the volume of the reaction vessel
d) increasing the temperature
e) adding H_2(g)

Answer: e

45. In which of the following equilibrium systems will an increase in the pressure have NO effect on the concentrations of products and reactants?
a) $CaO(s) + CO_2(g) \rightleftharpoons CaCO_3(s)$
b) $N_2(g) + 3\ H_2(g) \rightleftharpoons 2\ NH_3(g)$
c) $H_2(g) + Cl_2(g) \rightleftharpoons 2\ HCl(g)$
d) $H_2(g) + CO_2(g) \rightleftharpoons CO(g) + H_2O(\ell)$
e) $2\ H_2O_2(g) \rightleftharpoons 2\ H_2O(g) + O_2(g)$

Answer: c

46. A flask contains the following chemical system at equilibrium.
$$Ca(OH)_2(s) \rightleftharpoons Ca^{2+}(aq) + 2\ OH^-(aq)$$
Addition of which of the following substances will increase the solubility of $Ca(OH)_2(s)$?
a) $HNO_3(aq)$ b) $Ca(OH)_2(s)$ c) $NaOH(s)$ d) $CaCl_2(s)$ e) $Ca(NO_3)_2(s)$

Answer: a

47. The symbol Q is called the ________.

Answer: reaction quotient

48. In 1913, the Haber-Bosch process was patented. The product of the Haber-Bosch process is ________.

Answer: ammonia (or NH_3)

49. If a stress is applied to an equilibrium system, the system will respond is such a way as to relieve that stress. This is a statement of ________ principle.

Answer: Le Chatelier's

50. When 2.0 mole of acetic acid is diluted with water to a volume of 1.0 L, a small percentage of the acid ionizes to form acetate ion and hydronium ion.
$$CH_3CO_2H(aq) + H_2O(\ell) \rightleftharpoons CH_3CO_2^-(aq) + H_3O^+(aq)$$
At equilibrium, 0.30% of the acetic acid ionizes at 25 °C. What percentage of the acid ionizes when 0.50 mole of acetic acid is diluted with water to 1.0 L at 25 °C?

Answer: 0.60%

51. The standard enthalpy of formation of ammonia is -46.1 kJ/mol.

$$1/2\ N_2(g) + 3/2\ H_2(g) \rightleftharpoons NH_3(g)$$

Commercially, the reaction is carried out at high temperatures. Using your knowledge of kinetics and equilibrium, explain an advantage and a disadvantage of synthesizing ammonia at high temperatures.

Answer: The rate of formation of ammonia is negligible at room temperature. The advantage of high temperatures is that reaction rates increase. The disadvantage is that the equilibrium constant for this (exothermic) reaction decreases as the temperature of the reaction increases.

Diagnostic Quiz Questions

Goal #1: Understand the nature and characteristics of chemical equilibria

1. Which of the following statements correctly describe a chemical system that has reached equilibrium?
    1. The forward and reverse rates of reaction are equal.
    2. No further macroscopic changes in the system are observed.
    3. The concentrations of reactants and products remain constant.

a) 1 only
b) 2 only
c) 3 only
d) 1, 2, and 3

Answer: d

Goal #2: Understand the significance of the equilibrium constant, K.

2. What is the equilibrium constant expression, K_c, for the following chemical reaction?

$$H_2(g) + Cl_2(g) \rightleftharpoons 2\ HCl(g)$$

a) $K_c = \dfrac{[HCl]}{[H_2][Cl_2]}$

b) $K_c = \dfrac{[HCl]^2}{[H_2][Cl_2]}$

c) $K_c = \dfrac{2 \times [HCl]}{[H_2][Cl_2]}$

d) $K_c = \dfrac{[H_2][Cl_2]}{[HCl]^2}$

Answer: b

3. What is the equilibrium constant expression, K_p, for the following chemical reaction?

$$N_2(g) + 3\,H_2(g) \rightleftharpoons 2\,NH_3(g)$$

a) $K_p = \dfrac{P_{NH_3}}{P_{N_2}P_{H_2}}$

b) $K_p = \dfrac{P_{NH_3}}{P^{0.5}_{N_2}P^{1.5}_{H_2}}$

c) $K_p = \dfrac{P_{N_2}P^3_{H_2}}{P^2_{NH_3}}$

d) $K_p = \dfrac{P^2_{NH_3}}{P_{N_2}P^3_{H_2}}$

Answer: d

4. At 298 K, K_p is 5.85×10^{-3} for the decomposition of N_2O_4 to NO_2. Determine K_c for this reaction.

$$N_2O_4(g) \rightleftharpoons 2\,NO_2(g)$$

a) 2.39×10^{-4}
b) 5.85×10^{-3}
c) 0.143
d) 171

Answer: a

5. Write the equilibrium constant expression for the following chemical reaction.

$$Ca(OH)_2(s) \rightleftharpoons Ca^{2+}(aq) + 2\,OH^-(aq)$$

a) $K = \left[Ca^{2+}\right]\left[OH^-\right]$

b) $K = \left[Ca^{2+}\right]\left[OH^-\right]^2$

c) $K = \dfrac{\left[Ca^{2+}\right]\left[OH^-\right]}{\left[Ca(OH)_2\right]}$

d) $K = \dfrac{\left[Ca^{2+}\right]\left[OH^-\right]^2}{\left[Ca(OH)_2\right]}$

Answer: b

6. What is the chemical reaction that is described by the equilibrium constant expression below?

$$K = \frac{[NH_3][H_3O^+]}{[NH_4^+]}$$

a) $NH_4^+(aq) + H_2O(\ell) \rightleftharpoons NH_3(aq) + H_3O^+(aq)$
b) $NH_3(aq) + H_2O(\ell) \rightleftharpoons NH_4^+(aq) + H_3O^+(aq)$
c) $NH_4^+(aq) + OH^-(aq) \rightleftharpoons NH_3(aq) + H_2O(\ell)$
d) $NH_3(aq) + H_3O^+(aq) \rightleftharpoons NH_4^+(aq) + H_2O(\ell)$

Answer: a

Goal #3: Understand how to use K in quantitative studies of chemical equilibria.

7. The equilibrium constant for the dissolution of AgI(s) in water is 1.5×10^{-16}.

$$AgI(s) \rightleftharpoons Ag^+(aq) + I^-(aq)$$

A solution is prepared by diluting 2.0 mole AgI, 1.0×10^{-8} mole Ag^+, and 3×10^{-8} mole I^- to a volume of 1.0 L. Calculate the reactant quotient and determine whether a reaction occurs?
a) $Q = K$; no reaction occurs.
b) $Q < K$; AgI(s) dissolves until equilibrium is established.
c) $Q > K$; AgI(s) dissolves until equilibrium is established.
d) $Q > K$; Ag^+ and I^- combine to form AgI(s) until equilibrium is established.

Answer: d

8. For the following reaction,

$$2\ SO_3(g) \rightleftharpoons 2\ SO_2(g) + O_2(g)$$

the equilibrium constant, K_p, is 1.32 at 627 °C. What is the equilibrium constant for the reaction below?

$$2\ SO_2(g) + O_2(g) \rightleftharpoons 2\ SO_3(g)$$

a) -1.32
b) 0.758
c) 1.32
d) 2.64

Answer: b

9. For the following reaction,

$$N_2(g) + O_2(g) \rightleftharpoons 2\,NO(g)$$

the equilibrium constant, K, is 6.7×10^{-10} at 900 K. What is the equilibrium constant for the reaction below?

$$NO(g) \rightleftharpoons 1/2\,N_2(g) + 1/2\,O_2(g)$$

a) 6.7×10^{-10}
b) 3.9×10^{4}
c) 2.6×10^{-5}
d) 1.5×10^{9}

Answer: b

10. The following equilibrium constants are given at 25 °C:

$$AgBr(s) \rightleftharpoons Ag^{+}(aq) + Br^{-}(aq) \quad K = 5.4 \times 10^{-13}$$

$$Ag^{+}(aq) + 2\,NH_3(aq) \rightleftharpoons Ag(NH_3)_2^{+}(aq) \quad K = 1.6 \times 10^{7}$$

What is the value of K for the reaction below?

$$Ag(NH_3)_2^{+}(aq) + Br^{-}(aq) \rightleftharpoons AgBr(s) + 2\,NH_3(aq)$$

a) 8.6×10^{-6}
b) 3.4×10^{-20}
c) 1.2×10^{5}
d) 3.0×10^{19}

Answer: c

11. Calculate the equilibrium constant, K_p, for the following chemical reaction,

$$2\,SO_2(g) + O_2(g) \rightleftharpoons 2\,SO_3(g)$$

if the equilibrium partial pressures are 0.15 atm SO_2, 0.023 atm O_2, and 0.38 atm SO_3 at 1000 K.

a) 9.1×10^{-3}
b) 1.1×10^{2}
c) 2.8×10^{2}
d) 3.6×10^{-3}

Answer: c

12. The reaction

$$2\ NO_2(g) \rightleftharpoons N_2O_4(g)$$

is studied at 298 K. Initially, 0.401 mol NO_2 is placed in an empty 1.0-L flask. At equilibrium, the concentration of N_2O_4 is 0.184 mol/L. Calculate K_c for the reaction.
a) 3.9
b) 1.7×10^2
c) 5.6
d) 5.9×10^{-3}

Answer: b

13. If 0.10 mol/L formic acid is put in water, 4.15% dissociates to form formate ion and hydronium ion.

$$HCOOH(aq) + H_2O(\ell) \rightleftharpoons HCOO^-(aq) + H_3O^+(aq)$$

Calculate the equilibrium constant, K, for this reaction.
a) 4.3×10^{-2}
b) 5.6×10^3
c) 1.7×10^2
d) 1.8×10^{-4}

Answer: d

14. The reaction

$$PCl_5(g) \rightleftharpoons PCl_3(g) + Cl_2(g)$$

has an equilibrium constant, K_c, of 1.2 at 250 °C. If 0.300 mol PCl_5 is placed in a 1.0-L flask and allowed to reach equilibrium, what is the concentration of Cl_2?
a) 0.36 M
b) 0.16 M
c) 0.25 M
d) 1.2 M

Answer: c

15. Hydrogen and iodine react to form hydrogen iodide.

$$H_2(g) + I_2(g) \rightleftharpoons 2\ HI(g)$$

K_c is 0.504 at 25 °C. If initial concentrations of 0.170 M H_2 and 0.170 M I_2 are allowed to equilibrate, what is the equilibrium concentration of HI?
a) 0.0445 M
b) 0.0891 M
c) 0.0684 M
d) 0.0706 M

Answer: b

16. Hypobromous acid reacts with water to give H_3O^+ and hypobromite ion.

$$HOBr(aq) + H_2O(\ell) \rightleftharpoons H_3O^+(aq) + OBr^-(aq)$$

The equilibrium constant is 2.5×10^{-9}. If 0.50 mol/L HOBr is allowed to equilibrate with water, what is the equilibrium concentration of OBr^-?

a) 1.3×10^{-9} M
b) 7.1×10^{-5} M
c) 3.5×10^{-5} M
d) 5.0×10^{-5} M

Answer: c

17. Assume that the following endothermic chemical reaction is at equilibrium.

$$C(s) + H_2O(g) \rightleftharpoons H_2(g) + CO(g) \quad \Delta H^\circ = 131.3 \text{ kJ}$$

Which one of the following changes will cause the reaction to shift to the right, creating more products?

a) decreasing the temperature.
b) increasing the pressure.
c) removal of a gaseous product.
d) addition of solid carbon.

Answer: c

CHAPTER 17: Principles of Reactivity: The Chemistry of Acids and Bases

17.1 Acids, Bases, and the Equilibrium Concept

1. Acids that ionize extensively in solution are referred to as
   a) dissociators.
   b) weak electrolytes.
   c) Arrhenius acids.
   d) Brønsted-Lowry acids.
   e) strong acids.

   Answer: e

17.2 The Brønsted-Lowry concepts of Acids and Bases

2. According to the Brønsted-Lowry definition, an acid
   a) increases the H_3O^+ concentration in an aqueous solution.
   b) increases the OH^- concentration in an aqueous solution.
   c) is a proton acceptor.
   d) is a proton donor.
   e) is an electron pair-acceptor.

   Answer: d

3. According to the Brønsted-Lowry definition, a base
   a) increases the H_3O^+ concentration in an aqueous solution.
   b) increases the OH^- concentration in an aqueous solution.
   c) is a proton acceptor.
   d) is a proton donor.
   e) is an electron-pair donor.

   Answer: c

4. Which of the following is never a Brønsted-Lowry acid in an aqueous solution?
   a) hydrogen chloride, $HCl(g)$
   b) dihydrogen sulfide, $H_2S(g)$
   c) ammonium chloride, $NH_4Cl(s)$
   d) hydrogen fluoride, $HF(g)$
   e) magnesium oxide, $MgO(s)$

   Answer: e

5. Which of the following is never a Brønsted-Lowry base in an aqueous solution?
   a) hydrogen bromide, $HBr(g)$
   b) sodium dihydrogen phosphate, $NaH_2PO_4(s)$
   c) sodium phosphate, $Na_3PO_4(s)$
   d) ammonia, $NH_3(g)$
   e) sodium bicarbonate, $NaHCO_3(s)$

   Answer: a

6. In the following reaction

   $$HCO_3^-(aq) + H_2O(\ell) \rightleftharpoons CO_3^{2-}(aq) + H_3O^+(aq)$$

   a) HCO_3^- is an acid and H_2O is its conjugate base.
   b) HCO_3^- is an acid and CO_3^{2-} is its conjugate base.
   c) H_3O^+ is an acid and HCO_3^- is its conjugate base.
   d) H_2O is an acid and CO_3^{2-} is its conjugate base.
   e) H_3O^+ is an acid and CO_3^{2-} is its conjugate base.

   Answer: b

7. In the following reaction

   $$HF(aq) + HPO_4^{2-}(aq) \rightleftharpoons F^-(aq) + H_2PO_4^-(aq)$$

   a) HF is an acid and F^- is its conjugate base.
   b) HF is an acid and HPO_4^{2-} is its conjugate base.
   c) HPO_4^{2-} is an acid and $H_2PO_4^-$ is its conjugate base.
   d) $H_2PO_4^-$ is an acid and F^- is its conjugate base.
   e) HPO_4^{2-} is an acid and HF is its conjugate base.

   Answer: a

8. The conjugate base of OH^- is ________.
   a) H_3O^+ b) H_2O c) OH^- d) O^{2-} e) H^-

   Answer: d

9. What is the conjugate base of $[Cr(H_2O)_6]^{3+}(aq)$?
   a) H_3O^+
   b) $[Cr(H_2O)_5OH]^{2+}$
   c) $[Cr(H_2O)_5H_3O]^{4+}$
   d) $[Cr(H_2O)_6]^{2+}$
   e) $[Cr(H_2O)_5]^{3+}$

   Answer: b

10. The conjugate acid of HSO_4^- is ________.
a) OH^- b) SO_4^{2-} c) H_2SO_4 d) H_2O e) H_3O^+

Answer: c

11. Which acid is a potentially dangerous component of rhubarb leaves?
a) oxalic b) arsenous c) arsenic d) nitric e) selenic

Answer: a

12. Molecules or ions that can alternately behave as either a Brønsted-Lowry acid or base are called
a) weak electrolytes.
b) hydronium ions.
c) polyprotic acids or bases.
d) conjugate acids or bases.
e) amphiprotic.

Answer: e

13. Which is NOT an amphiprotic species in water?
a) $HC_2O_4^-$ b) $CH_3CO_2^-$ c) HPO_4^{2-} d) $H_2PO_4^-$ e) HCO_3^-

Answer: b

14. Which of the following molecules or ions is amphiprotic in water?
a) NH_4^+ b) HF c) CH_3CO_2H d) CN^- e) HCO_3^-

Answer: e

17.3 Water and the pH Scale

15. At 15 °C, the water ionization constant, K_w, is 4.5×10^{-15}. What is the H_3O^+ concentration in neutral water at this temperature?
a) 2.0×10^{-29} M b) 4.5×10^{-15} M c) 6.7×10^{-8} M d) 1.5×10^{-7} M e) 2.2 M

Answer: c

16. A solution is prepared by diluting 0.16 mol HNO_3 with water to a volume of 1.5 L. What is the pH of this solution?
a) -0.80 b) 0.80 c) 0.97 d) 1.83 e) 2.24

Answer: c

17. At 25 °C, what is the H_3O^+ concentration in 0.044 M NaOH(aq)? ($K_w = 1.0 \times 10^{-14}$)
a) 4.4×10^{-16} M b) 2.3×10^{-13} M c) 4.4×10^{-7} M d) 1.36 M e) 12.6 M

Answer: b

18. Which one of the following aqueous solutions will have a pH of 10.00 at 25 °C? ($K_w = 1.0 \times 10^{-14}$)
a) 1.0×10^{1} M HCl
b) 1.0×10^{2} M HNO_3
c) 1.0×10^{-4} M HBr
d) 1.0×10^{-4} M NaOH
e) 1.0×10^{2} M KOH

Answer: d

19. What is the pH of 0.28 M HNO_3(aq) at 25 °C? ($K_w = 1.0 \times 10^{-14}$)
a) -0.55 b) 0.28 c) 0.55 d) 1.27 e) 13.45

Answer: c

20. What is the pH of 2.1×10^{-5} M NaOH(aq) at 25 °C? ($K_w = 1.0 \times 10^{-14}$)
a) -4.68 b) 4.68 c) 3.23 d) 9.32 e) 10.77

Answer: d

21. What is the H_3O^+ concentration of an aqueous solution with a pH of 12.17?
a) 6.8×10^{-13} M b) 1.9×10^{-9} M c) 5.2×10^{-6} M d) 1.5×10^{-2} M e) 1.1 M

Answer: a

22. What is the OH^- concentration of an aqueous solution with a pH of 11.45? ($K_w = 1.0 \times 10^{-14}$)
a) 3.5×10^{-12} M b) 9.4×10^{-10} M c) 1.1×10^{-5} M d) 2.8×10^{-3} M e) 7.8×10^{-2} M

Answer: d

23. What is the OH^- concentration of an aqueous solution with a pH of -0.88? ($K_w = 1.0 \times 10^{-14}$)
a) 1.3×10^{-15} M b) 7.6×10^{-14} M c) 3.4×10^{-7} M d) 0.13 M e) 7.6 M

Answer: a

24. An aqueous solution with a pH of 3.00 is diluted from 2.0 L to 4.0 L. What is the pH of the diluted solution?
a) 1.50 b) 1.73 c) 3.00 d) 3.30 e) 6.00

Answer: d

25. An aqueous solution with a pH of 10.60 is diluted from 1.0 L to 1.5 L. What is the pH of the diluted solution?
a) 4.84 b) 7.07 c) 9.60 d) 10.42 e) 10.78

Answer: d

17.4 Equilibrium Constants for Acids and Bases

26. Which of the following chemical equations corresponds to the acid ionization constant, K_a, for formic acid (HCO_2H)?
a) $HCO_2H(aq) + H_2O(\ell) \rightleftharpoons HCO_2^-(aq) + H_3O^+(aq)$
b) $HCO_2H(aq) + H_3O^+(aq) \rightleftharpoons HCO_2H_2^+(aq) + H_2O(\ell)$
c) $HCO_2^-(aq) + H_2O(\ell) \rightleftharpoons HCO_2H(aq) + OH^-(aq)$
d) $HCO_2^-(aq) + H_3O^+(aq) \rightleftharpoons HCO_2H\ (aq) + H_2O(\ell)$
e) $HCO_2H(aq) + OH^-(aq) \rightleftharpoons HCO_2^-(aq) + H_2O(\ell)$

Answer: a

27. Which of the following chemical equations corresponds to the base ionization constant, K_b, for nitrite ion (NO_2^-)?
a) $HNO_2(aq) + H_2O(\ell) \rightleftharpoons NO_2^-(aq) + H_3O^+(aq)$
b) $NO_2^-(aq) + H_2O(\ell) \rightleftharpoons HNO_2(aq) + OH^-(aq)$
c) $HNO_2(aq) + OH^-(aq) \rightleftharpoons NO_2^-(aq) + H_2O(\ell)$
d) $NO_2^-(aq) + OH^-(aq) \rightleftharpoons HNO_3^{2-}(aq)$
e) $NO_2^-(aq) + H_3O^+(aq) \rightleftharpoons HNO_2(aq) + H_2O(\ell)$

Answer: b

28. Assuming equal initial concentrations of the given species, which of the following is the strongest acid in an aqueous solution?
a) benzoic acid, $K_a = 6.3 \times 10^{-5}$
b) formic acid, $K_a = 1.8 \times 10^{-4}$
c) hydrogen phosphate ion, $K_a = 3.6 \times 10^{-13}$
d) hydrogen sulfite ion, $K_a = 6.2 \times 10^{-8}$
e) nitrous acid, $K_a = 4.5 \times 10^{-4}$

Answer: e

29. Assuming equal initial concentrations of the given species, which of the following weak acids has the strongest conjugate base in an aqueous solution?
   a) acetic acid, $K_a = 1.8 \times 10^{-5}$
   b) formic acid, $K_a = 1.8 \times 10^{-4}$
   c) hydrogen sulfite ion, $K_a = 6.2 \times 10^{-8}$
   d) nitrous acid, $K_a = 4.5 \times 10^{-4}$
   e) phosphoric acid, $K_a = 7.5 \times 10^{-3}$

   Answer: c

30. Boric acid has a pK_a value of 9.14. What is the value of K_b for sodium benzoate?
   a) 7.2×10^{-10} b) 1.4×10^{-5} c) 1.1×10^{-4} d) -4.86 e) 4.86

   Answer: b

31. Which salt produces a pH-neutral aqueous solution at 25 °C?
   a) KBr b) NH_4Cl c) $NaNO_2$ d) KF e) $FeCl_3$

   Answer: a

32. Which salt forms a 0.10 M aqueous solution with the highest pH?
   a) $NaNO_3$ b) NH_4Cl c) $FeCl_3$ d) $Ca(ClO_4)_2$ e) NaF

   Answer: e

33. Which salt forms a 0.10 M aqueous solution with the lowest pH?
   a) $MgCO_3$ b) NaCl c) NaF d) NH_4Cl e) K_3PO_4

   Answer: d

17.5 Equilibrium Constants and Acid-Base Reactions

34. If you mix equal molar quantities of HF ($K_a = 7.2 \times 10^{-4}$) and KCN ($K_b = 2.5 \times 10^{-5}$), the resulting solution will be
   a) acidic because K_a of HF is greater than K_b of CN^-.
   b) basic because K_a of HF is greater than K_a of HCN.
   c) basic because K_a of HF is greater than K_b of CN^-.
   d) basic because K_b of F^- is less than K_b of CN^-.
   e) neutral because the weak acid neutralizes the weak base.

   Answer: a

17.6 Types of Acid-Base Reactions

35. Given the following acid dissociation constants,

K_a (HF) = 7.2×10^{-4}

K_a (NH_4^+) = 5.6×10^{-10}

determine the equilibrium constant for the reaction below at 25 °C.

$HF(aq) + NH_3(aq) \rightleftharpoons NH_4^+(aq) + F^-(aq)$

a) 4.0×10^{-13} b) 1.3×10^{-8} c) 7.8×10^{-7} d) 1.3×10^{6} e) 2.5×10^{12}

Answer: d

36. Given the following equilibrium constants,

K_a (NH_4^+) = 5.6×10^{-10}

K_b (NO_2^-) = 2.2×10^{-11}

$K_w = 1.00 \times 10^{-14}$

determine the equilibrium constant for the reaction below at 25 °C.

$NH_4^+(aq) + NO_2^-(aq) \rightleftharpoons HNO_2(aq) + NH_3(aq)$

a) 1.2×10^{-34} b) 2.5×10^{-13} c) 1.2×10^{-6} d) 8.1×10^{5} e) 3.9×10^{12}

Answer: c

17.7 Calculations with Equilibrium Constants

37. What is the H_3O^+ concentration in 0.45 M HCN(aq)? (K_a of HCN = 4.0×10^{-10})

a) 4.0×10^{-10} M b) 1.8×10^{-10} M c) 1.3×10^{-5} M d) 2.0×10^{-5} M e) 0.45 M

Answer: c

38. What is the OH^- concentration in 0.51 M $CH_3CO_2^-$(aq)? (K_b of $CH_3CO_2^-$ = 5.6×10^{-10})

a) 2.8×10^{-10} M b) 5.6×10^{-10} M c) 5.9×10^{-10} M d) 1.7×10^{-5} M e) 1.8×10^{-5} M

Answer: d

39. What is the pH of 0.015 M aqueous benzoic acid? (K_a of $C_5H_6CO_2H$ = 6.3×10^{-5})

a) 1.82 b) 3.03 c) 4.20 d) 6.02 e) 12.18

Answer: b

40. What is the pH of 9.2×10^{-3} M aqueous cyanide ion? (K_b of CN^- = 2.5×10^{-5})

a) 2.04 b) 3.33 c) 4.60 d) 9.40 e) 10.67

Answer: e

41. A solution is made by diluting 0.50 mol NaClO to a volume of 3.0 L with water. What is the pH of the solution? (K_b of $ClO^- = 2.9 \times 10^{-7}$)
a) 3.66 b) 7.46 c) 10.34 d) 10.58 e) 13.22

Answer: c

42. The pH of aqueous 0.050 M trimethylamine is 11.24. What is the K_b of this base?
a) 6.6×10^{-22} b) 5.8×10^{-12} c) 6.3×10^{-5} d) 1.7×10^{-3} e) 2.76

Answer: c

43. The pH of aqueous 0.30 M hypochlorous acid, HClO, is 3.99. What is the pK_a of this acid?
a) 6.54 b) 7.46 c) 10.01 d) 12.80 e) 13.30

Answer: b

44. What is the pH of the solution which results from mixing 25 mL of 0.30 M HF(aq) and 25 mL of 0.30 M NaOH(aq) at 25 °C? (K_a of HF $= 7.2 \times 10^{-4}$)
a) 5.84 b) 8.16 c) 8.31 d) 10.86 e) 12.02

Answer: b

45. What is the pH of the solution which results from mixing 35 mL of 0.50 M NH_3(aq) and 35 mL of 0.50 HCl(aq) at 25 °C? (K_b for $NH_3 = 1.8 \times 10^{-5}$)
a) 0.60 b) 2.67 c) 4.74 d) 4.93 e) 9.26

Answer: d

17.8 Polyprotic Acids and Bases

46. Which of the following chemical equations corresponds to K_{a3} for phosphoric acid?
a) $HPO_4^{2-}(aq) + H_2O(\ell) \rightleftharpoons PO_4^{3-}(aq) + H_3O^+(aq)$
b) $PO_4^{3-}(aq) + H_2O(\ell) \rightleftharpoons HPO_4^{2-}(aq) + OH^-(aq)$
c) $H_3PO_4(aq) + H_2O(\ell) \rightleftharpoons H_2PO_4^-(aq) + H_3O^+(aq)$
d) $H_3PO_4(aq) + 3\,H_2O(\ell) \rightleftharpoons PO_4^{3-}(aq) + 3\,H_3O^+(aq)$
e) $H_2PO_4^-(aq) + H_2O(\ell) \rightleftharpoons HPO_4^{2-}(aq) + H_3O^+(aq)$

Answer: a

47. Which of the following chemical equations corresponds to K_{b1} for SO_3^{2-}?
a) $HSO_3^-(aq) + H_2O(\ell) \rightleftharpoons H_2SO_3(aq) + OH^-(aq)$
b) $SO_3^{2-}(aq) + H_3O^+(aq) \rightleftharpoons HSO_3^-(aq) + H_2O(\ell)$
c) $H_2SO_3(aq) + OH^-(aq) \rightleftharpoons HSO_3^-(aq) + H_2O(\ell)$
d) $HSO_3^-(aq) + OH^-(aq) \rightleftharpoons SO_3^{2-}(aq) + H_2O(\ell)$
e) $SO_3^{2-}(aq) + H_2O(\ell) \rightleftharpoons HSO_3^-(aq) + OH^-(aq)$

Answer: e

48. What is the pH of 0.80 M Na_3PO_4(aq) at 25 °C? ($K_{a1} = 7.5 \times 10^{-3}$, $K_{a2} = 6.2 \times 10^{-8}$, and $K_{a3} = 3.6 \times 10^{-13}$)
a) 0.87 b) 6.27 c) 7.73 d) 8.01 e) 13.13

Answer: e

17.9 The Lewis Concept of Acids and Bases

49. Of the following list of molecules and ions, which can behave as Lewis bases: BH_3, NH_3, Cl^-, Al^{3+}?
a) BH_3 only b) NH_3 and Cl^- c) Al^{3+} and NH_3 d) Cl^- and BH_3 e) Cl^- and Al^{3+}

Answer: b

50. All of the following species can behave as Lewis acids EXCEPT ________.
a) F^- b) Al^{3+} c) SO_3 d) $Al(OH)_3$ e) CO_2

Answer: a

17.10 Molecular Structure, Bonding, and Acid-Base Behavior

51. All of the following compounds are acids containing chlorine. Which compound is the weakest acid?
a) HCl b) HClO c) $HClO_2$ d) $HClO_3$ e) $HClO_4$

Answer: b

52. Which of the following molecules or ions is the strongest acid?
a) $CH_3CO_2^-$ b) CH_3CO_2H c) $CBrH_2CO_2H$ d) CBr_2HCO_2H e) CBr_3CO_2H

Answer: e

Short Answer Questions

53. When a Lewis acid combines with a Lewis base, the base supplies both electrons to the bond. This type of chemical bond is called a ________ covalent bond.

    Answer: coordinate

54. The acid strength of an oxoacid increases as the number of oxygen atoms attached to the central atom increases. For example, HNO_3 is a stronger acid than HNO_2. One explanation is that the electrons in the O-H bond are attracted by nearby electronegative atoms. The extent to which nearby atoms attract electrons from other parts of a molecule is called the ________ effect.

    Answer: inductive

55. The chemical equations below show the reaction of $Al(OH)_3$ as a Lewis acid and as a Lewis base, respectively.

    $$Al(OH)_3(s) + OH^-(aq) \rightleftharpoons Al(OH)_4^-(aq)$$
    $$Al(OH)_3(s) + 3\ H_3O^+(aq) \rightleftharpoons Al^{3+}(aq) + 3\ H_2O(\ell)$$

    Substances that can behave as either Lewis acids or bases are called ________ substances.

    Answer: amphoteric

56. Write a net ionic equation for the neutralization reaction of ammonium chloride and sodium hydroxide. Identify the spectator ions.

    Answer: Net ionic equation: $NH_4^+(aq) + OH^-(aq) \rightleftharpoons NH_3(aq) + H_2O(\ell)$; spectator ions: Cl^- and Na^+.

57. Which is the stronger Brønsted-Lowry acid, $Fe(H_2O)_6^{2+}$ or $Fe(H_2O)_6^{3+}$? Explain.

    Answer: $Fe(H_2O)_6^{3+}$. When a proton dissociates from the oxygen atom, the oxygen is left with an extra electron. This negative charge is stabilized by the interaction of the oxygen with the iron cation. Since the Fe^{3+} is more positively charged than the Fe^{2+}, it can more effectively stabilize the negative charge.

Diagnostic Quiz Questions

Goal #1: Use the Brønsted-Lowry and Lewis theories of acids and bases

1. According to the Brønsted-Lowry definition, an acid
   a) increases the H^+ concentration in water.
   b) increases the OH^- concentration in water.
   c) is a proton acceptor.
   d) is a proton donor.

   Answer: d

2. Which of the following compounds is a weak acid?
   a) HCl
   b) HNO_3
   c) HF
   d) H_2SO_4

   Answer: c

3. Which of the following molecules or ions is amphiprotic?
   a) H_3PO_4
   b) NH_4^+
   c) HCO_3^-
   d) HBr

   Answer: c

4. In the following reaction

   $$H_2S(aq) + H_2O(\ell) \rightleftharpoons HS^-(aq) + H_3O^+(aq)$$

   a) H_2O is an acid and H_2S is its conjugate base.
   b) H_2S is an acid and H_2O is its conjugate base.
   c) H_2O is an acid and H_3O^+ is its conjugate base.
   d) H_2S is an acid and HS^- is its conjugate base.

   Answer: d

5. What are the hydronium ion and hydroxide ion concentrations in 3.5×10^{-2} M HNO_3(aq) at 25 °C?
   a) $[H_3O^+] = 3.5 \times 10^{-2}$ M, $[OH^-] = 2.9 \times 10^{-13}$ M
   b) $[H_3O^+] = 3.5 \times 10^{-2}$ M, $[OH^-] = 3.5 \times 10^{-2}$ M
   c) $[H_3O^+] = 3.5 \times 10^{-2}$ M, $[OH^-] = 3.5 \times 10^{12}$ M
   d) $[H_3O^+] = 2.9 \times 10^{-13}$ M, $[OH^-] = 3.5 \times 10^{-2}$ M

   Answer: a

6. What is the pH of 4.5×10^{-3} M NaOH(aq) at 25 °C?
   a) 2.35
   b) 5.40
   c) 8.60
   d) 11.65

   Answer: d

Goal #2: Apply the principles of chemical equilibrium to acids and bases in aqueous solution

7. Which of the following compounds will give rise to an acidic solution in water?
   a) KNO_3
   b) Na_2CO_3
   c) NH_4Cl
   d) $NaCH_3CO_2$

   Answer: c

8. Which of the following species will give rise to a basic solution in water?
   a) NH_4Br
   b) $Al(H_2O)_6^{3+}$
   c) KCN
   d) $NaHSO_4$

   Answer: c

9. The K_a for acetic acid is 1.8×10^{-5} at 25 °C. What is the pK_b for acetate ion?
   a) 1.80
   b) 4.74
   c) 9.26
   d) 12.20

   Answer: c

10. The pH of a 0.351 M CN^- solution is 11.47. What is the K_b for CN^-?
    a) 3.3×10^{-23}
    b) 9.7×10^{-12}
    c) 2.5×10^{-5}
    d) 8.4×10^{-3}

    Answer: c

11. Which of the following chemical reactions corresponds to the acid ionization constant, K_a, for hydrogen carbonate ion (HCO_3^-)?
   a) $HCO_3^-(aq) + OH^-(aq) \rightleftharpoons CO_3^{2-}(aq) + H_2O(\ell)$
   b) $HCO_3^-(aq) + H_3O^+(aq) \rightleftharpoons CO_2(g) + 2\,H_2O(\ell)$
   c) $HCO_3^-(aq) + H_3O^+(aq) \rightleftharpoons H_2CO_3(aq) + H_2O(\ell)$
   d) $HCO_3^-(aq) + H_2O(\ell) \rightleftharpoons CO_3^{2-}(aq) + H_3O^+(aq)$

Answer: d

12. What is the pH of 0.50 M $CH_3CO_2^-$ at 25 °C? The K_a for acetic acid, CH_3CO_2H, is 1.8×10^{-5}.
   a) 2.52
   b) 4.78
   c) 9.22
   d) 11.48

Answer: c

13. What is the pH of 5.0×10^{-3} M HF? The K_a for hydrofluoric acid is 7.2×10^{-4}.
   a) 2.72
   b) 2.80
   c) 4.60
   d) 5.44

Answer: b

14. What is the pH of 0.20 M H_3PO_4? The K_a values for phosphoric acid are $K_{a1} = 7.5 \times 10^{-3}$, $K_{a2} = 6.2 \times 10^{-8}$, and $K_{a3} = 3.6 \times 10^{-13}$.
   a) 1.45
   b) 1.41
   c) 2.12
   d) 2.82

Answer: a

Goal #3: Predict the outcome of reactions of acids and bases.

15. Which of the following chemical equations is a net ionic equation for the reaction between HCl and a weak base?
    a) $2\ HCl(aq) + Sr(OH)_2(aq) \rightarrow SrCl_2(aq) + 2\ H_2O(\ell)$
    b) $HCl(aq) + NaF(aq) \rightarrow NaCl(aq) + HF(aq)$
    c) $H_3O^+(aq) + F^-(aq) \rightarrow HF(aq) + H_2O(\ell)$
    d) $H_3O^+(aq) + OH^-(aq) \rightarrow 2\ H_2O(\ell)$

    Answer: c

16. Calculate the pH after mixing 25 mL of 0.70 M HCl and 25 mL of 0.70 M $NaNO_2$. The K_a for nitrous acid is 4.5×10^{-4}.
    a) 0.46
    b) 1.75
    c) 1.91
    d) 2.21

    Answer: c

Goal #4: Understand the influence of structure and bonding on acid-base properties

17. Which of the following molecules or ions is a Lewis acid?
    a) CH_3NH_2
    b) NH_3
    c) Fe^{3+}
    d) OH^-

    Answer: c

18. Which of the following acids has the largest K_a?
    a) CH_3CO_2H
    b) $ClCH_2CO_2H$
    c) Cl_2CHCO_2H
    d) Cl_3CCO_2H

    Answer: d

CHAPTER 18: Principles of Reactivity: Other Aspects of Aqueous Equilibria

18.1 The Common Ion Effect

1. What is the effect of adding 10 mL of 0.1 M NaOH(aq) to 100 mL of 0.2 M NH_4^+(aq)?
   1. The pH will decrease.
   2. The concentration of NH_3 will increase.
   3. The concentration of NH_4^+ will decrease.

   a) 1 only b) 2 only c) 3 only d) 2 and 3 e) 1, 2, and 3

   Answer: d

2. Given a solution of 0.10 M NH_3(aq), what is the effect of adding NH_4Cl(s) to this solution?
   1. The pH will decrease.
   2. The concentration of NH_3 will increase.
   3. The concentration of H_3O^+ will increase.

   a) 1 only b) 2 only c) 3 only d) 1 and 3 e) 1, 2, and 3

   Answer: e

3. What is the pH of a solution that results from adding 25 mL of 0.15 M HCl to 25 mL of 0.52 M NH_3? (K_b of $NH_3 = 1.8 \times 10^{-5}$)

   a) 2.74 b) 4.35 c) 9.65 d) 11.26 e) 11.41

   Answer: c

4. What is the pH of a solution that results from adding 25 mL of 0.50 M NaOH to 75 mL of 0.50 M CH_3CO_2H? (K_a of $CH_3CO_2H = 1.8 \times 10^{-5}$)

   a) 2.67 b) 3.17 c) 4.44 d) 5.04 e) 5.35

   Answer: c

5. What is the pH of an aqueous solution of 0.30 M HF and 0.15 M F^-? (K_a of HF $= 7.2 \times 10^{-4}$)

   a) 1.83 b) 2.84 c) 3.14 d) 3.44 e) 10.86

   Answer: b

6. What is the pH of an aqueous solution containing 0.50 M CO_3^{2-} and 0.20 M HCO_3^-? (K_b of $CO_3^{2-} = 2.1 \times 10^{-4}$)

   a) 3.28 b) 3.68 c) 10.32 d) 10.72 e) 12.01

   Answer: d

18.2: Controlling pH: Buffer Solutions

7. Which of the following combinations would be best to buffer an aqueous solution at a pH of 2.0?
   a) H_3PO_4 and $H_2PO_4^-$, $K_{a1} = 7.5 \times 10^{-3}$
   b) HNO_2 and NO_2^-, $K_a = 4.5 \times 10^{-4}$
   c) CH_3CO_2H and CH_3COO^-, $K_a = 1.8 \times 10^{-5}$
   d) $H_2PO_4^-$ and HPO_4^{2-}, $K_{a2} = 6.2 \times 10^{-8}$
   e) NH_4^+ and NH_3, $K_a = 5.7 \times 10^{-10}$

   Answer: a

8. Which of the following combinations would be the best to buffer an aqueous solution at a pH of 5.0?
   a) H_3PO_4 and $H_2PO_4^-$, $K_{a1} = 7.5 \times 10^{-3}$
   b) HNO_2 and NO_2^-, $K_a = 4.5 \times 10^{-4}$
   c) CH_3CO_2H and CH_3COO^-, $K_a = 1.8 \times 10^{-5}$
   d) $H_2PO_4^-$ and HPO_4^{2-}, $K_{a2} = 6.2 \times 10^{-8}$
   e) NH_4^+ and NH_3, $K_a = 5.7 \times 10^{-10}$

   Answer: c

9. Each of the following mixtures can produce an effect buffer solution EXCEPT
   a) $HClO_4$ and $NaClO_4$.
   b) HF and NaF.
   c) $NaHCO_3$ and Na_2CO_3.
   d) Na_2HPO_4 and Na_3PO_4.
   e) NH_4Cl and NH_3.

   Answer: a

10. All of the following statements concerning buffers are true EXCEPT
    a) buffers are resistant pH changes upon addition of small quantities of strong acids or bases.
    b) buffers are used as colored indicators in acid-base titrations.
    c) the pH of a buffer is close to the pK_a of the weak acid from which it is made.
    d) buffers contain appreciable quantities of a weak acid and its conjugate base.
    e) buffers are resistant to changes in pH when diluted with water.

    Answer: b

11. Which of the following mathematical expressions is the Henderson-Hasselbalch equation?

a) $pK_a = pH + \log\left(\frac{[\text{conjugate base}]}{[\text{acid}]}\right)$

b) $pH = pK_a + \log\left(\frac{[OH^-]}{[H_3O^+]}\right)$

c) $pH = pK_a + \log\left(\frac{[\text{acid}]}{[\text{conjugate base}]}\right)$

d) $pK_a = pH - \log\left(\frac{[\text{acid}]}{[\text{conjugate base}]}\right)$

e) $pH = pK_a + \log\left(\frac{[\text{conjugate base}]}{[\text{acid}]}\right)$

Answer: e

12. What is the pH of the buffer that results when 4.0 g of NH_3 and 8.0 g of NH_4Cl are diluted with water to a volume of 0.50 L? (K_a of $NH_4^+ = 5.6 \times 10^{-10}$)

a) 8.95 b) 9.06 c) 9.25 d) 9.45 e) 9.55

Answer: d

13. What is the pH of the buffer that results when 11 g of $NaCH_3CO_2$ is mixed with 85 mL of 1.0 M CH_3CO_2H and diluted with water to 1.0 L? (K_a of $CH_3CO_2H = 1.8 \times 10^{-5}$)

a) 2.91 b) 3.86 c) 4.55 d) 4.74 e) 4.94

Answer: e

14. What is the pH of the buffer that results when 12.5 g of NaH_2PO_4 and 22.0 g Na_2HPO_4 are mixed and diluted with water to 0.500 L? (The acid dissociation constants for phosphoric acid are $K_{a1} = 7.5 \times 10^{-3}$, $K_{a2} = 6.2 \times 10^{-8}$, and $K_{a3} = 3.6 \times 10^{-13}$)

a) 2.30 b) 7.04 c) 7.38 d) 12.27 e) 12.62

Answer: c

15. What is the pH of a buffer that results when 0.50 mole of Na_3PO_4 is mixed with 0.75 mole of HCl and diluted with water to 1.00 L? (The acid dissociation constants for phosphoric acid are $K_{a1} = 7.5 \times 10^{-3}$, $K_{a2} = 6.2 \times 10^{-8}$, and $K_{a3} = 3.6 \times 10^{-13}$)

a) 7.03 b) 7.21 c) 7.38 d) 12.27 e) 12.44

Answer: b

16. What is the pH of a buffer that results when 33 g $NaHCO_3$ is mixed with 100.0 mL of 2.50 M NaOH and diluted with water to 250 mL? (K_a for $HCO_3^- = 4.8 \times 10^{-11}$)
a) 8.20 b) 10.07 c) 10.12 d) 10.32 e) 10.56

Answer: e

17. What mass of solid KF (molar mass = 58.1 g/mol) should be added to 2.0 L of 0.25 M HF to make a buffer with a pH of 3.14? (pK_a for HF = 3.14)
a) 7.3 g b) 15 g c) 29 g d) 58 g e) 120 g

Answer: c

18. If the ratio of acid to base in a buffer increases by a factor of 10, the pH of the buffer
a) decreases by 1.
b) decreases by 10.
c) increases by 10.
d) increases by 1.
e) remains unchanged.

Answer: a

19. The K_a of hypochlorous acid, HClO, is 3.5×10^{-8}. What $[ClO^-]/[HClO]$ ratio is necessary to make a buffer with a pH of 7.71?
a) 2.0×10^{-8} b) 0.25 c) 0.56 d) 1.8 e) 3.9

Answer: d

20. A buffer is prepared by combining 250 mL of 0.25 M NaOH and 250 mL of a 0.600 M weak acid, HA. If the pH of the buffer is 6.33, what is the pK_a of the acid?
a) 5.95 b) 6.18 c) 6.33 d) 6.48 e) 6.71

Answer: d

21. A buffer is composed of 0.250 mol $H_2PO_4^-$ and 0.250 mol HPO_4^{2-} diluted with water to a volume of 1.00 L. The pH of the buffer is 7.210. How many moles of NaOH must be added to increase the pH to 8.210?
a) 0.045 mol b) 0.125 mol c) 0.205 mol d) 0.250 mol e) 0.500 mol

Answer: c

22. How many moles of HCl must be added to 1.00 L of 0.72 M NH_3 to make a buffer with a pH of 9.50? (K_a of $NH_4^+ = 5.6 \times 10^{-10}$)
a) 0.26 mol b) 0.31 mol c) 0.41 mol d) 0.46 mol e) 1.3 mol

Answer: a

23. A buffer contains 1.0 mol CH_3CO_2H and 1.0 mol $CH_3CO_2^-$ diluted with water to 1.0 L. How many moles of NaOH are required to increase the pH of the buffer to 5.10? (pK_a of CH_3CO_2H = 4.74)
a) 0.36 mol b) 0.39 mol c) 0.44 mol d) 0.61 mol e) 0.64 mol

Answer: b

24. What volume of 0.50 M NaOH should be added to 2.0 L of 0.25 M HCO_3^- to make a buffer with a pH of 10.02? (pK_a of HCO_3^- = 10.32)
a) 0.17 mL b) 83 mL c) 2.5×10^2 mL d) 3.3×10^2 mL e) 5.0×10^2 mL

Answer: d

18.3 Acid-Base Titrations

25. Which one of the following conditions is always met at the equivalence point of the titration of a monoprotic weak base with a strong acid?
a) The pH of the solution is equal to 7.00.
b) The volume of acid added from the buret equals the volume of base titrated.
c) The molarity of the acid equals the initial molarity of the weak base.
d) The percent ionization of the acid equals the percent ionization of the base.
e) The moles of acid added from the buret equals the initial moles of weak base.

Answer: e

26. Which one of the following conditions is always true for a titration of a weak acid with a strong base?
a) A colored indicator with a pK_a less than 7 should be used.
b) If a colored indicator is used, it must change color rapidly in the weak acid's buffer region.
c) Equal volumes of weak acid and strong base are required to reach the equivalence point.
d) The equivalence point occurs at a pH equal to 7.
e) The equivalence point occurs at a pH greater than 7.

Answer: e

27. A volume of 25.0 mL of 0.100 M $HCO_2H(aq)$ is titrated with 0.100 M NaOH(aq). What is the pH after the addition of 12.5 mL of NaOH? (K_a for $HCO_2H = 1.8 \times 10^{-4}$)
a) 2.52 b) 3.74 c) 4.74 d) 7.00 e) 10.26

Answer: b

28. A 25.0 mL sample of 0.200 M HClO(aq) is titrated with 0.250 M NaOH(aq). What is the pH of a solution when 85.0% of the acid is titrated? (K_a of HClO = 3.5×10^{-8})
a) 6.70 b) 7.39 c) 7.46 d) 7.55 e) 8.21

Answer: e

29. A 25.0 mL sample of 0.0200 M $NH_3(aq)$ is titrated with 0.0100 M HCl(aq). What is the pH at the equivalence point? (K_b of $NH_3 = 1.8 \times 10^{-5}$)
a) 3.46 b) 5.48 c) 5.72 d) 8.25 e) 10.54

Answer: c

30. A 50.0 mL sample of 0.0240 M $NH_3(aq)$ is titrated with aqueous hydrochloric acid. What is the pH after the addition of 15.0 mL of 0.0600 M HCl(aq)? (K_b of $NH_3 = 1.8 \times 10^{-5}$)
a) 8.78 b) 8.86 c) 9.25 d) 9.38 e) 9.73

Answer: a

31. A 20.0 mL sample of 0.20 M sodium acetate is titrated with 0.11 M HCl(aq). What is the pH after the addition of 50.0 mL HCl(aq)? (K_b of $CH_3CO_2^- = 5.6 \times 10^{-10}$)
a) 1.10 b) 1.67 c) 2.82 d) 9.11 e) 9.51

Answer: b

32. Potassium hydrogen phthalate (molar mass = 204.2 g/mol) is used to standardize sodium hydroxide. If 26.37 mL of NaOH(aq) is required to titrate 0.7719 g KHP to the equivalence point, what is the concentration of the NaOH(aq)?

$$HC_8H_4O_4^-(aq) + OH^-(aq) \rightleftharpoons C_8H_4O_4^{2-}(aq) + H_2O(\ell)$$

a) 0.02036 M b) 0.02937 M c) 0.09968 M d) 0.1433 M e) 5.977 M

Answer: d

33. A 50.0 mL sample of vinegar is titrated with 0.774 M NaOH(aq). If the titration requires 41.6 mL of NaOH(aq), what is the concentration of acetic acid in the vinegar?
a) 0.0921 M b) 0.429 M c) 0.644 M d) 0.930 M e) 0.967 M

Answer: c

34. An impure sample of sodium carbonate, Na_2CO_3, is titrated with 0.150 M HCl according to the reaction below.

$$2\,HCl(aq) + Na_2CO_3(aq) \rightleftharpoons CO_2(g) + H_2O(\ell) + 2\,NaCl(aq)$$

What is the percent of Na_2CO_3 in a 0.927 g sample if the titration requires 27.3 mL of HCl? The molar mass of Na_2CO_3 is 106.0 g/mol.
a) 0.221% b) 23.4% c) 46.8% d) 93.7% e)104%

Answer: b

35. Which is the best colored indicator to use in the titration of 0.0010 M $CH_3CO_2^-(aq)$ with HCl(aq)? Why? (K_b of $CH_3CO_2^- = 5.6 \times 10^{-10}$)

Indicator	pK_a
Bromocresol green	4.7
Phenol Red	7.8
Phenolphthalein	9.0

a) Bromocresol green. The pH at the equivalence point is less than 7.0.
b) Phenol Red. The pK_b of acetate ion and the pK_b of the indicator are similar.
c) Phenol Red. The equivalence point of an acid-base titration occurs at a pH of 7.0.
d) Phenolphthalein. The pK_b of acetate ion and the pK_b of the indicator are similar.
e) Phenolphthalein. The pH at the equivalence point is greater than 7.0.

Answer: a

36. What color change is exhibited by phenolphthalein during a titration of aqueous acetic acid with aqueous sodium hydroxide?
a) colorless to pink
b) pink to colorless
c) green to yellow
d) yellow to blue
e) blue to yellow

Answer: a

18.4 Solubility of Salts

37. Which of the following equations is the solubility product for magnesium iodate, $Mg(IO_3)_2$?
a) $K_{sp} = [Mg^{2+}][I^-]^2[O^{2-}]^6$
b) $K_{sp} = [Mg^{2+}][I^-]^2[3O^{2-}]^2$
c) $K_{sp} = [Mg^{2+}][IO_3^-]$
d) $K_{sp} = [Mg^{2+}]^2[IO_3^-]$
e) $K_{sp} = [Mg^{2+}][IO_3^-]^2$

Answer: e

38. The solubility of $SrSO_4$ in water is 0.107 g in 1.0 L at 25 °C. What is the value of K_{sp} for $SrSO_4$?
a) 3.4×10^{-7} b) 5.8×10^{-4} c) 1.2×10^{-3} d) 1.1×10^{-2} e) 2.1×10^{-1}

Answer: a

39. The solubility of $PbCl_2$ in water is 4.4 g/L at 25 °C. What is the value of K_{sp} for $PbCl_2$?
a) 4.0×10^{-6} b) 1.6×10^{-5} c) 2.5×10^{-4} d) 5.0×10^{-4} e) 3.4×10^{2}

Answer: b

40. The K_{sp} of $PbBr_2$ is 6.6×10^{-6} at 25 °C. What is the concentration of Br^- in a saturated solution of $PbBr_2$(aq)?
a) 2.6×10^{-3} M b) 1.2×10^{-2} M c) 1.9×10^{-2} M d) 2.4×10^{-2} M e) 3.8×10^{-2} M

Answer: d

41. The K_{sp} of $CaSO_4$ is 4.9×10^{-5} at 25 °C. What is the concentration of Ca^{2+} in a saturated solution of $CaSO_4$?
a) 2.4×10^{-9} M b) 2.5×10^{-5} M c) 4.9×10^{-5} M d) 7.0×10^{-3} M e) 3.7×10^{-2} M

Answer: d

42. The K_{sp} of AgBr is 5.4×10^{-13} at 25 °C. Calculate the molar solubility of AgBr in 0.015 M NaBr(aq) at 25 °C.
a) 8.1×10^{-15} mol/L
b) 3.6×10^{-11} mol/L
c) 1.1×10^{-8} mol/L
d) 7.3×10^{-7} mol/L
e) 4.9×10^{-5} mol/L

Answer: b

43. What is the molar solubility of Ag_2CrO_4 in 0.0050 M K_2CrO_4 at 25 °C? The value of K_{sp} for Ag_2CrO_4 is 5.4×10^{-12} at 25 °C.
a) 1.6×10^{-5} mol/L
b) 2.2×10^{-5} mol/L
c) 3.3×10^{-5} mol/L
d) 6.6×10^{-5} mol/L
e) 1.1×10^{-4} mol/L

Answer: a

44. At 25 °C, only 1.04 mg $Cu(NO_3)_2$ will dissolve per liter of a solution that is buffered at pH 6.80. What is the value of K_{sp} for $Cu(OH)_2$? The molar mass of $Cu(NO_3)_2$ is 187.6 g/mol.
a) 2.2×10^{-20} b) 1.4×10^{-19} c) 4.1×10^{-18} d) 3.5×10^{-13} e) 8.8×10^{-13}

Answer: a

45. What is the molar solubility of $Fe(OH)_3(s)$ in a solution that is buffered at pH 2.75 at 25 °C? The K_{sp} of $Fe(OH)_3$ is 6.3×10^{-38} at 25 °C.
a) 1.1×10^{-29} mol/L
b) 1.1×10^{-26} mol/L
c) 2.0×10^{-15} mol/L
d) 2.2×10^{-10} mol/L
e) 3.5×10^{-4} mol/L

Answer: e

18.5 Precipitation Reactions

46. At what pH will an aqueous solution of 0.210 M Mn^{2+} begin to precipitate as $Mn(OH)_2$ at 25 °C? The K_{sp} for $Mn(OH)_2$ is 1.9×10^{-13} at 25 °C.
a) 3.94 b) 6.02 c) 7.98 d) 9.86 e) 10.06

Answer: c

47. The concentration of Ca^{2+} in an aqueous solution is 2.2×10^{-3} M. What concentration of CO_3^{2-} is required to begin precipitating $CaCO_3$? The K_{sp} of $CaCO_3$ is 3.4×10^{-9}.
a) 7.5×10^{-12} M b) 1.5×10^{-6} M c) 5.8×10^{-5} M d) 7.0×10^{-4} M e) 2.7×10^{-2} M

Answer: b

48. What mass of Na_2SO_4 (molar mass = 142.0 g/mol) must be added to 350 mL of 0.16 M Ag^+ to initiate precipitation of Ag_2SO_4? The K_{sp} of Ag_2SO_4 is 1.2×10^{-5}. Assume no volume change occurs upon addition of Na_2SO_4.
a) 4.7×10^{-4} g b) 3.7×10^{-3} g c) 1.1×10^{-2} g d) 2.3×10^{-2} g e) 6.7×10^{-2} g

Answer: d

49. An aqueous solution contains 0.010 M Br^- and 0.010 M I^-. If Ag^+ is added until AgBr(s) just begins to precipitate, what are the concentrations of Ag^+ and I^-? (K_{sp} of AgBr = 5.4×10^{-13}, K_{sp} of AgI = 8.5×10^{-17})
a) $[Ag^+] = 5.4 \times 10^{-11}$ M, $[I^-] = 1.0 \times 10^{-2}$ M
b) $[Ag^+] = 8.5 \times 10^{-15}$ M, $[I^-] = 1.0 \times 10^{-2}$ M
c) $[Ag^+] = 5.4 \times 10^{-11}$ M, $[I^-] = 1.6 \times 10^{-6}$ M
d) $[Ag^+] = 8.5 \times 10^{-15}$ M, $[I^-] = 6.4 \times 10^{1}$ M
e) $[Ag^+] = 8.5 \times 10^{-15}$ M, $[I^-] = 1.6 \times 10^{-6}$ M

Answer: c

50. The following anions can be separated by precipitation as silver salts: Cl^-, Br^-, I^-, CrO_4^{2-}. If Ag^+ is added to a solution containing the four anions, each at a concentration of 0.10 M, in what order will they precipitate?

Compound	K_{sp}
AgCl	1.8×10^{-10}
Ag_2CrO_4	1.1×10^{-12}
AgBr	5.4×10^{-13}
AgI	8.5×10^{-17}

a) $AgCl \rightarrow Ag_2CrO_4 \rightarrow AgBr \rightarrow AgI$
b) $AgI \rightarrow AgBr \rightarrow Ag_2CrO_4 \rightarrow AgCl$
c) $Ag_2CrO_4 \rightarrow AgCl \rightarrow AgBr \rightarrow AgI$
d) $Ag_2CrO_4 \rightarrow AgI \rightarrow AgBr \rightarrow AgCl$
e) $AgI \rightarrow AgBr \rightarrow AgCl \rightarrow Ag_2CrO_4$

Answer: e

18.6 Solubility and Complex Ions

51. Consider the reaction

$Zn(OH)_2(s) + 2\ OH^-(aq) \rightleftharpoons Zn(OH)_4^{2-}(aq)$ $\qquad K = 8.7 \times 10^{-2}$

If K_{sp} for $Zn(OH)_2$ is 3.0×10^{-17}, what is the value of the formation constant, K_{form}, for the reaction below?

$Zn^{2+}(aq) + 4\ OH^-(aq) \rightleftharpoons Zn(OH)_4^{2-}(aq)$

a) 2.6×10^{-18} b) 3.4×10^{-16} c) 2.9×10^{15} d) 3.3×10^{16} e) 3.8×10^{17}

Answer: c

52. Given the following reactions,

$AgCl(s) \rightleftharpoons Ag^+(aq) + Cl^-(aq)$ $\qquad K_{sp} = 1.8 \times 10^{-10}$

$Ag^+(aq) + 2\ NH_3(aq) \rightleftharpoons Ag(NH_3)_2^+(aq)$ $\qquad K_{form} = 1.6 \times 10^7$

determine the equilibrium constant for the reaction below.

$AgCl(s) + 2\ NH_3(aq) \rightleftharpoons Ag(NH_3)_2^+(aq) + Cl^-(aq)$

a) 1.1×10^{-17} b) 2.9×10^{-3} c) 3.5×10^{2} d) 1.6×10^{7} e) 8.9×10^{16}

Answer: b

Short Answer Questions

53. The buffer that controls the pH of blood is predominantly composed of a mixture of HCO_3^- and ________.

Answer: H_2CO_3

54. For a monoprotic acid titration, the ________ point in a titration is the point where the moles of strong base added equals the moles of acid initially present.

Answer: equivalence

55. To increase the pH of a buffer by 1.00, you must increase the ratio of base to acid by a factor of _____ .

Answer: 10

56. To make a buffer with a pH of 6.00, you should use a weak acid with a K_a close to _______.

Answer: 1.0×10^{-6}

57. Hyperventilation can cause your blood pH to rise. One way to lower your blood pH is to breath into a paper bag, thus recycling the air you exhale. Why does this procedure lower your blood pH?

Answer: By breathing recycled air, you increase your CO_2 intake. Carbon dioxide is an acidic oxide; it reacts with water in your blood to form carbonic acid, H_2CO_3, which helps neutralize the excess base in your blood stream.

58. Why do salts containing basic anions have a greater solubility in water than predicted from calculations using K_{sp} values? Examples of such salts are $CaCO_3$, PbF_2, $Ca_3(PO_4)_2$.

Answer: Basic anions react with water to form their conjugate acids, such as HCO_3^-, HF, and HPO_4^{2-}. These reactions reduce the concentration of the basic anion. Additional salt dissociates to replace the basic anions.

Diagnostic Quiz Questions

Goal #1: Understand the common ion effect.

1. What is the effect of adding sodium hydroxide to an aqueous acetic acid solution?
   1. The pH increases.
   2. The concentration of acetate ion increases.
   3. The concentration of H_3O^+ decreases.

   a) 1 only
   b) 2 only
   c) 3 only
   d) 1, 2, and 3

   Answer: d

Goal #2: Understand the control of pH in aqueous solutions with buffers

2. What is the pH of a solution that results from adding 25 mL of 0.10 M NaOH to 25 mL of 0.25 M $H_2PO_4^-$? (K_a for dihydrogen phosphate ion is 6.2×10^{-8})
   a) 6.81
   b) 7.03
   c) 7.38
   d) 7.61

   Answer: b

3. Which of the following combinations of acids and conjugate bases would be the best to create a pH 7.0 buffer?
   a) H_3PO_4 ($K_a = 7.5 \times 10^{-3}$) and $H_2PO_4^-$
   b) HNO_2 ($K_a = 4.5 \times 10^{-4}$) and NO_2^-
   c) $H_2PO_4^-$ ($K_a = 6.2 \times 10^{-8}$) and HPO_4^{2-}
   d) NH_4^+ ($K_a = 5.7 \times 10^{-10}$) and NH_3

   Answer: c

4. Which of the following statements concerning buffers are correct?
   1. Buffers are composed of a weak acid diluted with water.
   2. Buffers are resistant to changes in pH when diluted with water.
   3. Buffers are used as colored indicators for acid-base titrations.

   a) 1 only
   b) 2 only
   c) 3 only
   d) 1 and 2

   Answer: b

5. What is the pH of the buffer that results when 5.0 g of NH_3 and 5.0 g of NH_4Cl are diluted with water to a volume of 250 mL? (K_a for $NH_4^+ = 5.6 \times 10^{-10}$)
   a) 4.25
   b) 8.75
   c) 9.25
   d) 9.75

   Answer: d

6. What mass of solid KF (molar mass = 58.1 g/mol) should be added to 1.00 L of 0.250 M HF to make a buffer of pH = 3.209? (K_a for HF = 7.20×10^{-4})
   a) 9.18 g
   b) 12.5 g
   c) 14.5 g
   d) 16.9 g

   Answer: d

7. A buffer contains 0.50 mol CH_3CO_2H and 0.50 mol $CH_3CO_2^-$ diluted with water to 1.0 L. How much NaOH (in moles) is required to increase the pH of the buffer to 5.01? (pK_a for CH_3CO_2H = 4.74)
   a) 0.15 mol
   b) 0.19 mol
   c) 0.23 mol
   d) 0.33 mol

   Answer: a

Goal #3: Evaluate the pH in the course of acid-base titrations

8. The pH at the equivalence point of a titration of a strong acid with a strong base is
   a) greater than 7.
   b) greater than or equal to 7.
   c) less than 7.
   d) equal to 7.

   Answer: d

9. Calculate the pH after 35.0 mL of 0.100 M HCl has been added to 70.0 mL of 0.100 M NH_3? The K_a for NH_4^+ is 5.6×10^{-10}.
   a) 6.15
   b) 7.00
   c) 8.95
   d) 9.25

   Answer: d

10. A 25.0 mL sample of 0.160 M NH_3 is titrated with 0.100 M HCl. What is the pH at the equivalence point? The K_a for NH_4^+ is 5.6×10^{-10}.
    a) 5.02
    b) 5.23
    c) 7.00
    d) 9.25

    Answer: b

11. A 25.0 mL sample of 0.100 M HF is titrated with 0.100 M NaOH. What is the pH after 30.0 mL of NaOH is added? The K_a for HF is 7.2×10^{-4}.
   a) 7.92
   b) 10.86
   c) 11.96
   d) 13.00

   Answer: c

Goal #4: Apply chemical equilibrium concepts to the solubility of ionic compounds.

12. Calcium fluoride dissolves in water to the extent of 18.5 mg per liter of solution. What is the K_{sp} for CaF_2?
   a) 5.32×10^{-11}
   b) 5.61×10^{-8}
   c) 2.37×10^{-4}
   d) 1.33×10^{-2}

   Answer: a

13. The K_{sp} for $PbCl_2$ is 1.7×10^{-6}. What is the concentration of Cl^- in a saturated solution of $PbCl_2$?
   a) 3.8×10^{-3} M
   b) 7.5×10^{-3} M
   c) 9.5×10^{-3} M
   d) 1.5×10^{-2} M

   Answer: d

14. The K_{sp} for AgI is 8.5×10^{-17}. What is the molar solubility of AgI in 0.050 M NaI?
   a) 1.7×10^{-15} mol/L
   b) 1.9×10^{-12} mol/L
   c) 9.2×10^{-9} mol/L
   d) 4.1×10^{-8} mol/L

   Answer: a

15. At what pH will a solution 0.15 M Cu^{2+} begin to precipitate as $Cu(OH)_2$? The K_{sp} for $Cu(OH)_2$ is 1.6×10^{-19}.
   a) 5.01
   b) 7.53
   c) 7.83
   d) 8.99

   Answer: a

16. A solution contains 0.10 M Cl^- and 0.10 M Br^-. If Ag^+ is added until AgCl just begins to precipitate, what are the concentrations of Ag^+ and Br^-? (K_{sp} for AgCl = 1.8×10^{-10}, K_{sp} for AgBr = 3.3×10^{-13})
    a) $[Ag^+] = 1.8 \times 10^{-9}$ M, $[Br^-] = 3.3 \times 10^{-12}$ M
    b) $[Ag^+] = 1.8 \times 10^{-9}$ M, $[Br^-] = 1.8 \times 10^{-4}$ M
    c) $[Ag^+] = 3.3 \times 10^{-12}$ M, $[Br^-] = 1.0 \times 10^{-1}$ M
    d) $[Ag^+] = 3.3 \times 10^{-12}$ M, $[Br^-] = 1.8 \times 10^{-4}$ M

    Answer: b

17. Consider the reaction

    $Ni(OH)_2(s) + 4\ CN^-(aq) \rightleftharpoons Ni(CN)_4^{2-}(aq) + 2\ OH^-(aq) \qquad K = 5.5 \times 10^{15}$

    If K_{sp} for $Ni(OH)_2$ is 5.5×10^{-16}, what is the value of the formation constant, K_{form}, for the reaction below?

    $Ni^{2+}(aq) + 4\ CN^-(aq) \rightleftharpoons Ni(CN)_4^{2-}(aq)$

    a) 1.0×10^{-31}
    b) 3.3×10^{-1}
    c) 3.0×10^{0}
    d) 1.0×10^{31}

    Answer: d

CHAPTER 19: Principles of Reactivity: Entropy and Free Energy

19.1 Spontaneous Change and Equilibrium

1. Reactants A and B are placed in an empty flask. If the reaction A(g) + B(g) $\rightleftharpoons$ 2 C(g) has $K_p > 1$, which of the following statements are correct?
   1. The forward reaction is spontaneous.
   2. The forward rate of reaction is fast.
   3. The reaction is reactant-favored.

   a) 1 only　　b) 2 only　　c) 3 only　　d) 1 and 2　　e) 1, 2, and 3

   Answer: a

19.2 Heat and Spontaneity

2. The following processes occur spontaneously at 25 °C. Which of these processes are endothermic?
   1. the combustion of methane to produce water and carbon dioxide
   2. the expansion of an ideal gas into a vacuum
   3. the melting of ice

   a) 1 only　　b) 2 only　　c) 3 only　　d) 1 and 2　　e) 2 and 3

   Answer: c

19.3 Dispersal of Energy and Matter

3. A statement of the second law of thermodynamics is that
   a) spontaneous reactions are always exothermic.
   b) energy is conserved in a chemical reaction.
   c) the Gibbs free energy is a function of both enthalpy and entropy.
   d) $\Delta S = - \Delta H$ for any chemical reaction.
   e) in a spontaneous process, the entropy of the universe increases.

   Answer: e

19.4 Entropy and the Second Law of Thermodynamics

4. As defined by Ludwig Boltzmann, the third law of thermodynamics states that
   a) in a spontaneous process, the entropy of the universe increases.
   b) there is no disorder in a perfect crystal at 0 K.
   c) the total entropy of the universe is always increasing.
   d) the total energy of the universe is constant.
   e) mass and energy are conserved in all chemical reactions.

   Answer: b

5. Which of the following linear chain alcohols is likely to have the highest standard entropy in the liquid state?
a) CH_3OH
b) CH_3CH_2OH
c) $CH_3CH_2CH_2OH$
d) $CH_3CH_2CH_2CH_2OH$
e) $CH_3CH_2CH_2CH_2CH_2OH$

Answer: e

6. Which of the following processes are reversible?
   1. the expansion of an ideal gas into a vacuum
   2. the melting of ice at 0 °C
   3. the combustion of carbon to produce carbon dioxide

a) 1 only b) 2 only c) 3 only d) 1 and 2 e) 2 and 3

Answer: b

7. If a chemical reaction has a positive change in entropy then
a) the change in enthalpy is negative.
b) the reaction is exothermic.
c) heat goes from the system into the surroundings.
d) the disorder of the system increases.
e) the reaction is spontaneous.

Answer: d

8. Which of the following processes involves a decrease in entropy?
a) the decomposition of $NH_3(g)$ into $H_2(g)$ and $N_2(g)$ gas
b) the dissolution of NaCl in water
c) the condensation of steam to liquid water
d) the evaporation of ethanol
e) the sublimation of dry ice (i.e., $CO_2(s)$)

Answer: c

9. All of the following statements concerning entropy are true EXCEPT
a) entropy values for substances are greater than or equal to zero.
b) entropy is a state function.
c) a positive change in entropy denotes a change toward greater disorder.
d) entropy is zero for elements under standard conditions.
e) the entropy of a substance in the gas phase is greater than the solid phase.

Answer: d

10. All of the following processes lead to an increase in entropy EXCEPT
    a) decreasing the temperature of a gas.
    b) melting a solid
    c) chemical reactions that increase the number of moles of gas.
    d) forming mixtures from pure substances.
    e) evaporating a liquid

    Answer: a

11. Which reaction is likely to have a negative change in entropy?
    a) $2\ NH_3(g) \rightarrow N_2(g) + 3\ H_2(g)$
    b) $CaCO_3(s) \rightarrow CaO(s) + CO_2(g)$
    c) $NaCl(s) \rightarrow Na^+(aq) + Cl^-(aq)$
    d) $N_2O_4(g) \rightarrow 2\ NO_2(g)$
    e) $2\ CO(g) \rightarrow 2\ C(s) + O_2(g)$

    Answer: e

12. Calculate the standard entropy change for the combustion of methanol at 25 °C.

$$2\ CH_3OH(\ell) + 3\ O_2(g) \rightarrow 2\ CO_2(g) + 4\ H_2O(g)$$

Species	$S°$ (J/K·mol)
$CH_3OH\ (\ell)$	127.2
$O_2(g)$	205.1
$CO_2(g)$	213.7
$H_2O(g)$	188.8

a) -2052.3 J/K b) -312.9 J/K c) +70.2 J/K d) +312.9 J/K e) +2052.3 J/K

Answer: d

13. Calculate the standard entropy change for the following reaction,

$$2\ HgO(s) \rightarrow 2\ Hg(\ell) + O_2(g)$$

given $S°[HgO] = 70.3$ J/K·mol, $S°[Hg(\ell)] = 76.0$ J/K·mol, and $S°[O_2(g)] = 205.1$ J/K·mol.

a) -216.5 J/K b) +210.8 J/K c) +216.5 J/K d) +351.4 J/K e) +497.7 J/K

Answer: c

14. The standard entropy of formation of $PCl_3(g)$ is -33.2 J/K·mol.

$$1/4\ P_4(s) + 3/2\ Cl_2(g) \rightarrow PCl_3(g)$$

Calculate the standard molar entropy of $PCl_3(g)$ given $S°[P_4(s)] = 41.1$ J/K·mol and $S°[Cl_2(g)] = 223.1$ J/K·mol.

a) -378.1 J/K·mol
b) -297.4 J/K·mol
c) +212.2 J/K·mol
d) +231.0 J/K·mol
e) +311.7 J/K·mol

Answer: e

19.5 Entropy Changes and Spontaneity

15. For the following reaction at 25.0 °C,

$$2\ C(s) + O_2(g) \rightarrow 2\ CO(g)$$

calculate ΔS°_{univ} given $\Delta S^{\circ}_{sys} = 179.1$ J/K and $\Delta H^{\circ}_{sys} = -221.1$ kJ.

a) -9023 J/K b) -562.3 J/K c) -167.7 J/K d) +170.2 J/K e) +920.7 J/K

Answer: e

16. Use the following thermodynamic data

Species	$\Delta H°$ (kJ/mol)	$S°$ (J/K·mol)
$H_2O_2(\ell)$	-187.78	109.6
$H_2O(\ell)$	-285.83	69.91
$O_2(g)$	0	205.14

to calculate ΔS°_{univ} for the decomposition of hydrogen peroxide at 25.0 °C.

$$2\ H_2O_2(\ell) \rightarrow 2\ H_2O(\ell) + O_2(g)$$

a) -783.5 J/K b) –531.9 J/K c) +494.2 J/K d) +783.5 J/K e) +3741 J/K

Answer: d

19.6 Gibbs Free Energy

17. Predict the signs of ΔH and ΔS for the condensation of steam at 85 °C.

a) $\Delta H < 0$ and $\Delta S < 0$
b) $\Delta H < 0$ and $\Delta S > 0$
c) $\Delta H > 0$ and $\Delta S < 0$
d) $\Delta H > 0$ and $\Delta S > 0$
e) $\Delta H = 0$ and $\Delta S < 0$

Answer: a

18. Hydrogen gas is a non-polluting fuel. Predict the signs of ΔH, ΔS, and ΔG for the combustion of hydrogen gas at 150 °C.

$$2\ H_2(g) + O_2(g) \rightarrow 2\ H_2O(g)$$

a) $\Delta H < 0$, $\Delta S > 0$, $\Delta G < 0$
b) $\Delta H < 0$, $\Delta S < 0$, $\Delta G < 0$
c) $\Delta H < 0$, $\Delta S > 0$, $\Delta G < 0$
d) $\Delta H > 0$, $\Delta S < 0$, $\Delta G < 0$
e) $\Delta H > 0$, $\Delta S < 0$, $\Delta G > 0$

Answer: b

19. Predict the signs of ΔH, ΔS, and ΔG for the evaporation of water 25 °C.
a) $\Delta H > 0$, $\Delta S < 0$, $\Delta G < 0$
b) $\Delta H > 0$, $\Delta S > 0$, $\Delta G > 0$
c) $\Delta H < 0$, $\Delta S > 0$, $\Delta G < 0$
d) $\Delta H < 0$, $\Delta S > 0$, $\Delta G > 0$
e) $\Delta H < 0$, $\Delta S < 0$, $\Delta G < 0$

Answer: b

20. If $\Delta G^\circ < 0$ for a reaction at all temperatures, then ΔH° is ________ and ΔS° is ________.
a) negative, positive
b) positive, negative
c) negative, negative
d) positive, positive
e) positive, either positive or negative

Answer: a

21. The dissolution of ammonium nitrate occurs spontaneously in water at 25 °C. As NH_4NO_3 dissolves, the temperature of the water decreases. What are the signs of ΔH, ΔS, and ΔG for this process?
a) $\Delta H > 0$, $\Delta S < 0$, $\Delta G > 0$
b) $\Delta H > 0$, $\Delta S > 0$, $\Delta G > 0$
c) $\Delta H > 0$, $\Delta S > 0$, $\Delta G < 0$
d) $\Delta H < 0$, $\Delta S < 0$, $\Delta G < 0$
e) $\Delta H < 0$, $\Delta S > 0$, $\Delta G > 0$

Answer: c

22. Diluting concentrated sulfuric acid with water can be dangerous. The temperature of the solution can increase rapidly. What are the signs of ΔH, ΔS, and ΔG for this process?
    a) $\Delta H < 0$, $\Delta S > 0$, $\Delta G < 0$
    b) $\Delta H < 0$, $\Delta S < 0$, $\Delta G < 0$
    c) $\Delta H < 0$, $\Delta S > 0$, $\Delta G > 0$
    d) $\Delta H > 0$, $\Delta S > 0$, $\Delta G < 0$
    e) $\Delta H > 0$, $\Delta S < 0$, $\Delta G > 0$

    Answer: a

23. At what temperatures will a reaction be spontaneous if $\Delta H = +158$ kJ and $\Delta S = +411$ J/K?
    a) All temperatures below 384 K
    b) Temperatures between 158 K and 411 K
    c) All temperatures above 384 K
    d) The reaction will be spontaneous at any temperature.
    e) The reaction will never be spontaneous.

    Answer: c

24. At what temperatures will a reaction be spontaneous if $\Delta H = -76.0$ kJ and $\Delta S = +231$ J/K?
    a) All temperatures below 329 K
    b) Temperatures between 0 K and 231 K
    c) All temperatures above 329 K
    d) The reaction will be spontaneous at any temperature.
    e) The reaction will never be spontaneous.

    Answer: d

25. For a reaction, $\Delta H = +62.9$ kJ and $\Delta S = +132.9$ J/K. At what temperature will $\Delta G = 0.00$ kJ?
    a) 2.11 K
    b) 70.0 K
    c) 473 K
    d) 8.36×10^6 K
    e) ΔG is greater than 0.00 kJ at any temperature.

    Answer: c

26. If a process is exothermic and spontaneous, which of the following must be true?
    a) $\Delta G > 0$ and $\Delta H < 0$
    b) $\Delta G < 0$ and $\Delta H < 0$
    c) $\Delta G < 0$ and $\Delta S > 0$
    d) $\Delta H < 0$ and $\Delta S > 0$
    e) $\Delta H > 0$ and $\Delta S < 0$

    Answer: b

27. Calculate ΔG°_{rxn} for the reaction below at 25.0 °C

$$2\ H_2O(g) + S(s) \rightarrow 2\ H_2S(g) + O_2(g)$$

given $\Delta H^\circ_{rxn} = +442.4$ kJ and $\Delta S^\circ_{rxn} = +206.9$ J/K.

a) +380.7 kJ b) +504.1 kJ c) +649.3 kJ d) +1277.7 kJ e) $+6.125 \times 10^4$ kJ

Answer: a

28. Calculate ΔG°_{rxn} for the reaction below at 25.0 °C

$$2\ Ca(s) + O_2(g) \rightarrow 2\ CaO$$

given $\Delta H^\circ_{rxn} = -1270.2$ kJ and $\Delta S^\circ_{rxn} = -211.9$ J/K.

a) -1058.3 kJ b) -1207.0 kJ c) -1269.5 kJ d) -1333.4 kJ e) -1482.1 kJ

Answer: b

29. Calculate ΔG°_{rxn} for the reaction below at 25.0 °C

$$CO(g) + H_2O(\ell) \rightarrow H_2(g) + CO_2(g)$$

given ΔG°_f [CO(g)] = -137.2 kJ/mol, ΔG°_f [$H_2O(\ell)$] = -237.2 kJ/mol, ΔG°_f [$H_2(g)$] = 0.0 kJ/mol, and ΔG°_f [$CO_2(g)$] = -394.4 kJ/mol.

a) -768.8 kJ b) -294.4 kJ c) -20.0 kJ d) +20.0 kJ e) +768.8 kJ

Answer: c

30. The ΔG°_{rxn} for the following reaction is +70.9 kJ.

$$SO_3(g) \rightarrow SO_2(g) + \tfrac{1}{2}\ O_2(g)$$

Given ΔG°_f [$SO_3(g)$] = -371.0 kJ/mol, calculate ΔG°_f [$SO_2(g)$].

a) -300.1 kJ/mol b) -5.23 kJ/mol c) +5.23 kJ/mol d) +300.1 kJ/mol e) +441.9 kJ/mol

Answer: a

31. Calculate ΔG° for the reaction below at 25.0 °C.

$$2\ Fe(s) + 3\ Cl_2(g) \rightarrow 2\ FeCl_3(s)$$

Species	ΔH°_f (kJ/mol)	S°_f (J/K·mol)
Fe(s)	0	27.78
$Cl_2(g)$	0	223.08
$FeCl_3(s)$	-344.0	142.3

a) -212.7 kJ b) -311.6 kJ c) -376.4 kJ d) -556.8 kJ e) -819.3 kJ

Answer: d

32. Calculate ΔG° for the reaction below at 25.0 °C.

$P_4(s) + 6\ H_2O(\ell) \rightarrow 4\ H_3PO_4(\ell)$

Species	ΔH_f° (kJ/mol)	S_f° (J/K·mol)
$P_4(s)$	0	22.80
$H_2O(\ell)$	-285.8	69.95
$H_3PO_4(\ell)$	-1279.0	110.5

a) -993.2 kJ b) -998.5 kJ c) -3137 kJ d) -3252 kJ e) -3401 kJ

Answer: e

33. Calculate ΔG° for the reaction below at 125 °C.

$C_2H_5OH(g) + 3\ O_2(g) \rightarrow 2\ CO_2(g) + 3\ H_2O(g)$

Species	ΔH_f° (kJ/mol)	S_f° (J/K·mol)
$C_2H_5OH(g)$	-235.3	282.7
$O_2(g)$	0	205.1
$CO_2(g)$	-393.5	213.7
$H_2O(g)$	-241.8	188.8

a) -366.0 kJ b) -1277.1 kJ c) -1289.1 kJ d) -1315.2 kJ e) -3.94×10^4 kJ

Answer: d

34. All of the following substances have a standard free energy of zero EXCEPT
a) $Br_2(\ell)$. b) $I_2(g)$. c) $S_8(s)$. d) $Cl_2(g)$. e) $Hg(\ell)$.

Answer: b

35. Given that

$C(s) + O_2(g) \rightarrow CO_2(g) \qquad \Delta G^\circ = -394.4$ kJ

$CO(g) + \frac{1}{2}O_2(g) \rightarrow CO_2(g) \qquad \Delta G^\circ = -257.2$ kJ

calculate ΔG° for the following reaction.

$2\ C(s) + O_2(g) \rightarrow 2\ CO(g)$

a) -651.6 kJ b) -274.4 kJ c) -137.2 kJ d) +137.2 kJ e) +274.4 kJ

Answer: b

36. Estimate the boiling point of carbon tetrachloride given the following thermodynamic parameters.

	$CCl_4(\ell)$	$CCl_4(g)$
ΔH_f° (kJ/mol)	-128.4	-96.0
ΔS° (J/K·mol)	214.4	309.7
ΔG_f° (kJ/mol)	-57.6	-53.6

a) -272 °C b) 25 °C c) 67 °C d) 69 °C e) 109 °C

Answer: c

19.7 ΔG°, K, and Product Favorability

37. Thermodynamics can be used to determine all of the following EXCEPT
 a) the rate of reaction.
 b) the extent to which a reaction occurs.
 c) the direction in which a reaction is spontaneous.
 d) the temperature at which a reaction is spontaneous.
 e) the entropy change of a reaction.

Answer: a

38. For a chemical reaction, if $\Delta G^\circ = 0$, then
 a) $K > 1$ b) $K = 0$ c) $K < 1$ d) $K < 0$ e) $K = 1$

Answer: e

39. All of the following relationships are true EXCEPT
 a) $\Delta G^\circ_{sys} = \Delta H^\circ_{sys} - T\Delta S^\circ_{sys}$
 b) $\Delta G^\circ_{sys} = - RT\ln(K)$
 c) $\Delta S^\circ_{univ} = \Delta S^\circ_{sys} + \Delta S^\circ_{surr}$
 d) $\Delta H = \Delta H^\circ_{sys} + RT\ln(K)$
 e) $\Delta G^\circ_{sys} = - T\Delta S^\circ_{univ}$

Answer: d

40. For a chemical system, ΔG° and ΔG are equal when
   a) the system is in equilibrium.
   b) the reactants and products are in standard state conditions.
   c) the equilibrium constant, K, equals 0.
   d) the reaction quotient, Q, is less than 1.
   e) the reactants and products are in the gas phase.

   Answer: b

41. The standard free energy change for a chemical reaction is -18.3 kJ/mole. What is the equilibrium constant for the reaction at 87 °C? (R = 8.314 J/K·mol)
   a) 2.2×10^{-3} b) 1.0 c) 6.1 d) 4.5×10^{2} e) 1.3×10^{6}

   Answer: d

42. The standard free energy change for the formation of $N_2H_4(\ell)$ from elements is +149 kJ/mol at 25 °C. What is the equilibrium constant for the reaction? (R = 8.314 J/K·mol)
   a) 7×10^{-61} b) 8×10^{-27} c) 1 d) 1×10^{26} e) 1×10^{60}

   Answer: b

43. What is the equilibrium constant for reaction below at 25 °C? (R = 8.314 J/K·mol)
   $$2\ H_2O_2(\ell) \rightleftharpoons 2\ H_2O(\ell) + O_2(g)$$
   given $\Delta G^\circ_f\ [H_2O_2(\ell)]$ = -120.3 kJ/mol, $\Delta G^\circ_f\ [H_2O(\ell)]$ = -237.1 kJ/mol, and $\Delta G^\circ_f\ [O_2(g)]$ = 0.0 kJ/mol.
   a) 5×10^{-95} b) 1×10^{-41} c) 1 d) 3×10^{20} e) 9×10^{40}

   Answer: e

44. The equilibrium constant for a reaction at 25 °C is 4.7×10^{-8}. What is ΔG°? (R = 8.314 J/K·mol)
   a) +1.52 kJ b) +3.51 kJ c) +6.81 kJ d) +18.2 kJ e) +41.8 kJ

   Answer: e

45. Calculate ΔG° for the following reaction at 425 °C,
   $$H_2(g) + I_2(g) \rightleftharpoons 2\ HI(g)$$
   given K = 56. (R = 8.314 J/K·mol)
   a) -23.4 kJ b) -14.2 kJ c) -10.1 kJ d) -6.18 kJ e) +14.2 kJ

   Answer: a

Short Answer Questions

46. The total entropy of the universe is always increasing. This is a statement of the ________ law of thermodynamics.

    Answer: second

47. A chemical reaction with an equilibrium constant less than one is said to be ________-________.

    Answer: reactant-favored

48. The change in entropy for any process is not dependent upon the pathway by which the process occurs. In other words, the change in entropy for any process is a ________ function.

    Answer: state

49. For any process, the change in entropy of the universe equals the sum of the entropy changes for the ________ and for the surroundings.

    Answer: system

50. At what temperature is the entropy of a pure crystal 0.0 J/K.

    Answer: 0 K

51. Does the formation of complex molecules such as proteins and nucleic acids from more simple molecules contradict the second law of thermodynamics?

    Answer: No. The formation of complex molecules involves a decrease in entropy locally. The local decrease is offset by an increase in entropy to the universe.

52. Use the standard entropies of vaporization for liquid water and steam to estimate the enthalpy of vaporization of water at its normal boiling point. $\Delta S°$ $[H_2O(\ell)] = 69.9$ J/K·mol and $\Delta S°$ $[H_2O(g)] = 188.8$ J/K·mol.

    Answer: 44.4 kJ/mol

Diagnostic Quiz Questions

Goal #1: Understand the concept of entropy and how it relates to spontaneity.

1. Which one of the following processes leads to a decrease in entropy?
   a) evaporating a liquid
   b) forming mixtures from pure substances
   c) chemical reactions that increase the number of gas particles
   d) freezing a liquid

   Answer: d

2. Of the following product-favored processes, which are endothermic?
   1. the combustion of methane to produce water and carbon dioxide
   2. the reaction of a strong acid with a strong base
   3. the melting of ice

   a) 1 only
   b) 2 only
   c) 3 only
   d) 2 and 3

   Answer: c

3. Calculate the standard molar entropy change for the combustion of 1.00 mole of methane.
   Unbalanced reaction: $CH_4(g) + O_2(g) \rightarrow CO_2(g) + H_2O(g)$

Species	$S°$ (J/K·mol)
$CH_4(g)$	186.3
$O_2(g)$	205.1
$CO_2(g)$	213.7
$H_2O(g)$	188.8

   a) –5.2 J/K
   b) +11.1 J/K
   c) -11.1 J/K
   d) +1187.8 J/K

   Answer: a

4. Which of the following molecules is likely to have the highest standard entropy in the liquid state?
   a) CH_4
   b) C_2H_6
   c) C_3H_8
   d) C_4H_{10}

   Answer: d

5. Which reaction is likely to have a negative change in entropy?
   a) $2\ C_2H_6(g) + 7\ O_2(g) \rightarrow 4\ CO_2(g) + 6\ H_2O(g)$
   b) $N_2O_4(g) \rightarrow 2\ NO_2(g)$
   c) $2\ NH_3(g) \rightarrow N_2(g) + 3\ H_2(g)$
   d) $CaO(s) + CO_2(g) \rightarrow CaCO_3(s)$

   Answer: d

Goal #2: Predict whether a reaction is spontaneous.

6. Predict the signs of ΔH and ΔS for the evaporation of water.
   a) $\Delta H > 0$ and $\Delta S > 0$
   b) $\Delta H > 0$ and $\Delta S < 0$
   c) $\Delta H < 0$ and $\Delta S > 0$
   d) $\Delta H < 0$ and $\Delta S < 0$

   Answer: a

7. For a given reaction, ΔH is +165 kJ and ΔS is +421 J/K. At what temperatures is the reaction spontaneous?
   a) None. The reaction is never spontaneous.
   b) The reaction is spontaneous at all temperatures.
   c) $T < 392$ K
   d) $T > 392$ K

   Answer: d

Goal #3: Understand and use a new thermodynamic function, free energy.

8. Predict signs of ΔH, ΔS, and ΔG for the combustion of pentane, $C_5H_{12}(\ell)$, producing $CO_2(g)$ and $H_2O(g)$?
   a) $\Delta H < 0, \Delta S < 0, \Delta G < 0$
   b) $\Delta H < 0, \Delta S < 0, \Delta G > 0$
   c) $\Delta H < 0, \Delta S > 0, \Delta G < 0$
   d) $\Delta H > 0, \Delta S > 0, \Delta G < 0$

   Answer: c

9. Calculate ΔG° for the reaction below at 25.0 °C.

$$4\,Fe(s) + 3\,O_2(g) \rightarrow 2\,Fe_2O_3(s)$$

Species	ΔH_f° (kJ/mol)	S_f° (J/K·mol)
$Fe(s)$	0	27.78
$O_2(g)$	0	205.14
$Fe_2O_3(s)$	-824.2	87.40

a) -1483.9 kJ
b) -1156.8 kJ
c) -780.8 kJ
d) +1156.8 kJ

Answer: a

10. The ΔG°_{rxn} for the following reaction is –70.9 kJ.

$$SO_2(g) + \tfrac{1}{2}\,O_2(g) \rightarrow SO_3(g)$$

Given ΔG_f° [$SO_2(g)$] = -300.2 kJ/mole, calculate ΔG_f° [$SO_3(g)$].

a) -371.1 kJ/mole
b) -229.3 kJ/mole
c) +229.3 kJ/mole
d) +371.1 kJ/mole

Answer: a

Goal #4: Understand the relationship of a free energy change for a reaction, its equilibrium constant, and whether the reaction is product- or reactant- favored.

11. For a chemical reaction, if $\Delta G^\circ < 0$, then

a) $K < 0$
b) $K > 1$
c) $K < 1$
d) $K = 1$

Answer: b

12. The standard free energy change for the formation of carbon monoxide is -137.2 kJ/mol at 25 °C. What is the equilibrium constant for this reaction? ($R = 8.314$ J/K·mol)

a) 9×10^{-20}
b) 5.5×10^{-2}
c) 2.9×10^{4}
d) 1×10^{24}

Answer: d

13. Calculate $\Delta G°$ for the following reaction at 298 K,

$$N_2O_4(g) \rightleftharpoons 2\ NO_2(g)$$

given $K = 0.15$. ($R = 8.314$ J/K·mol)

a) 2.04 kJ/mol
b) 2.88 kJ/mol
c) 4.70 kJ/mol
d) 27.8 kJ/mol

Answer: c

CHAPTER 20: Principles of Reactivity: Electron Transfer Reactions

20.1 Oxidation-Reduction Reactions

1. In the following reaction,
$$2\ Fe^{3+}(aq) + Zn(s) \rightarrow 2\ Fe^{2+}(aq) + Zn^{2+}(aq)$$
a) $Fe^{3+}(aq)$ is the reducing agent and Zn(s) is the oxidizing agent.
b) Zn(s) is the reducing agent and $Fe^{3+}(aq)$ is the oxidizing agent.
c) $Fe^{3+}(aq)$ is the reducing agent and $Fe^{2+}(aq)$ is the oxidizing agent.
d) Zn(s) is the reducing agent and $Zn^{2+}(aq)$ is the oxidizing agent.
e) Zn(s) is the reducing agent and $Fe^{2+}(aq)$ is the oxidizing agent.

Answer: b

2. Assuming the following reaction proceeds in the forward direction,
$$Ni^{2+}(aq) + Cr(s) \rightarrow Ni(s) + Cr^{2+}(aq)$$
a) $Ni^{2+}(aq)$ is oxidized and Cr(s) is reduced.
b) $Ni^{2+}(aq)$ is oxidized and Ni(s) is reduced.
c) Ni(s) is oxidized and $Ni^{2+}(aq)$ is reduced.
d) Cr(s) is oxidized and $Cr^{2+}(aq)$ is reduced.
e) Cr(s) is oxidized and $Ni^{2+}(aq)$ is reduced.

Answer: e

3. The following reaction occurs spontaneously.
$$2\ H^{+}(aq) + Ca(s) \rightarrow Ca^{2+}(aq) + H_2(g)$$
Write the balanced reduction half-reaction.
a) $2\ H^{+}(aq) + 2\ e^{-} \rightarrow H_2(g)$
b) $2\ H^{+}(aq) \rightarrow H_2(g) + 2\ e^{-}$
c) $H_2(g) \rightarrow 2\ H^{+}(aq) + 2\ e^{-}$
d) $Ca(s) + 2\ e^{-} \rightarrow Ca^{2+}(aq)$
e) $Ca(s) \rightarrow Ca^{2+}(aq) + 2\ e^{-}$

Answer: a

4. The following reaction occurs spontaneously.
$$3\ Cu^{2+}(aq) + 2\ Fe(s) \rightarrow 2\ Fe^{3+}(aq) + 3\ Cu(s)$$
Write the balanced oxidation half-reaction.
a) $2\ Fe(s) \rightarrow 2\ Fe^{3+}(aq) + 6\ e^{-}$
b) $2\ Fe(s) + 6\ e^{-} \rightarrow 2\ Fe^{3+}(aq)$
c) $3\ Cu^{2+}(aq) + 6\ e^{-} \rightarrow 3\ Cu(s)$
d) $3\ Cu^{2+}(aq) \rightarrow 3\ Cu(s) + 6\ e^{-}$
e) $3\ Cu(s) \rightarrow 3\ Cu^{2+}(aq) + 6\ e^{-}$

Answer: a

5. Write a balanced half-reaction for the reduction of hydrogen peroxide to water in an acidic solution.

a) $2\ H_2O_2(\ell) \rightarrow 2\ H_2O(\ell) + O_2(g)$

b) $2\ H_2O_2(\ell) + 2e^- \rightarrow 2\ H_2O(\ell) + O_2(g)$

c) $H_2O_2(\ell) + 2\ H^+(aq) + 2\ e^- \rightarrow 2\ H_2O(\ell)$

d) $H_2O_2(\ell) + 4\ H^+(aq) + 2\ e^- \rightarrow 2\ H_2O(\ell) + H_2(g)$

e) $H_2O_2(\ell) + 2\ H^+(aq) + 4\ e^- \rightarrow 2\ H_2(g) + O_2(g)$

Answer: c

6. Write a balanced half-reaction for the reduction of $ClO_3^-(aq)$ to $Cl_2(g)$ in an acidic solution.

a) $2\ ClO_3^-(aq) + 6\ H^+(aq) + 10\ e^- \rightarrow Cl_2(g) + 6\ OH^-(aq)$

b) $2\ ClO_3^-(aq) + 12\ H^+(aq) + 5\ e^- \rightarrow Cl_2(g) + 6\ H_2O(\ell)$

c) $2\ ClO_3^-(aq) + 10\ e^- \rightarrow Cl_2(g) + 6\ H_2O(\ell) + 3\ O_2(g)$

d) $2\ ClO_3^-(aq) + 12\ H^+(aq) + 10\ e^- \rightarrow Cl_2(g) + 6\ H_2O(\ell)$

e) $2\ ClO_3^-(aq) + 18\ H^+(aq) \rightarrow Cl_2(g) + 6\ H_3O^+(aq)$

Answer: d

7. Write a balanced half-reaction for the reduction of permanganate ion, MnO_4^-, to MnO_2 in a basic solution.

a) $MnO_4^-(aq) + 4\ OH^-(aq) + 3\ e^- \rightarrow MnO_2(s) + 2\ H_2O(\ell) + 2\ O_2(g)$

b) $MnO_4^-(aq) + 2\ OH^-(aq) + 3\ e^- \rightarrow MnO_2(s) + 2\ HO_2(aq)$

c) $MnO_4^-(aq) + 3\ e^- \rightarrow MnO_2(s) + O_2(g)$

d) $MnO_4^-(aq) + 2\ H^+(aq) + 3\ e^- \rightarrow MnO_2(s) + 2\ OH^-(aq)$

e) $MnO_4^-(aq) + 2\ H_2O(\ell) + 3\ e^- \rightarrow MnO_2(s) + 4\ OH^-(aq)$

Answer: e

8. Write a balanced chemical equation for the following reaction in an acidic solution.

$$Cr_2O_7^{2-}(aq) + Br^-(aq) \rightarrow Cr^{3+}(aq) + Br_2(aq)$$

a) $Cr_2O_7^{2-}(aq) + 2\ Br^-(aq) \rightarrow 2\ Cr^{3+}(aq) + Br_2(aq)$

b) $Cr_2O_7^{2-}(aq) + 6\ Br^-(aq) + 14\ H^+(aq) \rightarrow 3\ Br_2(aq) + 2\ Cr^{3+}(aq) + 7\ H_2O(\ell)$

c) $Cr_2O_7^{2-}(aq) + 2\ Br^-(aq) + 14\ H^+(aq) \rightarrow 2\ Cr^{3+}(aq) + Br_2(aq) + 7\ H_2O(\ell)$

d) $Cr_2O_7^{2-}(aq) + 2\ Br^-(aq) + 7\ H^+(aq) \rightarrow 2\ Cr^{3+}(aq) + Br_2(aq) + 7\ OH^-(aq)$

e) $Cr_2O_7^{2-}(aq) + 6\ Br^-(aq) + 7\ H^+(aq) \rightarrow 2\ Cr^{3+}(aq) + 3\ Br_2(aq) + 7\ OH^-(aq)$

Answer: b

9. Write a balanced chemical equation for the oxidation of Fe(s) by concentrated nitric acid. Two products of the reaction are $NO_2(g)$ and $Fe^{3+}(aq)$.
a) $HNO_3(aq) + Fe(s) \rightarrow NO_2(g) + Fe^{3+}(aq) + OH^-(aq)$
b) $6\ HNO_3(aq) + 2\ Fe(s) \rightarrow 6\ NO_2(g) + 2\ Fe^{3+}(aq) + 6\ OH^-(aq)$
c) $3\ HNO_3(aq) + Fe(s) + 3\ H^+(aq) \rightarrow 3\ NO_2(g) + Fe^{3+}(aq) + 3\ H_2O(\ell)$
d) $3\ HNO_3(aq) + Fe(s) \rightarrow 3\ NO_2(g) + Fe^{3+}(aq) + 3\ OH^-(g)$
e) $2\ HNO_3(aq) + Fe(s) + H_2O(\ell) \rightarrow 2\ NO_2(g) + Fe^{3+}(aq) + H_2(g) + O_2(g)$

Answer: c

10. Write a balanced chemical equation for the following reaction in a basic solution.
$ClO_3^-(aq) + Mn(OH)_2(s) \rightarrow Cl^-(aq) + MnO_2(s)$
a) $2\ ClO_3^-(aq) + Mn(OH)_2(s) \rightarrow 2\ Cl^-(aq) + MnO_2(s) + 2\ H_2O(\ell) + 2\ O_2(g)$
b) $2\ ClO_3^-(aq) + 3\ Mn(OH)_2(s) \rightarrow 2\ Cl^-(aq) + 3\ MnO_2(s) + 6\ OH^-(aq)$
c) $2\ ClO_3^-(aq) + 3\ Mn(OH)_2(s) + 6H^+(aq) \rightarrow 2\ Cl^-(aq) + 3\ MnO_2(s) + 6\ H_2O(\ell)$
d) $ClO_3^-(aq) + 3\ Mn(OH)_2(s) \rightarrow Cl^-(aq) + 3\ MnO_2(s) + 3\ H_2O(\ell)$
e) $ClO_3^-(aq) + 2\ Mn(OH)_2(s) \rightarrow Cl^-(aq) + 2\ MnO_2(s) + 2\ OH^-(aq) + H_2O(\ell)$

Answer: d

20.2 Simple Voltaic Cells

11. All of the following statements concerning voltaic cells are true EXCEPT
a) a salt bridge allows cations and anions to move between the half-cells.
b) electrons flow from the cathode to the anode.
c) reduction occurs at the cathode.
d) a voltaic cell can be used as a source of energy.
e) a voltaic cell consists of two-half cells.

Answer: b

12. What is the correct cell notation for a voltaic cell based on the reaction below?
$Ni^{2+}(aq) + Zn(s) \rightarrow Ni(s) + Zn^{2+}(aq)$
a) $Zn(s)\ |\ Zn^{2+}(aq)\ ||\ Ni^{2+}(aq)\ |\ Ni(s)$
b) $Zn(s)\ ||\ Zn^{2+}(aq), Ni^{2+}(aq)\ |\ Ni(s)$
c) $Ni(s)\ ||\ Ni^{2+}(aq), Zn^{2+}(aq)\ ||\ Zn(s)$
d) $Ni(s)\ |\ Zn^{2+}(aq)\ ||\ Ni^{2+}(aq)\ |\ Zn(s)$
e) $Ni(s)\ |\ Ni^{2+}(aq)\ ||\ Zn^{2+}(aq)\ |\ Zn(s)$

Answer: a

13. What is the correct cell notation for a voltaic cell based on the reaction below?

$$PbCl_2(s) + 2\ Li(s) \rightarrow Pb(s) + 2\ Li^+(aq) + 2\ Cl^-(aq)$$

a) $Pb(s)\ |\ Pb^{2+}(aq)\ ||\ Cl^-(aq), Li^+(aq)\ ||\ Li(s)$
b) $Pb(s), PbCl_2(s)\ |\ Cl^-(aq)\ ||\ Li^{2+}(aq)\ ||\ Li(s)$
c) $Li(s)\ |\ Li^+(aq), Cl^-(aq)\ ||\ PbCl_2(s)\ |\ Pb(s)$
d) $Li(s)\ |\ Li^+(aq), Cl^-(aq)||\ PbCl_2(s)\ ||\ Pb(s)$
e) $Li(s)\ |\ Li^+(aq)\ ||\ Cl^-(aq)\ |\ PbCl_2(s)\ |\ Pb(s)$

Answer: e

14. Write a balanced chemical equation for the overall reaction represented by the cell notation below.

$$Pt\ |\ Sn^{2+}(aq), Sn^{4+}(aq)\ ||\ SO_4^{2-}(aq)\ |\ PbSO_4(s)\ |\ Pb(s)$$

a) $Pb(s) + Sn^{4+}(aq) \rightarrow Pb^{2+}(s) + Sn^{2+}(aq)$
b) $Pb(s) + Sn^{4+}(aq) + SO_4^{2-}(aq) \rightarrow PbSO_4(s) + Sn^{2+}(aq)$
c) $PbSO_4(s) + Sn^{2+}(aq) \rightarrow Pb(s) + Sn^{4+}(aq) + SO_4^{2-}(aq)$
d) $PbSO_4(s) + Sn^{4+}(aq) \rightarrow Pb(s) + Sn^{2+}(aq) + SO_4^{2-}(aq)$
e) $Pb^{2+}(s) + Sn^{2+}(aq) \rightarrow Pb(s) + Sn^{4+}(aq)$

Answer: c

15. Write a balanced net ionic equation for the overall reaction represented by the cell notation below.

$$Zn\ |\ ZnCl_2(aq)\ ||\ HCl(aq)\ |\ H_2(g)\ |\ Pt$$

a) $2\ H^+(aq) + Zn(s) \rightarrow H_2(g) + Zn^{2+}(aq)$
b) $H_2(g) + Zn^{2+}(aq) \rightarrow 2\ H^+(aq) + Zn(s)$
c) $H_2(g) + ZnCl_2(aq) \rightarrow 2\ HCl(aq) + Zn(s)$
d) $Pt(s) + Zn^{2+}(aq) \rightarrow Pt^{2+}(aq) + Zn(s)$
e) $Pt(s) + ZnCl_2(aq) \rightarrow PtCl_2(aq) + Zn(s)$

Answer: a

20.3 Commercial Voltaic Cells

16. The fuel cells used aboard NASA's Space Shuttles are electrochemical cells that generate electricity from which overall chemical reaction?

a) $Fe_2O_3(s) + Al(s) \rightarrow Fe(s) + Al_2O_3(s)$
b) $Pb(s) + PbO_2(s) + 2\ H_2SO_4(aq) \rightarrow 2\ PbSO_4(s) + 2\ H_2O(\ell)$
c) $2\ NiO(OH)(s) + Cd(s) + H_2O(\ell) \rightarrow 2\ Ni(OH)_2(s) + Cd(OH)_2(s)$
d) $2\ H_2(g) + O_2(g) \rightarrow 2\ H_2O(\ell)$
e) $N_2H_4(\ell) + O_2(g) \rightarrow N_2(g) + 2\ H_2O(\ell)$

Answer: d

20.4 Standard Electrochemical Potentials

17. Which of the following half-reactions has been assigned a standard reduction potential, $E°$, of 0.00 Volts?

a) $2\ H_2O(\ell) + 2\ e^- \rightarrow H_2(g) + 2\ OH^-(aq)$
b) $O_2(g) + 4\ e^- \rightarrow 2\ O^{2-}(aq)$
c) $Hg_2Cl_2(s) + 2\ e^- \rightarrow 2\ Hg(\ell) + 2\ Cl^-(aq)$
d) $Li^+(aq) + e^- \rightarrow Li(s)$
e) $2\ H^+(aq) + 2\ e^- \rightarrow H_2(g)$

Answer: e

18. Use the standard reduction potentials below to determine which element or ion is the best oxidizing agent?

$Br_2(\ell) + 2\ e^- \rightarrow 2\ Br^-(aq)$ $E° = +1.08$ V
$Hg^{2+}(aq) + e^- \rightarrow Hg(\ell)$ $E° = +0.86$ V
$Ni^{2+}(aq) + 2\ e^- \rightarrow Ni(s)$ $E° = -0.25$ V

a) Br_2 b) Br^- c) Hg^{2+} d) Ni^{2+} e) Ni

Answer: a

19. Use the standard reduction potentials below to determine which element or ion is the best reducing agent?

$Pd^{2+}(aq) + 2\ e^- \rightarrow Pd(s)$ $E° = +0.90$ V
$Sn^{2+}(aq) + 2\ e^- \rightarrow Sn(s)$ $E° = -0.14$ V
$Cr^{2+}(aq) + 2\ e^- \rightarrow Cr(s)$ $E° = -0.91$ V

a) Pd^{2+} b) Pd(s) c) Sn^{2+} d) Cr^{2+} e) Cr(s)

Answer: e

20. Which of the following species are likely to behave as oxidizing agents: K^+, MnO_4^-, $Cr_2O_7^{2-}$, and I^-?

a) K^+ only
b) MnO_4^- and $Cr_2O_7^{2-}$
c) MnO_4^-, $Cr_2O_7^{2-}$, and I^-
d) $Cr_2O_7^{2-}$ and I^-
e) I^- only

Answer: b

21. Consider the following half-reactions:

$Cu^{2+}(aq) + 2\,e^- \rightarrow Cu(s)$ $E° = +0.34$ V
$Sn^{2+}(aq) + 2\,e^- \rightarrow Sn(s)$ $E° = -0.14$ V
$Fe^{2+}(aq) + 2\,e^- \rightarrow Fe(s)$ $E° = -0.44$ V
$Al^{3+}(aq) + 3\,e^- \rightarrow Al(s)$ $E° = -1.66$ V
$Mg^{2+}(aq) + 2\,e^- \rightarrow Mg(s)$ $E° = -2.37$ V

Which of the above metals or metal ions will reduce $Fe^{2+}(aq)$?
a) Cu(s) and Sn(s)
b) $Cu^{2+}(aq)$ and $Sn^{2+}(aq)$
c) $Al^{3+}(aq)$ and $Mg^{2+}(aq)$
d) Al(s) and Mg(s)
e) Sn(s) and $Al^{3+}(aq)$

Answer: d

22. Consider the following half-reactions:

$F_2(g) + 2\,e^- \rightarrow 2\,F^-(aq)$ $E° = +2.87$ V
$Cu^{2+}(aq) + 2\,e^- \rightarrow Cu(s)$ $E° = +0.34$ V
$Sn^{2+}(aq) + 2\,e^- \rightarrow Sn(s)$ $E° = -0.14$ V
$Al^{3+}(aq) + 3\,e^- \rightarrow Al(s)$ $E° = -1.66$ V
$Na^+(aq) + e^- \rightarrow Na(s)$ $E° = -2.71$ V

Which of the above elements or ions will oxidize Sn(s)?
a) Cu(s) and $Al^{3+}(aq)$
b) $F^-(aq)$ and Cu(s)
c) $F_2(g)$ and $Cu^{2+}(aq)$
d) Al(s) and Na(s)
e) $Al^{3+}(aq)$ and $Na^+(aq)$

Answer: c

23. Given the following two half-reactions, determine which overall reaction is spontaneous and calculate its standard cell potential.

$Cd^{2+}(aq) + 2\,e^- \rightarrow Cd(s)$ $E° = -1.21$ V
$Sn^{2+}(aq) + 2\,e^- \rightarrow Sn(s)$ $E° = -0.14$ V

a) $Cd^{2+}(aq) + Sn(s) \rightarrow Cd(s) + Sn^{2+}(aq)$ $E^o_{cell} = +1.07$ V
b) $Cd^{2+}(aq) + Sn(s) \rightarrow Cd(s) + Sn^{2+}(aq)$ $E^o_{cell} = -1.35$ V
c) $Cd(s) + Sn^{2+}(aq) \rightarrow Cd^{2+}(aq) + Sn(s)$ $E^o_{cell} = +1.07$ V
d) $Cd(s) + Sn^{2+}(aq) \rightarrow Cd^{2+}(aq) + Sn(s)$ $E^o_{cell} = -1.07$ V
e) $Cd(s) + Sn^{2+}(aq) \rightarrow Cd^{2+}(aq) + Sn(s)$ $E^o_{cell} = +1.35$ V

Answer: c

24. Given the following two half-reactions, determine which overall reaction is spontaneous and calculate its standard cell potential.

$Sn^{4+}(aq) + 2\,e^- \rightarrow Sn^{2+}(aq)$ $E° = +0.15$ V

$Ag^+(aq) + e^- \rightarrow Ag(s)$ $E° = +0.80$ V

a) $Sn^{4+}(aq) + 2\,Ag(s) \rightarrow Sn^{2+}(aq) + 2\,Ag^+(aq)$ $E^{o}_{cell} = -0.65$ V

b) $Sn^{4+}(aq) + 2\,Ag(s) \rightarrow Sn^{2+}(aq) + 2\,Ag^+(aq)$ $E^{o}_{cell} = +0.95$ V

c) $Sn^{2+}(aq) + 2\,Ag^+(aq) \rightarrow Sn^{4+}(aq) + 2\,Ag(s)$ $E^{o}_{cell} = -0.65$ V

d) $Sn^{2+}(aq) + 2\,Ag^+(aq) \rightarrow Sn^{4+}(aq) + 2\,Ag(s)$ $E^{o}_{cell} = +0.65$ V

e) $Sn^{2+}(aq) + 2\,Ag^+(aq) \rightarrow Sn^{4+}(aq) + 2\,Ag(s)$ $E^{o}_{cell} = +0.95$ V

Answer: d

25. Calculate E^{o}_{cell} for the reaction below,

$Tl^+(aq) + 2\,Cl^-(aq) \rightarrow 2\,Tl(s) + Cl_2(g)$

given the following standard reduction potentials.

$Cl_2(g) + 2\,e^- \rightarrow 2\,Cl^-(aq)$ $E° = +1.36$ V

$Tl^+(aq) + e^- \rightarrow Tl(s)$ $E° = -0.34$ V

a) -2.04 V b) -1.70 V c) 1.02 V d) +1.70 V e) +2.04 V

Answer: b

26. Calculate E^{o}_{cell} for the reaction below,

$5\,Cd(s) + 2\,MnO_4^-(aq) + 16\,H^+(aq) \rightarrow 5\,Cd^{2+}(aq) + 2\,Mn^{2+}(aq) + 8\,H_2O(\ell)$

given the following standard reduction potentials.

$MnO_4^-(aq) + 8\,H^+(aq) + 5\,e^- \rightarrow Mn^{2+}(aq) + 4\,H_2O(\ell)$ $E° = +1.52$ V

$Cd^{2+}(aq) + 2\,e^- \rightarrow Cd(s)$ $E° = -0.40$ V

a) +1.04 V b) +1.12V c) +1.92 V d) +5.04 V e) +8.40 V

Answer: c

27. Calculate E^{o}_{cell} for the electrochemical cell below,

Ag(s) | AgCl(s) | Cl^-(aq, 1.0 M) || Fe^{3+}(aq, 1.0 M), Fe^{2+}(aq, 1.0 M) | Pt(s)

given the following reduction half-reactions.

$Ag^+(aq) + e^- \rightarrow Ag(s)$ $E° = +0.799$ V

$Fe^{3+}(aq) + e^- \rightarrow Fe^{2+}(aq)$ $E° = +0.771$ V

$AgCl(s) + e^- \rightarrow Ag(s) + Cl^-(aq)$ $E° = +0.222$ V

a) -0.250 V b) -0.028 V c) +0.549 V d) +0.993 V e) +1.570 V

Answer: c

28. Calculate E°_{cell} for the electrochemical cell below,

$Pb(s) \mid PbSO_4(s) \mid Pb^{2+}(aq) \parallel Ag^+(aq) \mid Ag(s)$

given the following standard reduction potentials.

$Ag^+(aq) + e^- \rightarrow Ag(s)$ $E° = +0.799$ V

$PbSO_4(s) + 2e^- \rightarrow Pb(s) + SO_4^{2-}(aq)$ $E° = -0.356$ V

a) -1.954 V b) -1.155 V c) -0.443 V d) +0.443 V e) +1.155 V

Answer: e

20.5 Electrochemical Cells under Nonstandard Conditions

29. Which of the following equations is a correct form of the Nernst equation?

a) $E = E^{\circ} - \left(\frac{RT}{nF}\right)\ln Q$

b) $E = E^{\circ} + \log\left(\frac{RT}{nF}\right)$

c) $E = E^{\circ} - \left(\frac{nF}{RT}\right)\log Q$

d) $E^{\circ} = E - \left(\frac{RT}{nF}\right)\ln Q$

e) $E = E^{\circ} - \left(\frac{nF}{RT}\right)\ln Q$

Answer: a

30. A Faraday, F, is defined as
a) the charge on a single electron.
b) the charge, in coulombs, carried by one mole of electrons.
c) the voltage required to reduce one mole of reactant.
d) the current required to reduce one mole of reactant.
e) the charge passed by one ampere of current in one second.

Answer: b

31. Calculate E for the following electrochemical cell at 25 °C

$Pt(s) \mid H_2(g, 1.00 \text{ atm}) \mid H^+(aq, 1.00 \text{ M}) \parallel Cu^{2+}(aq, 0.315 \text{ M}) \mid Cu(s)$

given the following standard reduction potentials.

$Cu^{2+}(aq) + 2e^- \rightarrow Cu(s)$ $E° = +0.337$ V

$2H^+(aq) + 2e^- \rightarrow H_2(g)$ $E° = 0.000$ V

a) +0.307 V b) +0.322 V c) +0.337 V d) +0.351 V e) +0.367 V

Answer: b

32. Calculate E for the following electrochemical cell at 25 °C

$Pt(s) \mid Sn^{2+}(aq, 0.50\ M), Sn^{4+}(aq, 0.50\ M) \parallel I^-(aq, 0.15\ M) \mid AgI(s) \mid Ag(s)$

given the following standard reduction potentials.

$AgI(s) + e^- \rightarrow Ag(s) + I^-(aq)$ $E° = -0.15\ V$

$Sn^{4+}(aq) + 2\ e^- \rightarrow Sn^{2+}(aq)$ $E° = +0.15\ V$

a) -0.35 V b) -0.32 V c) -0.30 V d) -0.25 V e) +0.05 V

Answer: d

33. Calculate the cell potential, at 25 °C, based upon the overall reaction

$3\ Cu^{2+}(aq) + 2\ Al(s) \rightarrow 3\ Cu(s) + 2\ Al^{3+}(aq)$

if $[Cu^{2+}] = 0.75$ M and $[Al^{3+}] = 0.0010$ M. The standard reduction potentials are as follows:

$Cu^{2+}(aq) + e^- \rightarrow Cu(s)$ $E° = +0.34\ V$

$Al^{3+}(aq) + 3\ e^- \rightarrow Al(s)$ $E° = -1.66\ V$

a) -2.11 V b) -1.26 V c) +1.94 V d) +2.06 V e) +2.11 V

Answer: d

34. The following electrochemical cell has a potential of +0.617 V at 25 °C.

$Pt \mid H_2(g, 1.00\ atm) \mid H^+(aq, 1.00\ M) \parallel Ag^+(aq) \mid Ag$

The standard reduction potential, $E°$, of $Ag^+ = +0.799$ V. What is the Ag^+ concentration?

a) 8.3×10^{-8} M b) 7.1×10^{-6} M c) 8.4×10^{-4} M d) 8.3×10^{-1} M e) 1.2×10^{3} M

Answer: c

35. What is the pH of the solution at the cathode if $E^\circ_{cell} = -0.174$ V for the following electrochemical cell at 25 °C?

$Pt \mid H_2(g, 1.0\ atm) \mid H^+(aq, 1.00\ M) \parallel H^+(aq) \mid H_2(g, 1.0\ atm) \mid Pt$

a) 2.94 b) 5.88 c) 6.77 d) 8.12 e) 13.54

Answer: a

36. Which factor will decrease the potential of the following electrochemical cell?

$Pt \mid Sn^{4+}(aq, 1.0\ M), Sn^{2+}(aq, 1.0\ M) \parallel Cu^{2+}(aq, 0.200\ M) \mid Cu$

a) switching from a platinum to a graphite anode
b) decreasing the size of the cathode
c) increasing the concentration of Cu^{2+}
d) decreasing the concentration of Sn^{4+}
e) increasing the temperature of the cell

Answer: e

20.6 Electrochemistry and Thermodynamics

37. E°_{cell} for the following galvanic cell is +0.254 V.

$$Hg_2^{2+}(aq) + 2\,I^-(aq) \rightarrow 2\,Hg(\ell) + I_2(s)$$

What is ΔG° for this reaction?

a) -49.0 kJ b) -24.5 kJ c) +24.5 kJ d) +49.0 kJ e) +197 kJ

Answer: a

38. Calculate ΔG° for the disproportionation reaction of Cu^+ at 25 °C,

$$2\,Cu^+(aq) \rightarrow Cu^{2+}(aq) + Cu(s)$$

given the following thermodynamic information.

$$Cu^+(aq) + e^- \rightarrow Cu(s) \qquad E^\circ = +0.518\text{ V}$$
$$Cu^{2+}(aq) + 2\,e^- \rightarrow Cu(s) \qquad E^\circ = +0.337\text{ V}$$

a) -165 kJ b) -135 kJ c) -34.9 kJ d) +17.5 kJ e) +135 kJ

Answer: c

39. If ΔG° for the following reaction is -324 kJ, calculate E°_{cell}.

$$Cr_2O_7^{2-}(aq) + 2\,Fe(s) + 14\,H^+(aq) \rightarrow 2\,Cr^{3+}(aq) + 2\,Fe^{3+}(aq) + 7\,H_2O(\ell)$$

a) -3.36 V b) -1.12 V c) +0.0201 V d) +0.560 V e) +1.68 V

Answer: d

40. Calculate the equilibrium constant for the following reaction at 25 °C,

$$Co^{2+}(aq) + 2\,Cr^{2+}(aq) \rightarrow Co(s) + 2\,Cr^{3+}(aq)$$

given the following thermodynamic information.

$$Co^{2+}(aq) + 2\,e^- \rightarrow Co(s) \qquad E^\circ = -0.28\text{ V}$$
$$Cr^{3+}(aq) + e^- \rightarrow Cr^{2+}(aq) \qquad E^\circ = -0.41\text{ V}$$

a) 5×10^{-24} b) 7×10^{-11} c) 4×10^{-5} d) 2×10^{4} e) 1×10^{10}

Answer: d

41. Given the following standard reduction potentials,

$$Fe^{2+}(aq) + 2\,e^- \rightarrow Fe(s) \qquad E^\circ = -0.440\text{ V}$$
$$FeS(s) + 2\,e^- \rightarrow Fe(s) + S^{2-}(aq) \qquad E^\circ = -1.010\text{ V}$$

determine the K_{sp} for FeS(s) at 25 °C.

a) 9×10^{-50} b) 5×10^{-20} c) 1×10^{-14} d) 2×10^{-10} e) 2×10^{20}

Answer: b

42. Calculate the standard reduction potential for the following reaction at 25 °C,

$$Ag(NH_3)_2^+(aq) + e^- \rightarrow Ag(s) + 2\ NH_3(aq)$$

given the following thermodynamic information.

$Ag^+(aq) + e^- \rightarrow Ag(s)$ $E° = 0.799$ V

$Ag^+(aq) + 2\ NH_3(aq) \rightarrow Ag(NH_3)_2^+(aq)$ $K_f = 1.7 \times 10^7$

a) -1.27 V b) -0.428 V c) +0.222 V d) +0.371 V e) +0.428 V

Answer: d

20.7 Electrolysis: Chemical Change using Electrical Energy

43. What charge, in coulombs, is required to deposit 0.301 g Cu(s) from a solution of $Cu^{2+}(aq)$?
a) 9.82×10^{-8} C b) 3.97×10^{-4} C c) 228 C d) 457 C e) 914 C

Answer: e

44. Al^{3+} is reduced to Al(s) at an electrode. If a current of 2.00 ampere is passed for 48 hours, what mass of aluminum is deposited at the electrode? Assume 100% current efficiency.
a) 3.58 g b) 32.2 g c) 48.3 g d) 96.6 g e) 2.90×10^2 g

Answer: b

45. One kind of battery used in watches contains mercury(II) oxide. As current flows, the mercury oxide is reduced to mercury.

$$HgO(s) + H_2O(\ell) + 2\ e^- \rightarrow Hg(\ell) + 2\ OH^-(aq)$$

If 2.5×10^{-5} amperes flows continuously 1095 days, what mass of $Hg(\ell)$ is produced?
a) 2.6×10^{-6} g b) 1.0×10^{-5} g c) 2.8×10^{-5} g d) 0.10 g e) 2.5 g

Answer: e

46. In an electrolytic cell, reduction occurs at the ________ and oxidation occurs at the ________.

Answer: cathode, anode

47. Gold and platinum are commonly used as inert electrodes in laboratory experiments. In commercial applications, such as batteries, ________ is more commonly used for inert electrodes because it is far less expensive.

Answer: graphite

48. What element serves as the anode in both a dry cell battery and an alkaline battery?

Answer: zinc

49. If electric current is passed through a solution of molten NaCl, the product at the cathode is ________.

Answer: Na(s)

50. When a secondary battery provides electrical energy, it is acting as a voltaic cell. When the battery is recharging, it is operating as a(n) ________ cell.

Answer: electrolytic

51. Write balanced reduction and oxidation half-reactions for the processes that occur at the cathode and anode of a fuel cell used aboard NASA's space shuttles.

Answer:

Cathode: $O_2(g) + 2\,H_2O(\ell) + 4\,e^- \rightarrow 4\,OH^-(aq)$

Anode: $H_2(g) \rightarrow 2\,H^+(aq) + 2\,e^-$

52. Explain the function of a salt bridge in a voltaic cell.

Answer: The salt bridge physically connects the redox couples in an electrochemical cell, thereby enabling electrical conduction through the cell. In addition, the salt bridge helps maintain electroneutrality as reduction and oxidation occur at the cathode and anode, respectively. As oxidation occurs, anions flow from the salt bridge toward the anode compartment and cations flow from the anode compartment into the salt bridge. Likewise, cations flow from the salt bridge toward the cathode compartment and anions flow from the cathode compartment into the salt bridge.

Diagnostic Quiz Questions

Goal #1: Balance equations for oxidation-reduction reactions in acidic or basic solutions using the half-reaction approach.

1. In the following reaction,

$$Sn^{2+}(aq) + Cu^{2+}(aq) \rightarrow Sn^{4+}(aq) + Cu(s)$$

a) Sn^{2+} is the reducing agent and Cu^{2+} is the oxidizing agent.
b) Sn^{2+} is the reducing agent and Sn^{4+} is the oxidizing agent.
c) Cu^{2+} is the reducing agent and Sn^{2+} is the oxidizing agent.
d) Cu^{2+} is the reducing agent and Cu(s) is the oxidizing agent.

Answer: a

2. Write a balanced half-reaction for the reduction of permanganate ion, MnO_4^-, to Mn^{2+} in an acidic solution.
   a) $MnO_4^-(aq) + 4\ H_2O(\ell) + 5\ e^- \rightarrow Mn^{2+}(aq) + 8\ OH^-(aq)$
   b) $MnO_4^-(aq) + 4\ H_2O(\ell) \rightarrow Mn^{2+}(aq) + 8\ OH^-(aq)$
   c) $MnO_4^-(aq) + 8\ H^+(aq) \rightarrow Mn^{2+}(aq) + 4\ H_2O(\ell)$
   d) $MnO_4^-(aq) + 8\ H^+(aq) + 5\ e^- \rightarrow Mn^{2+}(aq) + 4\ H_2O(\ell)$

   Answer: d

3. Write a balanced chemical equation for the following reaction in an acidic solution.
   $$Cr_2O_7^{2-}(aq) + I_2(aq) \rightarrow Cr^{3+}(aq) + IO_3^-(aq)$$
   a) $Cr_2O_7^{2-}(aq) + I_2(aq) + 2\ H^+(aq) \rightarrow 2\ Cr^{3+}(aq) + 2\ IO_3^-(aq) + H_2O(\ell)$
   b) $2\ Cr_2O_7^{2-}(aq) + 2\ I_2(aq) \rightarrow 4\ Cr^{3+}(aq) + 4\ IO_3^-(aq) + O_2(g)$
   c) $5\ Cr_2O_7^{2-}(aq) + 3\ I_2(aq) + 34\ H^+(aq) \rightarrow 10\ Cr^{3+}(aq) + 6\ IO_3^-(aq) + 17\ H_2O(\ell)$
   d) $2\ Cr_2O_7^{2-}(aq) + 2\ I_2(aq) + 4\ H^+(aq) \rightarrow 4\ Cr^{3+}(aq) + 4\ IO_3^-(aq) + 2\ H_2O(\ell)$

   Answer: c

4. Write a balanced chemical equation for the following reaction in a basic solution.
   $$MnO_2(s) + Cr(OH)_3(s) \rightarrow Mn(OH)_2(s) + CrO_4^{2-}(aq)$$
   a) $3MnO_2(s) + 2\ Cr(OH)_3(s) + 2\ H_2O(\ell) \rightarrow 3\ Mn(OH)_2(s) + 2\ CrO_4^{2-}(aq) + 4\ H^+(aq)$
   b) $MnO_2(s) + 2\ Cr(OH)_3(s) + 4\ OH^-(aq) \rightarrow Mn(OH)_2(s) + 2\ H_2CrO_4(aq) + 2\ H_2O(\ell)$
   c) $MnO_2(s) + Cr(OH)_3(s) + OH^-(aq) \rightarrow Mn(OH)_2(s) + CrO_4^{2-}(aq)$
   d) $3\ MnO_2(s) + 2\ Cr(OH)_3(s) + 4\ OH^-(aq) \rightarrow 3\ Mn(OH)_2(s) + 2\ CrO_4^{2-}(aq) + 2\ H_2O(\ell)$

   Answer: d

Goal #2: Understand the principles underlying voltaic cells.

5. Which of the following statements are correct?
   1. Oxidation occurs at the cathode.
   2. A voltaic cell consists of two half-cells.
   3. In electrochemical cells, electrons flow in the external circuit from the cathode to the anode.

   a) 1 only
   b) 2 only
   c) 3 only
   d) 1, 2, and 3

   Answer: b

6. What is the correct cell notation for a galvanic cell based on the reaction below?

$$Cu^{2+}(aq) + 2\ Fe(s) \rightarrow Cu(s) + 2\ Fe^{2+}(aq)$$

a) $Cu(s)\ |\ Cu^{2+}(aq)\ ||\ Fe^{2+}(aq)\ |\ Fe(s)$
b) $Cu^{2+}(aq)\ |\ Cu(s)\ ||\ Fe(s)\ |\ Fe^{2+}(aq)$
c) $Fe^{2+}(aq)\ |\ Fe(s)\ ||\ Cu(s)\ |\ Cu^{2+}(aq)$
d) $Fe(s)\ |\ Fe^{2+}(aq)\ ||\ Cu^{2+}(aq)\ |\ Cu(s)$

Answer: d

7. What type of battery has the following cathodic and anodic half-reactions?

Cathode: $2\ MnO_2(s) + H_2O(\ell) + 2\ e^- \rightarrow Mn_2O_3(s) + 2\ OH^-(aq)$

Anode: $Zn(s) + 2\ OH^-(aq) \rightarrow ZnO(s) + H_2O(\ell) + 2\ e^-$

a) lead storage battery
b) alkaline battery
c) nickel-cadmium battery
d) LeClanché battery

Answer: b

Goal #3: Understand how to use electrochemical potentials

8. Use the standard reduction potentials below to determine which species is the best oxidizing agent?

$Cl_2(g) + 2\ e^- \rightarrow 2\ Cl^-(aq)$ $E° = +1.36$ V
$Ag^+(aq) + e^- \rightarrow Ag(s)$ $E° = +0.80$ V
$Fe^{2+}(aq) + 2\ e^- \rightarrow Fe(s)$ $E° = -0.44$ V

a) $Cl_2(g)$
b) $Ag^+(aq)$
c) $Fe^{2+}(aq)$
d) $Fe(s)$

Answer: a

9. Use the standard reduction potentials below to determine which species is the best reducing agent?

$Au^{3+}(aq) + 3\ e^- \rightarrow Au(s)$ $E° = +1.50$ V
$Ag^+(aq) + e^- \rightarrow Ag(s)$ $E° = +0.80$ V
$Cd^{2+}(aq) + 2\ e^- \rightarrow Cd(s)$ $E° = -0.403$ V

a) Au^{3+}
b) Ag^+
c) Cd^{2+}
d) $Cd(s)$

Answer: d

10. Given the following two half-reactions, determine which overall reaction is spontaneous and calculate its standard cell potential.

$Mg^{2+}(aq) + 2\,e^- \rightarrow Mg(s) \qquad E° = -2.37\ V$
$Ni^{2+}(aq) + 2\,e^- \rightarrow Ni(s) \qquad E° = -0.25\ V$

a) $Mg^{2+}(aq) + Ni(s) \rightarrow Mg(s) + Ni^{2+}(aq)$; $E°_{cell} = -2.12\ V$
b) $Mg^{2+}(aq) + Ni^{2+}(aq) \rightarrow Mg(s) + Ni(s)$; $E°_{cell} = +2.62\ V$
c) $Mg(s) + Ni^{2+}(aq) \rightarrow Mg^{2+}(aq) + Ni(s)$; $E°_{cell} = +2.12\ V$
d) $Mg(s) + Ni^{2+}(aq) \rightarrow Mg^{2+}(aq) + Ni(s)$; $E°_{cell} = -2.12\ V$

Answer: c

11. Calculate E for the following electrochemical cell at 25 °C

$Pt \mid H_2(g, 1.00\ atm) \mid H^+(aq, 0.500\ M) \parallel Cu^{2+}(aq, 0.175\ M) \mid Cu$

given the following standard reduction potentials.

$Cu^{2+}(aq) + 2\,e^- \rightarrow Cu(s) \qquad E° = 0.337\ V$
$2\,H^+(aq) + 2\,e^- \rightarrow H_2(g) \qquad E° = 0.000\ V$

a) +0.346 V
b) +0.342 V
c) +0.332 V
d) +0.335 V

Answer: c

12. Calculate $\Delta G°$ for the following reaction

$Ni^{2+}(aq) + Sn(s) \rightarrow Ni(s) + Sn^{2+}(aq)$

given the following reduction half-reactions and standard reduction potentials.

$Sn^{2+}(aq) + 2\,e^- \rightarrow Sn(s) \qquad E° = -0.15\ V$
$Ni^{2+}(aq) + 2\,e^- \rightarrow Ni(s) \qquad E° = -0.25\ V$

a) -77 kJ
b) -19 kJ
c) +19 kJ
d) +77 kJ

Answer: c

13. Given the following standard reduction potentials,

$Pb^{2+}(aq) + 2\,e^- \rightarrow Pb(s)$ $E° = -0.126$ V

$PbSO_4(s) + 2\,e^- \rightarrow Pb(s) + SO_4^{2-}(aq)$ $E° = -0.356$ V

determine the K_{sp} for $PbSO_4(s)$ at 25 °C.
a) 5.0×10^{-17}
b) 1.7×10^{-8}
c) 2.0×10^{16}
d) 6.0×10^{7}

Answer: b

Goal #4: Explore electrolysis, the use of electrical energy to produce chemical change.

14. $Al^{3+}(aq)$ is reduced to Al(s) at an electrode. If a current of 1.22 ampere is passed for 12.0 hours, what mass of aluminum is deposited at the electrode? Assume 100 % current efficiency.
a) 1.36×10^{-3} g
b) 4.91 g
c) 14.7 g
d) 44.2 g

Answer: b

15. How much time (in seconds) is required to produce 5.00 g Zn(s) from $Zn(NO_3)_2$ by electrolysis using a constant current of 2.00 A?
a) 7.98×10^{2} s
b) 1.84×10^{3} s
c) 3.69×10^{3} s
d) 7.38×10^{3} s

Answer: d

CHAPTER 21: The Chemistry of the Main Group Elements

21.1 Element Abundances

1. Which three elements are most abundant in the Earth's crust?
   a) hydrogen, helium, and lithium
   b) sodium, calcium, and iron
   c) oxygen, nitrogen, and iron
   d) carbon, nitrogen, and oxygen
   e) oxygen, aluminum, and silicon

   Answer: e

2. Which two elements are most abundant in our solar system?
   a) hydrogen and helium
   b) hydrogen and oxygen
   c) helium and iron
   d) hydrogen and iron
   e) carbon and oxygen

   Answer: a

21.2 The Periodic Table: A Guide to the Elements

3. What is the charge on a nonmetallic, monatomic ion from Group 5A?
   a) 5+ b) 3+ c) 1+ d) 3- e) 5-

   Answer: d

4. What is the highest oxidation number of bromine?
   a) 1- b) 0 c) 5+ d) 7+ e) 8+

   Answer: d

5. Predict the chemical formula for hypoiodic acid.
   a) HI_3 b) HIO_4 c) HIO_3 d) HIO_2 e) HIO

   Answer: e

6. Use your knowledge of group properties to predict the chemical formula for selenic acid.
   a) $HSeO_2$ b) H_2SeO_3 c) H_2SeO_4 d) H_3SeO_3 e) H_3SeO_4

   Answer: c

21.3 Hydrogen

7. Approximately 300 billion liters of hydrogen is produced annually. The majority of the hydrogen is used in the production of ________.
   a) ammonia
   b) hydrochloric acid
   c) methane
   d) sodium hydroxide
   e) water

   Answer: a

8. Which of the following chemical equations shows the formation of synthesis gas?
   a) $2\ H_2O(g) \rightarrow 2\ H_2(g) + O_2(g)$
   b) $16\ CH_4(g) + 7\ O_2(g) \rightarrow 2\ C_8H_{18}(g) + 14\ H_2O(g)$
   c) $C(s) + 2\ H_2(g) \rightarrow CH_4(g)$
   d) $CO(g) + H_2O(g) \rightarrow CO_2(g) + H_2(g)$
   e) $C(s) + H_2O(g) \rightarrow H_2(g) + CO(g)$

   Answer: e

9. Metal hydrides are used to remove traces of water from nonpolar solvents. Write a balanced chemical equation for the reaction of lithium hydride with water.
   a) $LiH(s) + H_2O(\ell) \rightarrow Li(s) + H_3O^+(aq)$
   b) $LiH(s) + H_2O(\ell) \rightarrow LiOH(s) + H_2(g)$
   c) $2\ LiH(s) + H_2O(\ell) \rightarrow Li_2O(s) + 2\ H_2(g)$
   d) $2\ LiH(s) + 2\ H_2O(\ell) \rightarrow Li_2O_2(s) + O_2(g) + 6\ H^+(aq)$
   e) $2\ LiH(s) + 4\ H_2O(\ell) \rightarrow 2\ LiO_2(s) + 5\ H_2(g)$

   Answer: b

10. All of the following combinations of elements or compounds will react to produce hydrogen gas EXCEPT
    a) $Na(s)$ and $H_2O(\ell)$.
    b) $MgH_2(s)$ and $H_2O(\ell)$.
    c) $Ca(s) + HCl(aq)$.
    d) $Na_2S(s) + HCl(aq)$.
    e) $Al(s) + KOH(aq)$.

    Answer: d

21.4 The Alkali Metals, Group 1A

11. All of the following statements concerning sodium are correct EXCEPT
    a) most sodium compounds are water soluble.
    b) sodium is soft and easily cut with a knife.
    c) sodium is usually found in nature as the reduced metal or as a sulfide salt.
    d) sodium ions help regulate osmotic pressure in the human body.
    e) sodium is commercially prepared by electrolysis of molten NaCl.

    Answer: c

12. The principal product of the reaction of potassium and oxygen is not K_2O. The product is ________ and has the formula ________.
    a) potassium trioxide, K_2O_3
    b) potassium peroxide, K_2O_2
    c) potassium superoxide, KO_2
    d) potassium monoxide, KO
    e) potassium dioxide, K_3O_2

    Answer: c

13. Elemental sodium is commercially produced by electrolysis ________. What valuable by-product is simultaneously produced during the electrolysis?
    a) brine, $Cl_2(g)$
    b) NaH, $H_2(g)$
    c) soda water, NaOH(aq)
    d) Na_2O_2, $O_2(g)$
    e) Na_2S, $S_8(s)$

    Answer: a

14. Write a balanced chemical equation for the reaction of sodium and water.
    a) $Na(s) + H_2O(\ell) \rightarrow NaO(s) + H_2(g)$
    b) $4\ Na(s) + 2\ H_2O(\ell) \rightarrow 4\ NaH(s) + O_2(g)$
    c) $2\ Na(s) + 2\ H_2O(\ell) \rightarrow Na_2O_2(s) + 2\ H_2(g)$
    d) $2\ Na(s) + H_2O(\ell) \rightarrow Na_2O(s) + H_2(g)$
    e) $2\ Na(s) + 2\ H_2O(\ell) \rightarrow 2\ NaOH(aq) + H_2(g)$

    Answer: e

15. Which of the following statements are CORRECT?
    1. Chili saltpeter is a common name for $NaNO_3$.
    2. Trona is a common name for $Na_2CO_3 \cdot NaHCO_3 \cdot 2\,H_2O$.
    3. Soda ash is a common for Na_2CO_3.

a) 1 only b) 2 only c) 3 only d) 1 and 3 e) 1, 2, and 3

Answer: e

21.5 The Alkaline Earth Elements, Group 2A

16. Write a balanced chemical equation for the reaction of lime and water to form slaked lime.

a) $CaO(s) + 2\,H_2O(\ell) \rightarrow CaCO_3(s) + 2\,H_2(g)$

b) $CaO(s) + H_2O(\ell) \rightarrow Ca(OH)_2(s)$

c) $Ca_3(PO_4)_2(s) + 2\,H_2O(\ell) \rightarrow 2\,CaHPO_4(aq) + Ca(OH)_2(aq)$

d) $CaCO_3(s) + H_2O(\ell) \rightarrow Ca(s) + H_3O^+(aq) + CO_2(g)$

e) $CaCO_3(s) + H_2O(\ell) \rightarrow CaO(s) + H_2CO_3(aq)$

Answer: b

17. Which of the following statements are CORRECT?
    1. Diagnostic imaging of the human digestive tract often requires a patient to consume a slurry of BaSO4(s), a compound that contains the toxic Ba^{2+} ion.
    2. Calcium is the central element in chlorophyll.
    3. Beryllium is toxic. Exposure (by breathing) to beryllium can cause berylliosis.

a) 1 only b) 2 only c) 3 only d) 1 and 3 e) 1, 2, and 3

Answer: d

18. At high temperatures limestone decomposes according to which balanced chemical equation?

a) $Ca(OH)_2(s) \rightarrow CaO(s) + H_2O(g)$
b) $2\,CaO(s) \rightarrow 2\,Ca(s) + O_2(g)$
c) $CaCO_3(s) \rightarrow CaO(s) + CO_2(g)$
d) $Ca(OH)_2(s) \rightarrow CaH_2(s) + O_2(g)$
e) $2\,CaCO_3(s) \rightarrow 2\,Ca(s) + O_2(g) + CO_2(g)$

Answer: c

19. Oxides of the alkaline earth family form ________ upon reaction with water.
    a) basic solutions
    b) carbon dioxide gas
    c) alkaline earth hydrides
    d) oxygen gas
    e) hydrogen gas

    Answer: a

21.6 Boron, Aluminum, and the Group 3A Elements

20. Aluminum is resistant to corrosion because
    a) the surface of aluminum is coated with unreactive AlN.
    b) the surface of aluminum is coated with unreactive Al_2O_3.
    c) aluminum is a noble metal.
    d) aluminum contains trace amounts of magnesium and chromium.
    e) aluminum has a large positive standard reduction potential.

    Answer: b

21. Who developed the process for obtaining aluminum from bauxite ore?
    a) Antoine Lavoisier
    b) Fritz Haber
    c) Michael Faraday
    d) Herman Frasch
    e) Charles Hall

    Answer: e

22. In what way is cryolite (Na_3AlF_6) used in the production of aluminum?
    a) Cryolite is a catalyst, it increases the rate of aluminum reduction.
    b) Cryolite-bauxite mixtures melt at lower temperatures than pure bauxite.
    c) Cryolite makes aluminum more corrosion resistant.
    d) Cryolite is reduced to aluminum at the cathode.
    e) Cryolite is oxidized to $F_2(g)$ at the anode.

    Answer: b

23. Molten Al_2O_3 is reduced by electrolysis at low potentials and high currents. If 4.0×10^4 amperes of current is passed through molten Al_2O_3 for 6.0 hours, what mass of Al is produced? Assume 100% current efficiency.
    a) 1.3×10^4 g b) 9.0×10^3 g c) 8.1×10^4 g d) 2.4×10^5 g e) 7.2×10^5 g

    Answer: c

24 Which of the following statements are CORRECT?

1. More than 20 neutral boron hydrides (have formulas B_xH_y) are known to exist.
2. Boric acid, $B(OH)_3$, is used as a flame retardant for cellulose insulation.
3. Borax, $Na_2B_4O_7 \cdot 10\ H_2O$, is used as a low-melting flux in metallurgy.

a) 1 only b) 2 only c) 3 only d) 1 and 3 3) 1, 2, and 3

Answer: e

21.7 Silicon and the Group 4A Elements

25. The building block of silicate minerals is the
a) linear SiO unit.
b) the bent SiO_2 unit.
c) trigonal-planar SiO_3 unit.
d) tetrahedral SiO_4 unit.
e) the octahedral SiO_6 unit.

Answer: d

26. Quartz is the pure crystalline form of
a) silica, SiO_2.
b) sodium silicate, $NaSiO_4$.
c) calcium orthosilicate, Ca_2SiO_4.
d) silicon tetrachloride, $SiCl_4$.
e) silicon, Si.

Answer: a

27. The only acid that attacks silicon dioxide is
a) $HClO_4$ b) HF c) HNO_3 d) H_2SO_4 e) HIO_4

Answer: b

21.8 Nitrogen, Phosphorus, and the Group 5A Elements

28. Which one of the following metals is not oxidized by nitric acid?
a) K b) Zn c) Al d) Cu e) Mg

Answer: c

29. The Ostwald process is an industrial method of producing ________ from the oxidation of ________.
    a) nitric acid, ammonia
    b) ammonia, nitric acid
    c) ammonia, nitrogen
    d) nitric acid, nitrogen
    e) dinitrogen monoxide, nitrogen

    Answer: a

30. Dinitrogen monoxide, or laughing gas, can be made by decomposing ammonium nitrate at 250 °C. Write a balanced equation for this reaction.
    a) $NH_4NO_3(s) \rightarrow N_2O(g) + 2\ H_2(g) + O_2(g)$
    b) $NH_4NO_3(s) + 2\ N_2(g) \rightarrow 3\ N_2O(g) + 2\ H_2(g)$
    c) $NH_4NO_3(s) \rightarrow N_2O(g) + 2\ H_2O(g)$
    d) $2\ NH_4NO_3(s) \rightarrow N_2O(g) + 2\ HNO_2(g) + H_2O(g) + 2\ H_2(g)$
    e) $3\ NH_4NO_3(s) \rightarrow N_2O(g) + 2\ N_2(g) + 8\ H_2O(g)$

    Answer: c

31. The reaction of ammonia and bleach produces hydrazine. Write a balanced chemical equation for this reaction.
    a) $2\ NH_3(aq) + NaClO(aq) \rightarrow N_2H_4(aq) + NaCl(aq) + H_2O(\ell)$
    b) $2\ NH_3(aq) + 2\ NaClO(aq) \rightarrow N_2H_4(aq) + 2\ NaCl(aq) + H_2(g) + O_2(g)$
    c) $2\ NH_3(aq) + 2\ NaClO(aq) \rightarrow N_2H_4(aq) + 2\ HCl(aq) + Na_2O_2(aq)$
    d) $2\ NH_3(aq) + 2\ NaClO(aq) \rightarrow N_2H_4(aq) + 2\ Na(s) + 2\ HClO(aq)$
    e) $2\ NH_3(aq) + 2\ NaClO(aq) \rightarrow N_2H_4(aq) + Cl_2(aq) + Na_2O(aq) + H_2O(\ell)$

    Answer: a

32. Which of the following nitrogen oxides are free radicals: NO, N_2O, NO_2, and N_2O_3?
    a) NO and N_2O
    b) N_2O and NO_2
    c) NO_2 and N_2O_3
    d) NO and NO_2
    3) N_2O and N_2O_3

    Answer: d

21.9 Oxygen, Sulfur and the Group 6A Elements

33. In the United States, most sulfur is obtained from
    a) the reduction of cinnabar, HgS.
    b) S_8 deposits along the Gulf of Mexico.
    c) the reduction of iron pyrite, FeS_2.
    d) the reduction of copper sulfide, CuS.
    e) the oxidation of H_2S gas.

    Answer: b

34. Which compound is ranked #1 in terms of the amount produced annually in the United States?
    a) NH_3 b) N_2O c) $NaOH$ d) CaO e) H_2SO_4

    Answer: e

35. Pure oxygen can be produced by the catalyzed decomposition of potassium chlorate. Write a balanced chemical equation for this reaction.
    a) $KClO_3(s) \rightarrow KClO(s) + O_2(g)$
    b) $2\ KClO_3(s) \rightarrow 2\ K(s) + Cl_2(g) + 3\ O_2(g)$
    c) $2\ KClO_3(s) \rightarrow 2\ KCl(s) + 3\ O_2(g)$
    d) $2\ KClO_3(s) \rightarrow 2\ KClO_2(s) + O_2(g)$
    e) $KClO_3(s) \rightarrow K(s) + ClO(g) + O_2(g)$

    Answer: c

36. Write a balanced chemical equation for the reaction of sulfur trioxide and water.
    a) $SO_3(g) + H_2O(\ell) \rightarrow H_2SO_4(aq)$
    b) $2\ SO_3(g) + 2\ H_2O(\ell) \rightarrow H_2SO_4(aq) + H_2S(g) + 2\ O_2(g)$
    c) $SO_3(g) + H_2O(\ell) \rightarrow SO_4^{2-}(g) + H_2(g)$
    d) $SO_3(g) + H_2O(\ell) \rightarrow SO_4(aq) + H_2(g)$
    e) $2\ SO_3(g) + 2\ H_2O(\ell) \rightarrow 2\ H_2SO_3(aq) + O_2(g)$

    Answer: a

37. Write a balanced chemical equation for the reaction of CuS and hydrochloric acid.
a) $CuS(s) + 2\ HCl(aq) \rightarrow CuCl_2(aq) + H_2(g) + S(s)$
b) $CuS(s) + 2\ HCl(aq) \rightarrow Cu(s) + SCl_2(g) + H_2(g)$
c) $CuS(s) + 2\ HCl(aq) \rightarrow CuH_2(s) + SCl_2(g)$
d) $CuS(s) + 2\ HCl(aq) \rightarrow CuCl_2(aq) + H_2S(g)$
e) $CuS(s) + 2\ HCl(aq) \rightarrow Cu(s) + S(s) + H_2(g) + Cl_2(g)$

Answer: d

21.10 The Halogens, Group 7A

38. Ammonium perchlorate can decompose violently according to the chemical equation below.
$$2\ NH_4ClO_4(s) \rightarrow N_2(g) + Cl_2(g) + 2\ O_2(g) + 4\ H_2O(g)$$
What volume of gas is produced by the decomposition of 225 g NH_4ClO_4 at 252 °C and 1.0 atm? ($R = 0.08206$ L·atm/K·mol)
a) 4.0×10^1 L b) 1.6×10^2 L c) 3.3×10^2 L d) 6.6×10^2 L e) 8.3×10^2 L

Answer: c

39. What is the molecular formula of chlorous acid?
a) $HClO$ b) $HClO_2$ c) $HClO_3$ d) $HClO_4$ e) HCl

Answer: b

40. All of the following are commercial uses of chlorine or chlorine containing compounds EXCEPT
a) water purification.
b) pulp and paper manufacturing.
c) manufacturing pharmaceuticals.
d) making plastics.
e) reducing metal ions in waste water.

Answer: e

41. The most reactive element is ________. This element forms compounds with every other element except He and Ne.
a) oxygen b) fluorine c) lithium d) phosphorus e) carbon

Answer: b

42. Anhydrous hydrogen fluoride is used in all of the following applications EXCEPT
a) etching fluorescent light bulbs.
b) the production of refrigerants.
c) as a propellant in rocket fuels.
d) the production of plastics.
e) the production of herbicides.

Answer: c

43. Write a balanced chemical equation for the reaction of Al(s) and Cl_2(g).
a) Al(s) + Cl_2(g) $\rightarrow$ $AlCl_2$(s)
b) 2 Al(s) + Cl_2(g) $\rightarrow$ 2 AlCl(s)
c) 2 Al(s) + 3 Cl_2(g) $\rightarrow$ 2 $AlCl_3$(s)
d) 4 Al(s) + 3 Cl_2(g) $\rightarrow$ 2 Al_2Cl_3(s)
e) 4 Al(s) + Cl_2(g) $\rightarrow$ 2 Al_2Cl(s)

Answer: c

44. A ________ reaction is one where an element or compound is simultaneously oxidized and reduced.

Answer: disproportionation

45. Red phosphorus (P) and white phosphorus (P_4) are examples of two different forms of the same element. Two or more different forms of an element are called ________.

Answer: allotropes

46. Silicon is purified in a process called ________, where a narrow segment of a silicon rod is melted and allowed to slowly recrystallize.

Answer: zone refining

47. The primary product of the reaction of potassium and oxygen is ________. This compound is used in oxygen generators.

Answer: potassium superoxide or KO_2

48. Hydrogen is absorbed into the lattice of many metals. The resulting compounds are called ________ hydrides because the hydrogen resides in the spaces between the metal atoms.

Answer: interstitial

49. Concentrated nitric acid reacts with copper to form $NO_2(g)$. Dilute nitric acid reacts with copper to form NO(g). Why does the reaction involving the dilute acid give a different nitrogen oxide than the concentrated acid?

Answer: The oxidation states of the nitrogen in each compound are listed below.

Compound	Oxidation State
HNO_3	+5
NO_2	+4
NO	+2

The more dilute acid contains fewer nitrate ions that can react with the copper. To fully oxidize the copper, each nitrogen needs to gain more electrons. Likewise, because of the large number of nitrate ions present in the concentrated acid, each nitrogen only gains a single electron.

50. One commercial method of producing Cl_2 is by electrolysis of aqueous NaCl.

$$NaCl(aq) + 2\,H_2O(\ell) \rightarrow Cl_2(g) + 2\,NaOH(aq) + H_2(g)$$

Owing to environmental concerns, some manufacturers are shifting to the soda-lime method for NaOH production. Explain why this method is environmentally preferred. Write a balanced chemical equation for the production of NaOH from soda ash and lime.

Answer: The electrolysis method of manufacturing sodium hydroxide also produces chlorine. Chlorine and its by-products are considered environmentally unfriendly substances. The balanced equation for the soda-lime process is below.

$$Na_2CO_3(s) + CaO(s) + H_2O(\ell) \rightarrow 2\,NaOH(aq) \text{ and } CaCO_3(s)$$

Diagnostic Quiz Questions

Goal #1: Relate the formulas and properties of compounds to the periodic table

1. What is the highest oxidation state of iodine?
   a) -1
   b) +1
   c) +7
   d) +17

   Answer: c

2. Use your knowledge of group properties to predict the chemical formula for arsenous acid.
   a) H_2As
   b) H_3AsO_3
   c) H_2AsO_4
   d) H_3AsO_4

   Answer: b

Goal #2: Describe the chemistry of the main group or A-Group elements, particularly, H; Na and K; Mg and Ca; B and Al; Si; Na and P; O and S; and F and Cl.

3. Write a balanced chemical equation for the reaction of strontium and sulfur.
   a) $Sr(s) + S_8(s) \rightarrow SrS_8(s)$
   b) $4\ Sr(s) + S_8(s) \rightarrow 4\ SrS_2(s)$
   c) $16\ Sr(s) + 3\ S_8(s) \rightarrow 8\ Sr_2S_3(s)$
   d) $8\ Sr(s) + S_8(s) \rightarrow 8\ SrS(s)$

   Answer: d

4. Write a balanced chemical equation for the reaction of silicon and iodine.
   a) $Si(s) + 2\ I_2(g) \rightarrow SiI_4(g)$
   b) $Si(s) + I_2(g) \rightarrow SiI_2(g)$
   c) $4\ Si(s) + 3\ I_2(g) \rightarrow 2\ Si_2I_3(g)$
   d) $3\ Si(s) + I_2(g) \rightarrow Si_3I_2(g)$

   Answer: a

5. Which of the following chemical formulas is incorrect?
   a) Ca_3N_2
   b) BaO_2
   c) Al_2S_3
   d) KF

   Answer: b

6. Which element is most abundant in our solar system?
   a) hydrogen
   b) helium
   c) carbon
   d) iron

   Answer: a

7. What is the primary commercial use of hydrogen?
   a) the manufacture of ammonia
   b) combustion in power plants for electricity
   c) the production of methane from coal
   d) the reaction with alkali metals to form metal hydrides

   Answer: a

8. Write a balanced chemical equation for the primary reaction of sodium and oxygen and name the product.
   a) $Na(s) + O_2(g) \rightarrow NaO_2(s)$; sodium superoxide
   b) $4\ Na(s) + O_2(g) \rightarrow 2\ Na_2O(s)$; sodium oxide
   c) $2\ Na(s) + O_2(g) \rightarrow Na_2O_2(s)$; sodium peroxide
   d) $2\ Na(s) + O_2(g) \rightarrow Na_2O_2(s)$; disodium dioxide

   Answer: c

9. Which of the following statements are correct?
   1. Magnesium is the central element in chlorophyll.
   2. Slaked lime is a common name for calcium hydroxide.
   3. Beryllium compounds are carcinogenic.

   a) 1 only
   b) 2 only
   c) 3 only
   d) 1, 2, and 3

   Answer: d

10. Write a balanced chemical equation for the reaction of diborane with oxygen to produce boric oxide and water.
    a) $2\ B_2(g) + 3\ O_2(g) \rightarrow 2\ B_2O_3(s)$
    b) $4\ BH_3(g) + 7\ O_2(g) \rightarrow 4\ BO_2(s) + 6\ H_2O(g)$
    c) $B_2H_6(g) + 3\ O_2(g) \rightarrow B_2O_3(s) + 3\ H_2O(g)$
    d) $2\ B_2H_6(g) + 7\ O_2(g) \rightarrow 4\ BO_2(s) + 6\ H_2O(g)$

    Answer: c

11. Write a balanced chemical equation for the reaction of aluminum with HBr(g).
    a) $2\ Al(s) + 6\ HBr(g) \rightarrow Al_2Br_6(s) + 3\ H_2(g)$
    b) $Al(s) + 6\ HBr(g) \rightarrow Al_2Br_6(s) + 6\ H^+(g)$
    c) $Al(s) + 2\ HBr(g) \rightarrow AlBr_2(s) + H_2(g)$
    d) $Al(s) + 3\ HBr(g) \rightarrow AlBr_3(s) + 3\ H^+(g)$

    Answer: a

12. Which of the following Group 4A compounds is most likely to have covalent bonds?
    a) PbO_2
    b) $SnBr_4$
    c) $SiCl_4$
    d) GeI_2

    Answer: c

CHAPTER 22: The Chemistry of the Transition Elements

22.1 Properties of the Transition Elements

1. Which element has the highest melting point?
a) Ti b) Zr c) W d) Au e) Hg

Answer: c

2. ________ is a crucial component of vitamin B_{12}.
a) Fe b) Co c) Zn d) Cu e) Mo

Answer: b

3. Which of the following transition metal compounds is used as white paint pigment?
a) TiO_2 b) $KMnO_4$ c) Fe_2O_3 d) Cr_2O_3 e) CdS

Answer: a

4. Which ion produces the red color in rubies?
a) Zn^{2+} b) Fe^{2+} c) Cr^{3+} d) Zr^{4+} e) Ti^{4+}

Answer: c

5. Which of the following ions have an $[Ar]3d^5$ electron configuration: Mn^{2+}, Fe^{3+}, and Co^{2+}?
a) Mn^{2+} only b) Fe^{3+} only c) Co^{2+} only d) Mn^{2+} and Fe^{3+} e) Mn^{2+} and Co^{2+}

Answer: d

6. Identify the element with the ground state electron configuration $[Xe]4f^{14}5d^5 6s^2$.
a) Tc b) Ta c) Os d) At e) Re

Answer: e

7. What is the maximum oxidation state for chromium?
a) +3 b) +6 c) +7 d) +8 e) +10

Answer: b

8. Which element from the first transition series does NOT react with hydrochloric acid?
a) titanium b) chromium c) cobalt d) nickel e) copper

Answer: e

9. Which of the following statements are CORRECT?
   1. The atomic radii of transition metals decrease initially, then increase as one moves from left to right across a period of the periodic table.
   2. The atomic radii of the period 6 transition metals are approximately 25% larger than the atomic radii of the period 5 transition metals.
   3. The densities of the period 5 transition metals are greater than those from either the period 4 or period 6 transition metals.

   a) 1 only b) 2 only c) 3 only d) 1 and 2 e) 1, 2, and 3

   Answer: a

10. Given the following standard reduction potentials,

$$Fe^{3+}(aq) + e^- \rightarrow Fe^{2+}(aq) \qquad E° = +0.771\ V$$
$$2\ H^+(aq) + 2\ e^- \rightarrow H_2(g) \qquad E° = 0.000\ V$$
$$Fe^{2+}(aq) + 2\ e^- \rightarrow Fe(s) \qquad E° = -0.44\ V$$

predict which one of the following reactions will occur.

a) $Fe^{2+}(aq) + H_2(g) \rightarrow Fe(s) + 2\ H^+(aq)$
b) $2\ Fe^{3+}(aq) + 6\ H^+(aq) \rightarrow 2\ Fe(s) + 3\ H_2(g)$
c) $Fe^{2+}(aq) + 2\ H^+(aq) \rightarrow Fe^{3+}(aq) + H_2(g)$
d) $Fe(s) + 2\ H^+(aq) \rightarrow Fe^{2+}(aq) + H_2(g)$
e) $3\ Fe^{2+}(s) \rightarrow Fe(s) + 2\ Fe^{3+}(aq)$

Answer: d

21.2 Metallurgy

11. If an element can be obtained profitably from a mineral, the mineral is called a(n) ________.

    a) gangue b) ore c) slag d) roast e) alloy

    Answer: b

12. Ores are often found mixed with impurities such as sand and clay. In metallurgy, these impurities are called ________.

    a) gangue b) alloys c) slags d) roasts e) precipitates

    Answer: a

13. Which of the following groups of metals occur in nature as the free elements?

    a) Ni, Cu, Zn b) Sc, Y, La c) Cr, Mo, W d) Zn, Cd, Hg e) Cu, Pt, Au

    Answer: e

14. What element or compound is used to reduce Fe_2O_3 to Fe in a blast furnace?

    a) H_2 b) CaO c) $CaSiO_3$ d) CO e) CO_2

    Answer: d

15. Copper ores are enriched to increase the percentage of copper by a process called ________.
a) electrolysis b) roasting c) flotation d) tempering e) pyrometallurgy

Answer: c

22.3 Coordination Compounds

16. Ions such as $[Co(H_2O)_6]^{3+}$ and $[Ag(CN)_2]^-$ are called ________.
a) coordination complexes
b) Lewis bases
c) chelates
d) polydentates
e) ligands

Answer: a

17. Which molecule or ion can NOT act as a ligand?
a) OH^- b) NH_3 c) Br^- d) CH_4 e) CN^-

Answer: d

18. All of the following molecules or ions can act as polydentate ligands EXCEPT ________.
a) phenanthroline ($C_{12}H_8N_2$)
b) ethylenediaminetetraacetate ion ($(O_2CCH_2)_2NCH_2CH_2N(CH_2CO_2)_2^{4-}$)
c) oxalate ion ($C_2O_4^{2-}$)
d) ethylenediamine ($H_2NCH_2CH_2NH2$)
e) methylamine (CH_3NH_2)

Answer: e

19. What is the coordination number of the central metal ion in $K_3[Co(H_2O)_4Cl_2]$?
a) 2 b) 3 c) 4 d) 6 e) 9

Answer: d

20. What is the coordination number of the central metal ion in $[Ni(en)_2(OH)_2]$?
a) 2 b) 4 c) 6 d) 8 e) 10

Answer: c

21. What is the oxidation state of iron in $K_3[Fe(CN)_6]$?
a) -3 b) +2 c) +3 d) +4 e) +6

Answer: c

22. What is the oxidation state of chromium in $[Cr(NH_3)_2(H_2O)_2Cl_2]Cl$?
   a) +1 b) +2 c) +3 d) +4 e) +6

   Answer: c

23. What is the name of the compound having the formula $K_4[CrBr_6]$?
   a) potassium bromochromate(II)
   b) potassium hexabromochromate(II)
   c) potassium chromohexabromate(II)
   d) hexabromo tetrapotassiumchromate(II)
   e) tetrapotassium hexabromochromate(II)

   Answer: b

24. What is the name of the compound having the formula $[Cr(en)_2(H_2O)_2]SO_4$?
   a) dihydroxodiethlyenediaminechromate(II) sulfate
   b) diaquabis(ethylenediamine)chromium(II) sulfate
   c) bis(ethylenediamine)diaquachromium(II) sulfate
   d) diaquebis(ethylenediamine)sulfatochromium(II)
   e) diaquabis(ethylenediamine)sulfatochromate(II)

   Answer: b

25. What is the name of the compound having the formula $(NH_4)_3[Co(CN)_6]$?
   a) ammonium hexacyanocobaltate(III)
   b) triammonium hexacyanocobaltate(III)
   c) triamminehexacyanocobaltate(III)
   d) hexacyanotriamminecobaltate(II)
   e) hexacyanocobaltate(II) ammonia

   Answer: a

26. What is the name of the compound having the formula $[Fe(CO)_6]Br_2$?
   a) hexacarbonmonoxideiron(II) bromate
   b) hexacarbonmonoxideiron(III) bromide
   c) dibromohexacarbonylironate(II)
   d) hexacarbonyliron(III) bromide
   e) hexacarbonyliron(II) bromide

   Answer: e

27. What is the formula for dichlorobis(phenanthroline)chromium(III) chloride?
 a) $[Cr(phen)_2]Cl_3$
 b) $[CrCl_3(phen)_2]$
 c) $Cl[CrCl_2(phen)_2]$
 d) $[CrCl_2(phen)_2]Cl$
 e) $[CrCl\ (phen)_2]Cl_2$

Answer: d

28. What is the formula for sodium diaquatetraamminevanadate(III)?
 a) $Na_3[V(H_2O)_2(NH_3)_4]$
 b) $Na_2[V(H_2O)_2(NH_3)_4]$
 c) $[Na_3V(H_2O)_2(NH_3)_4]$
 d) $[V(H_2O)_2(NH_3)_4]Na_2$
 e) $[V(H_2O)_2(NH_3)_4Na_3]$

Answer: a

22.4 Structures of Coordination Compounds

29. What are the possible geometries of a metal complex with a coordination number of 4?
 1. square planar
 2. tetrahedral
 3. octahedral

 a) 1 only b) 2 only c) 3 only d) 1 and 2 e) 1, 2, and 3

Answer: d

30. Which of the following complex ions have geometric isomers: $[Ni(H_2O)_6]^{2+}$, $[Ni(H_2O)_5NH_3]^{2+}$, $[Ni(H_2O)_4(NH_3)_2]^{2+}$, and $[Ni(H_2O)_3(NH_3)_3]^{2+}$?
 a) $[Ni(H_2O)_6]^{2+}$ and $[Ni(H_2O)_5NH_3]^{2+}$
 b) $[Ni(H_2O)_5NH_3]^{2+}$ and $[Ni(H_2O)_4(NH_3)_2]^{2+}$
 c) $[Ni(H_2O)_4(NH_3)_2]^{2+}$ and $[Ni(H_2O)_3(NH_3)_3]^{2+}$
 d) $[Ni(H_2O)_6]^{2+}$ and $[Ni(H_2O)_4(NH_3)_2]^{2+}$
 e) $[Ni(H_2O)_5NH_3]^{2+}$ and $[Ni(H_2O)_3(NH_3)_3]^{2+}$

Answer: c

31. Which of the following compounds may have linkage isomers?
 a) $[Cr(H_2O)_4Cl_2]Cl$
 b) $[Cr(H_2O)_5(SCN)]SO_4$
 c) $[Cr(H_2O)_4Cl_2]SCN$
 d) $[Cr(H_2O)_5Cl](NO_2)_2$
 e) $[Cr(NH_3)_4Cl_2]Cl$

Answer: b

32. Which one of the following complex ions has an optical isomer?
   a) $[ZnCl_4]^{2-}$
   b) $[Zn(en)_2]^{2+}$
   c) $[Zn(NH_3)_2en]^{2+}$
   d) $[Co(H_2O)_4en]^{2+}$
   e) $[Co(phen)_3]^{2+}$

Answer: e

33. How many isomers are possible for the square-planar complex ion $[PtBr_2Cl_2]^{2-}$?
   a) 2 b) 3 c) 4 d) 5 e) 6

Answer: a

34. Which of the following statements are CORRECT?
   1. Complex ions with the structure $[ML_2]^{n\pm}$ are always linear.
   2. Complex ions with the structure $[ML_4]^{n\pm}$ are always square-planar.
   3. Complex ions with the structure $[ML_6]^{n\pm}$ are always octahedral.

   a) 1 only b) 2 only c) 3 only d) 1 and 3 e) 2 and 3

Answer: d

22.5 Bonding in Coordination Compounds

35. Which is NOT a correct *d*-orbital designation?
   a) d_{xy} b) d_{yz} c) d_{z^2} d) $d_{x^2-y^2}$ e) $d_{y^2+z^2}$

Answer: e

36. In an octahedral complex, electrons in d_{z^2} and $d_{x^2-y^2}$ orbitals experience a greater repulsion from the lone pairs of electrons on ligands because the d_{z^2} and $d_{x^2-y^2}$ orbitals
   a) contain more electrons than the other three *d*-orbitals.
   b) each contain one unpaired electron.
   c) are oriented directly toward the incoming ligand electron pairs.
   d) are located a greater distance from the metal ion nucleus.
   e) are roughly spherical in shape, much like an *s*-orbital.

Answer: c

37. All of the following statements concerning crystal field theory are true EXCEPT
   a) in low-spin complexes, electrons are concentrated in the d_{xy}, d_{yz}, and d_{xz} orbitals.
   b) ligand-metal bonding is ionic.
   c) high-spin complexes contain the maximum number of unpaired electrons.
   d) in low-spin complexes, the crystal field splitting is large.
   e) the energy difference between *d*-orbitals often corresponds to the energy of visible light.

   Answer: b

38. Determine the number of unpaired electrons in an octahedral, high-spin iron(II) complex.
   a) 0 b) 1 c) 2 d) 3 e) 4

   Answer: e

39. Determine the number of unpaired electrons in an octahedral, low-spin Mn(II) complex.
   a) 1 b) 2 c) 3 d) 4 e) 5

   Answer: a

40. If $[Co(CN)_6]^{3-}$ is diamagnetic, which of the following sets of conditions best describes the complex?
   a) low-spin, Δ_0 small
   b) low-spin, Δ_0 large
   c) high-spin, Δ_0 small
   d) high-spin, Δ_0 large
   e) none of the above

   Answer: b

41. As bound ligands, which of the following causes the largest splitting of *d*-orbitals?
   a) CN^- b) H_2O c) F^- d) Cl^- e) $C_2O_4^{2-}$

   Answer: a

42. Determine the number of unpaired electrons in the tetrahedral complex ion $[CoCl_4]^{2-}$.
   a) 0 b) 1 c) 2 d) 3 e) 4

   Answer: d

43. Determine the number of unpaired electrons in the low spin, square-planar complex ion $[Ni(CN)_4]^{2-}$.
   a) 0 b) 1 c) 2 d) 3 e) 4

   Answer: a

44. All of the following complexes will be colored (in the visible) EXCEPT ________.
a) $[Ni(H_2O)_6]^{2+}$ b) $[Fe(CN)_6]^{4-}$ c) $[Ti(en)_3]^{4+}$ d) $[Cr(NH_3)_6]^{3+}$ e) $[Cu(NH_3)_4]^{2+}$

Answer: c

45. The spectrochemical series
a) lists ligands in order from monodentate to hexadentate.
b) lists ligands in the order of their color.
c) lists coordination compounds in order of their color.
d) lists complex ions in order of their color.
e) lists ligands in order of their tendency to split *d*-orbitals.

Answer: e

46. The period 6 transition metals have greater densities than either the period 4 or 5 transition metals. An explanation concerns the 4*f*-orbitals, which are filled with electrons just prior to the 5*d*-orbitals. As the 4*f*-orbitals are filled, the radii of the elements decrease. The decrease in size is referred to as the ________ contraction.

Answer: lanthanide

47. One method of recovering metals from their ores involves aqueous solutions; a second method involves heat. The aqueous method is called ________.

Answer: hydrometallurgy

48. Phenanthroline ($C_{12}H_8N_2$) can bind to metal ions using both its ammine groups. It is referred to as a ________ ligand, meaning it is "two-toothed."

Answer: bidentate

49. $EDTA^{4-}$, a hexadentate ligand, is most effective at chelating with metal ions when the pH is greater than 12. Explain why lower pHs decrease the chelating ability of the ligand.

Answer: At lower pH, EDTA4- is protonated (in successive fashion) forming $HEDTA^{3-}$, H_2EDTA^{2-}, H_3EDTA^-, H_4EDTA (and even H_5EDTA^+ and H_6EDTA^{2+}). Once protonated, a chelating ligand site loses it ability to form a coordinate-covalent bond with a metal ion.

50. The complex $[NiCl_4]^{2-}$ is paramagnetic, but the complex $[Ni(NH_3)_4]^{2+}$ is diamagnetic. Why do the magnetic properties of these two nickel complexes differ?

Answer: $[NiCl_4]^{2-}$ must have a tetrahedral geometry. In this geometry, the *d*-orbitals are split into two energies, with 3 orbitals (d_{xy}, d_{yz}, and d_{xz}) higher than the other two orbitals (d_{z^2} and $d_{x^2-y^2}$). Two of nickel's 8 *d*-electrons are unpaired. The tetraammine complex must be square planar. In this geometry, the *d*-orbitals split into four energy levels. The highest energy orbital is unoccupied. The lower orbitals are filled.

Diagnostic Quiz Questions

Goal #1: Identify and explain the chemical and physical properties of the transition elements.

1. What is the correct electron configuration for osmium?
   a) $[Xe]5d^6 6s^2$
   b) $[Xe]5f^{14} 6d^6 6s^2$
   c) $[Xe]4f^{14} 5d^6 6s^2$
   d) $[Xe]6d^6 6s^2$

   Answer: c

2. What is the correct electron configuration of Au^+?
   a) $[Xe]4f^{14} 5d^9$
   b) $[Xe]4f^{14} 5d^{10}$
   c) $[Xe]4f^{14} 5d^{10} 6s^2$
   d) $[Xe]4f^{14} 5d^9 6s^2$

   Answer: b

3. Which of the following elements is not oxidized by hydrochloric acid?
   a) V
   b) Ni
   c) Zn
   d) Fe

   Answer: a

4. Given the following standard reduction potentials,

   $Fe^{3+}(aq) + e^- \rightarrow Fe^{2+}(aq)$ $E° = +0.771$ V
   $Fe^{2+}(aq) + 2\,e^- \rightarrow Fe(s)$ $E° = -0.44$ V
   $2\,H^+(aq) + 2\,e^- \rightarrow H_2(g)$ $E° = 0.000$ V

   predict which one of the following reactions is spontaneous.
   a) $Fe^{2+}(aq) + 2\,H^+(aq) \rightarrow Fe^{3+}(aq) + H_2(g)$
   b) $Fe^{3+}(aq) + 6\,H^+(aq) \rightarrow Fe(s) + 3\,H_2(g)$
   c) $Fe^{2+}(aq) + H_2(g) \rightarrow Fe(s) + 2\,H^+(aq)$
   d) $Fe(s) + 2\,H^+(aq) \rightarrow Fe^{2+}(aq) + H_2(g)$

   Answer: d

5. What substance is used to reduce Fe_2O_3 to Fe in a blast furnace?
   a) CO
   b) CO_2
   c) O_2
   d) $CaCO_3$

   Answer: a

Goal #2: Understand the composition, structure, and bonding in coordination compounds.

6. What is the coordination number of the central metal in the complex salt $K_3[CoCl_6]$?
   a) 0
   b) 1
   c) 3
   d) 6

   Answer: d

7. What is the oxidation state of iron in $K_4[Fe(CN)_6]$?
   a) +2
   b) +3
   c) +4
   d) +6

   Answer: a

8. What is the formula for dibromobis(ethylenediamine)chromium(III) bromide?
   a) $[Cr(en)_2Br_2]Br$
   b) $[Cr(en)_2Br_3]$
   c) $[Cr(en)Br_2]Br$
   d) $[Cr(en)Br]Br_2$

   Answer: a

9. What is the name of $(NH_4)_2[CrCl_4]$?
   a) tetrachlorochromate(II) ammonium
   b) ammonium tetrachlorochromate(II)
   c) diamminetetrachlorochromium(II)
   d) diamminetetrachlorochromate(II)

   Answer: b

10. Which one of the following complex ions has geometric isomers?
   a) $[Ni(NH_3)_6]^{2+}$
   b) $[Ni(NH_3)_5Br]^{+}$
   c) $[Ni(NH_3)_4en]^{2+}$
   d) $[Ni(NH_3)_2(en)_2]^{2+}$

   Answer: d

11. How many isomers are possible for the square planar complex ion $[Pt(NH_3)_2Cl_2]^{2-}$?
   a) 2
   b) 3
   c) 4
   d) none

   Answer: a

12. Determine the number of unpaired electrons in an octahedral, low-spin iron(II) complex.
   a) 0
   b) 1
   c) 4
   d) 5

   Answer: a

Goal #3: Show how bonding is used to explain the magnetism and spectra of coordination compounds.

13. If $[Fe(CN)_6]^{4-}$ is diamagnetic, is the complex ion high spin or low spin and how many unpaired electrons does it contain?
   a) high spin; no unpaired electrons
   b) high spin; four unpaired electrons
   c) low spin; no unpaired electrons
   d) low spin; four unpaired electrons

   Answer: c

14. Which of the following complex ions will most likely absorb visible light?
   a) $[Zn(en)_3]^{2+}$
   b) $[CrCl_6]^{3-}$
   c) $[Sc(H_2O)_6]^{3+}$
   d) $[Ti(NH_3)_6]^{4+}$

   Answer: b

CHAPTER 23: Nuclear Chemistry

23.1 Natural Radioactivity

1. The most penetrating type of radiation is a(n)
   a) alpha particle.
   b) beta particle.
   c) gamma ray.
   d) positron.
   e) cathode ray.

   Answer: c

2. Which symbol represents a beta particle?
   a) ${}^{0}_{+1}e$ b) ${}^{0}_{-1}e$ c) ${}^{1}_{0}n$ d) ${}^{0}_{0}\gamma$ e) ${}^{4}_{2}He$

   Answer: b

3. Which symbol represents a positron?
   a) ${}^{4}_{2}He$ b) ${}^{0}_{-1}\beta$ c) ${}^{1}_{0}n$ d) ${}^{0}_{0}\gamma$ e) ${}^{0}_{+1}\beta$

   Answer: e

23.2 Nuclear Reactions and Radioactive Decay

4. Which of the following reactions is an example of positron emission?
   a) ${}^{232}_{90}Th \rightarrow {}^{228}_{88}Ra + {}^{4}_{2}He$
   b) ${}^{7}_{4}Be + {}^{0}_{-1}e \rightarrow {}^{7}_{3}Li$
   c) ${}^{82}_{35}Br \rightarrow {}^{82}_{36}Kr + {}^{0}_{-1}\beta$
   d) ${}^{40}_{19}K \rightarrow {}^{40}_{18}Ar + {}^{0}_{+1}\beta$
   e) ${}^{58}_{26}Fe + {}^{1}_{0}n \rightarrow {}^{59}_{26}Fe$

   Answer: d

5. Which of the following reactions is an example of electron capture?
   a) ${}^{14}_{6}C \rightarrow {}^{14}_{7}N + {}^{0}_{-1}\beta$
   b) ${}^{227}_{89}Ac \rightarrow {}^{223}_{87}Fr + {}^{4}_{2}He$
   c) ${}^{58}_{26}Fe + {}^{1}_{0}n \rightarrow {}^{59}_{26}Fe$
   d) ${}^{101}_{45}Rh + {}^{0}_{-1}e \rightarrow {}^{101}_{44}Ru$
   e) ${}^{40}_{19}K \rightarrow {}^{40}_{18}Ar + {}^{0}_{+1}\beta$

   Answer: d

6. If a nucleus undergoes positron emission
   a) its atomic number decreases by one and its mass number is unchanged.
   b) its atomic number is unchanged and its mass number increases by one.
   c) both its atomic number and its mass number are unchanged.
   d) its atomic number increases by one and its mass number is unchanged.
   e) its atomic number is unchanged and its mass number decreases by one.

   Answer: a

7. If a nucleus emits an alpha particle
   a) its atomic number decreases by four and its mass number decreases by two.
   b) its atomic number decreases by two and its mass number decreases by four.
   c) its atomic number decreases by two and its mass number decreases by two.
   d) its atomic number decreases by four and its mass number is unchanged.
   e) its atomic number is unchanged and its mass number decreases by four.

   Answer: b

8. If a nucleus gains a neutron and then undergoes beta emission,
   a) its atomic number decreases by one and its mass number increases by one.
   b) its atomic number is unchanged and its mass number increases by one.
   c) both its atomic number and its mass number are unchanged.
   d) its atomic number increases by one and its mass number is unchanged.
   e) its atomic number increases by one and its mass number increases by one.

   Answer: e

9. If a nucleus decays by successive β, α, α emissions, its atomic number will
   a) decrease by eight.
   b) decrease by seven.
   c) decrease by four.
   d) decrease by three.
   e) decrease by two.

   Answer: d

10. If plutonium-244 decays by successive α, β, β, α emissions, what nucleus is produced?
    a) $^{236}_{88}Ra$ b) $^{236}_{89}Ac$ c) $^{236}_{90}Th$ d) $^{240}_{90}Th$ e) $^{236}_{92}U$

    Answer: e

11. If californium-251 decays by successive α, α, β emissions, what nucleus is produced?
    a) $^{243}_{95}Am$ b) $^{243}_{93}Np$ c) $^{242}_{94}Pu$ d) $^{244}_{94}Pu$ e) $^{247}_{92}Pa$

    Answer: a

12. If Ag-106 decays by electron capture, what is the product?
a) ${}^{105}_{46}Pd$ b) ${}^{106}_{46}Pd$ c) ${}^{105}_{47}Ag$ d) ${}^{106}_{48}Cd$ e) ${}^{107}_{47}Ag$

Answer: b

13. What nuclide is produced by the beta decay of In-127?
a) ${}^{127}_{48}Cd$ b) ${}^{126}_{50}Sn$ c) ${}^{127}_{50}Sn$ d) ${}^{126}_{48}Cd$ e) ${}^{128}_{49}In$

Answer: c

14. Complete the following nuclear reaction.

$${}^{97}_{45}Rh \rightarrow {}^{0}_{+1}\beta + \_\_\_$$

a) ${}^{98}_{46}Pd$ b) ${}^{97}_{46}Pd$ c) ${}^{96}_{44}Ru$ d) ${}^{97}_{44}Ru$ e) ${}^{98}_{44}Ru$

Answer: d

15. By what (single step) process does polonium-211 decay to lead-207?
a) α particle emission
b) β particle emission
c) positron emission
d) electron capture
e) neutron capture

Answer: a

16. By what (single step) process does radon-226 decay to francium-236?
a) α particle emission
b) β particle emission
c) positron emission
d) electron capture
e) neutron capture

Answer: b

<u>23.3 Stability of Atomic Nuclei</u>

17. A plot of the number of neutrons versus the number of protons in nuclei shows a narrow band of stable isotopes. Elements below this narrow line may attain stability by
a) beta emission.
b) neutron emission.
c) electron capture.
d) gamma ray emission.
e) alpha particle capture.

Answer: c

18. All of the following statements concerning nuclei are true EXCEPT
    a) only hydrogen-1 and helium-3 have more protons than neutrons.
    b) from He to Ca, stable nuclei have roughly equal numbers of protons and neutrons.
    c) most stable isotopes have odd numbers of both protons and neutrons.
    d) the neutron to proton ratio in stable nuclei increases as mass increases.
    e) beyond calcium, the neutron to proton ratio is always greater than one.

    Answer: c

19. The point of maximum stability in the binding energy curve occurs in the vicinity of which one of the following isotopes?
    a) ${}^{12}_{6}C$ b) ${}^{56}_{26}Fe$ c) ${}^{184}_{74}W$ d) ${}^{209}_{83}Bi$ e) ${}^{238}_{92}U$

    Answer: b

20. The molar nuclear mass of carbon-14 is 14.003241 g/mol. The molar mass of a proton is 1.007825 g/mol. The molar mass of a neutron is 1.008665 g/mol. Calculate the binding energy of C-14. ($c = 3.00 \times 10^8$ m/s)
    a) 1.13×10^{-1} J/mol
    b) 1.13×10^{-4} J/mol
    c) 1.02×10^{13} J/mol
    d) 1.00×10^{16} J/mol
    e) 1.00×10^{16} J/mol

    Answer: c

21. Calculate the energy released (per mole of tritium consumed) for the fusion reaction of tritium and deuterium,

$${}^{3}_{1}H + {}^{2}_{1}H \rightarrow {}^{4}_{2}He + {}^{1}_{0}n$$

    given the following molar masses of nucleons and nuclei. ($c = 3.00 \times 10^8$ m/s)

particle	mass (g/mol)
proton	1.007825
neutron	1.008665
deuterium	2.0140
tritium	3.01605
helium	4.00260

    a) -1.88×10^{-5} J/mol
    b) -1.88×10^{-2} J/mol
    c) -8.45×10^{11} J/mol
    d) -1.69×10^{12} J/mol
    e) -1.69×10^{15} J/mol

    Answer: d

23.4 Rates of Nuclear Decay

22. Strontium-90 has a half-life of 28.1 years. Starting with 2.08 mg of this isotope, how much would remain after 112.4 years?
a) 0.13 mg b) 0.26 mg c) 0.52 mg d) 1.04 mg e) 1.56 mg

Answer: a

23. The decay constant for the spontaneous decay of iodine-131 is 9.98×10^{-7} s^{-1}. How much of a 0.50 g sample of this isotope remains after 3.0 days?
a) 0.11 g b) 0.13 g c) 0.19 g d) 0.28 g e) 0.39 g

Answer: e

24. The half-life of iodine-123 is 13.2 hours. What percentage of iodine-123 remains in a sample after 36 hours?
a) 13% b) 15% c) 37% d) 85% e) 95%

Answer: b

25. The half-life of carbon-14 is 5730 years. What percentage of carbon-14 remains in a sample after 7.00 half-lives?
a) 0.78% b) 1.6% c) 3.1% d) 6.3% e) 14%

Answer: a

26. Uranium-235 has a half-life of 7.04×10^8 years. How many years will it take for 95% of a U-235 sample to decay?
a) 5.2×10^7 yr b) 6.7×10^8 yr c) 1.3×10^9 yr d) 1.6×10^9 yr e) 3.0×10^9 yr

Answer: e

27. What is the half-life of an isotope if the decay constant is 7.2×10^{-4} s^{-1}?
a) 1.0×10^{-3} s b) 1.4×10^{-3} s c) 2.4×10^{-3} s d) 4.2×10^2 s e) 9.6×10^2 s

Answer: e

28. $^{241}_{95}Am$ is used in many home smoke alarms. If 25% of the americium in a smoke detector decays in 180 years, what is the half-life of this isotope?
a) 1.9×10^2 yr b) 2.1×10^2 yr c) 3.6×10^2 yr d) 4.3×10^2 yr e) 1.0×10^3 yr

Answer: d

29. Carbon-14 dating is only accurate for samples less than 4.0×10^4 years old. If a sample initially contains 3.5 mg C-14, what mass remains after 4.0×10^4 years? The half-life of carbon-14 is 5730 years.
a) 1.4×10^{-8} g   b) 7.9×10^{-6} g   c) 2.8×10^{-5} g   d) 5.1×10^{-5} g   e) 5.0×10^{-4} g

Answer: c

30. A sample of fluorine-18 is found to have an activity of 980 disintegrations per second (dps). After 8.0 hours the activity has decreased to 45 dps. What is the half-life, in years, for the decay of fluorine-18?
a) 5.9×10^{-3} hr   b) 0.39 hr   c) 1.8 hr   d) 4.1 hr   e) 1.2×10^2 hr

Answer: c

23.5 Artificial Nuclear Reactions

31. Which of the following statements are CORRECT?
    1. Neptunium, plutonium, and americium do not exist, even in trace quantities, in nature. They must be produced in particle accelerators.
    2. Nuclear reactions involving neutron capture followed by gamma ray emission are referred to as (n, γ) reactions.
    3. Bombarding nuclei with low energy neutrons is an efficient means of creating heavier isotopes.

a) 1 only   b) 2 only   c) 3 only   d) 2 and 3   e) 1, 2, and 3

Answer: d

23.6 Nuclear Fission

32. Complete the following fission reaction.

$$^{1}_{0}n + ^{235}_{92}U \rightarrow ^{236}_{92}U \rightarrow \_\_\_\_ + ^{92}_{36}Kr + 3\ ^{1}_{0}n$$

a) $^{141}_{56}Ba$ b) $^{143}_{56}Ba$ c) $^{141}_{53}I$ d) $^{143}_{55}Cs$ e) $^{141}_{55}Cs$

Answer: a

33. Enriched uranium is uranium that has a greater proportion of ________.
a) lead-207   b) uranium-235   c) uranium-238   d) radium-226   e) plutonium-248

Answer: b

34. Which two nuclei are the two fissionable isotopes most commonly used in nuclear reactors?
   a) $^{235}_{92}U$ and $^{239}_{94}Pu$
   b) $^{238}_{92}U$ and $^{232}_{90}Th$
   c) $^{238}_{92}U$ and $^{243}_{95}Am$
   d) $^{235}_{92}U$ and $^{226}_{88}Ra$
   e) $^{226}_{88}Ra$ and $^{239}_{94}Pu$

   Answer: a

35. What role do the cadmium control rods play in a fission reactor?
   a) The rods control the rate of fission by absorbing neutrons.
   b) The cadmium combines with spent uranium fuel to produce a non-radioactive product.
   c) The rods focus the neutrons toward the center of the reactor.
   d) The cadmium acts as a catalyst, enabling fission to occur at lower temperatures.
   e) The rods move forward and backward, driving the pistons that turn the turbines.

   Answer: a

36. Which of the following elements undergoes nuclear fusion to provide the primary source of energy from the sun?
   a) helium b) uranium c) hydrogen d) carbon e) nitrogen

   Answer: c

37. What do scientists call the sequence of rapidly occurring reactions that results when a nuclear fission reaction produces enough neutrons to produce more fission reactions?
   a) nucleation
   b) chain reaction
   c) nuclear fusion
   d) neutron multi-emission
   e) binding energy

   Answer: b

<u>23.7 Nuclear Fusion</u>

38. Complete the following fusion reaction.

$$^{4}_{2}He + ^{9}_{4}Be \rightarrow ^{1}_{0}n + \_\_\_\_$$

   a) $^{0}_{+1}\beta$ b) $2\ ^{7}_{3}Li$ c) $^{12}_{7}N$ d) $^{13}_{6}C$ e) $^{12}_{6}C$

   Answer: e

39. What particle(s) is(are) produced in the following reaction?

$$^{238}_{92}\text{U} + ^{16}_{8}\text{O} \rightarrow ^{250}_{100}\text{Fm} + \_\_\_$$

a) $^{1}_{+1}p$ b) 2 $^{0}_{-1}\beta$ c) 4 $^{0}_{+1}\beta$ d) $^{4}_{2}\text{He}$ e) 4 $^{1}_{0}\text{n}$

Answer: e

40. What percentage of the world's electricity is supplied by nuclear fusion reactors?
a) 0% b) 3% c) 8% d) 17% e) 39%

Answer: a

41. At the high temperatures necessary for fusion, matter exists as a plasma, which is
a) a liquid form of a radioactive element.
b) a gaseous cloud of electrons and positrons.
c) a mixture of unbound nuclei and electrons.
d) an equal mixture of matter and anti-matter.
e) a cloud of neutrons.

Answer: c

23.8 Radiation and Human Health

42. For an average person, which of the following is likely to be the **most** significant source of exposure to radiation?
a) cosmic radiation
b) elements found naturally in human tissue
c) luminous watch dials
d) emissions from nuclear power plants
e) fallout from nuclear testing

Answer: a

43. In the United States, radiation is often measured in curies (Ci). A curie is
a) 1 disintegration per second.
b) 3.7×10^{10} disintegrations per second.
c) 0.01 J energy absorbed per kg of tissue.
d) 1 J energy absorbed per kg of tissue.
e) 1 calorie energy absorbed per kg of tissue.

Answer: b

23.9 Applications of Nuclear Chemistry

44. Which of the following isotopes is used in the treatment of thyroid disorders?
a) $^{67}_{31}Ga$ b) $^{201}_{81}Tl$ c) $^{99m}_{43}Tc$ d) $^{18}_{9}F$ e) $^{131}_{53}I$

Answer: e

Short Answer Questions

45. All isotopes of atomic number greater than ________ are unstable and radioactive.

Answer: 83

46. Beta particle emission is accompanied by the concurrent emission of a(n) ________, a massless, chargeless particle.

Answer: antineutrino

47. ________ emission tomography uses gamma rays produced by matter-antimatter annihilations to map regions of interest in the human body.

Answer: Positron

48. Neutron ________ analysis is a non-destructive process in which a sample is irradiated with neutrons. The neutrons react with nuclei to form isotopes with masses one unit higher than the original nuclei. The nuclei are formed in excited states and they emit gamma radiation that can be used to both identify the presence of an element and quantify how much is present.

Answer: activation

49. Explain the difference between 1 rad and 1 rem of radiation.

Answer: One radiation absorbed dose (rad) equals 0.01 J of energy absorbed per kilogram of tissue. One roentgen equivalent man (rem) equals one ram multiplied by the quality factor for the type of radiation involved. The quality factor accounts for variations in damaging power of different types of radiation.

50. Explain the treatment named boron neutron capture therapy (BNCT).

Answer: Boron-10 is injected into tumors and then bombarded with neutrons. Boron-10 is effective at capturing neutrons. The resulting boron-11 nuclei disintegrate to produce alpha particles and lithium-7 atoms. The alpha particles destroy tissue in the localized area where the boron was injected.

Diagnostic Quiz Questions

Goal #1: Identify radioactive elements and describe natural and artificial nuclear reactions

1. What is the least penetrating type of radiation?
   a) alpha particle
   b) beta particle
   c) gamma ray
   d) both alpha particles and beta particles

   Answer: a

2. Which one of the following symbols is used to represent a gamma ray?
   a) ${}^{1}_{0}\text{n}$
   b) ${}^{4}_{2}\alpha$
   c) ${}^{0}_{-1}\text{e}$
   d) γ

   Answer: d

3. Which of the following reactions is an example of beta particle emission?
   a) ${}^{58}_{26}\text{Fe} + {}^{1}_{0}\text{n} \rightarrow {}^{59}_{26}\text{Fe}$
   b) ${}^{40}_{19}\text{K} \rightarrow {}^{40}_{18}\text{Ar} + {}^{0}_{+1}\beta$
   c) ${}^{227}_{89}\text{Ac} \rightarrow {}^{223}_{87}\text{Fr} + {}^{4}_{2}\text{He}$
   d) ${}^{14}_{6}\text{C} \rightarrow {}^{14}_{7}\text{N} + {}^{0}_{-1}\beta$

   Answer: d

4. Which of the following reactions is an example of electron capture?
   a) ${}^{40}_{19}\text{K} \rightarrow {}^{40}_{18}\text{Ar} + {}^{0}_{+1}\beta$
   b) ${}^{101}_{45}\text{Rh} + {}^{0}_{-1}\text{e} \rightarrow {}^{101}_{44}\text{Ru}$
   c) ${}^{177}_{79}\text{Au} \rightarrow {}^{173}_{77}\text{Ir} + {}^{4}_{2}\alpha$
   d) ${}^{105}_{45}\text{Rh} \rightarrow {}^{105}_{46}\text{Pd} + {}^{0}_{-1}\beta$

   Answer: b

5. If a nucleus decays by successive α, β, α emissions,
   a) the atomic number will decrease by 2 and the mass number will decrease by 4.
   b) the atomic number will decrease by 3 and the mass number will decrease by 4.
   c) the atomic number will decrease by 3 and the mass number will decrease by 8.
   d) the atomic number will decrease by 5 and the mass number will decrease by 8.

   Answer: c

6. Complete the following fission reaction.

$$^{1}_{0}n + ^{235}_{92}U \rightarrow ^{146}_{57}La + \_\_\_ + 3\ ^{1}_{0}n$$

   a) $^{87}_{35}Br$
   b) $^{87}_{35}Kr$
   c) $^{85}_{35}Br$
   d) $^{89}_{35}Kr$

   Answer: a

Goal #2: Calculate the binding energy and binding energy per nucleon for a particular isotope.

7. The mass of beryllium-9 is 9.012182 g/mol. The mass of a proton is 1.007825 g/mol. The mass of a neutron is 1.008665 g/mol. Calculate the binding energy per mole of Be-9. (c = 3.00×10^8 m/s)
   a) 4.13×10^{11} J/mol
   b) 5.54×10^{12} J/mol
   c) 5.62×10^{12} J/mol
   d) 8.11×10^{17} J/mol

   Answer: c

8. Calculate the heat released (per mole of He-4 produced) for the following fusion reaction,

$$^{2}_{1}H + ^{2}_{1}H \rightarrow ^{4}_{2}He$$

   given,

particle	mass (g/mol)
deuterium	2.0140
helium-4	4.00260

   a) -9.11×10^{11} J/mol
   b) -2.29×10^{12} J/mol
   c) -6.87×10^{13} J/mol
   d) -1.79×10^{14} J/mol

   Answer: b

Goal #3: Understand rates of radioactive decay.

9. Indium-122 has a half-life of 10.1 seconds. What is the first-order rate constant for the decay of I-122?
   a) $6.86 \times 10^{-2}\ s^{-1}$
   b) $9.90 \times 10^{-2}\ s^{-1}$
   c) $0.571\ s^{-1}$
   d) $1.62\ s^{-1}$

   Answer: a

10. Strontium-90 has a half-life of 28.1 years. Starting with 3.20 mg of this isotope, how much would remain after 112.4 years?
    a) 0.20 mg
    b) 0.40 mg
    c) 0.80 mg
    d) 1.60 mg

    Answer: a

11. The decay constant for the spontaneous decay of iodine-131 is $9.98 \times 10^{-7}\ s^{-1}$. How much of a 5.00 g sample of this isotope remains after 25.0 days?
    a) 0.116 g
    b) 0.349 g
    c) 0.578 g
    d) 0.625 g

    Answer: c

12. The half-life of carbon-14 is 5730 years. What percentage of carbon-14 remains in a sample after 2.50×10^4 years?
    a) 2.7%
    b) 4.9%
    c) 21%
    d) 23%

    Answer: b

Goal #4: Understand artificial nuclear reactions

13. Which of the following statements are correct?
    1. Only a small fraction of the currently known radioactive isotopes occur naturally.
    2. Neutrons are accelerated to high velocities using particles accelerators.
    3. High speed collisions between neutrons and nuclei more favorable for the formation of heavier nuclei than low speed collisions.

    a) 1 only
    b) 2 only
    c) 3 only
    d) 1 and 2

    Answer: a

Goal #5: Understand issues of health and safety with respect to radioactivity.

14. Which of the following natural sources of radiation is likely to be the greatest source of exposure to the average person?
    a) inhaled radioactive isotopes of oxygen and nitrogen
    b) radioactive elements found in human tissues
    c) radioactive elements in building materials, such as concrete.
    d) cosmic radiation

    Answer: d

15. Which of the following radiation units is a measure of the energy absorbed per kg of tissue?
    a) curie
    b) becquerel
    c) rem
    d) rad

    Answer: d

Goal #5: Be aware of some uses of radioactive isotopes in science and medicine.

16. Which radioactive isotope is used for the treatment of hyperthyroidism?
    a) Tl-73
    b) F-18
    c) I-123
    d) Ga-67

    Answer: c